Bounce

Game Histories

edited by Henry Lowood and Raiford Guins

Debugging Game History: A Critical Lexicon, edited by Henry Lowood and Raiford Guins, 2016

Zones of Control: Perspectives on Wargaming, edited by Pat Harrigan and Matthew Kirschenbaum, 2016

Gaming the Iron Curtain: How Teenagers and Amateurs in Communist Czechoslovakia Claimed the Medium of Computer Games, Jaroslav Švelch, 2018

The Elusive Shift: How Role-Playing Games Forged Their Identity, Jon Peterson, 2020

Homebrew Gaming and the Beginnings of Vernacular Digitality, Melanie Swalwell, 2021

Game Wizards: The Epic Battle for Dungeons & Dragons, Jon Peterson, 2021

Arcade Britannia: A Social History of the British Amusement Arcade, Alan Meades, 2022

Building SimCity: How to Put the World in a Machine, Chaim Gingold, 2024

The Game That Never Ends: How Lawyers Shape the Video Game Industry, Julien Mailland, 2024

Playing at the World, 2E: Volume 1: The Invention of Dungeons & Dragons, Jon Peterson, 2024

Playing at the World, 2E: Volume 2: Three Pillars of Dungeons & Dragons, Jon Peterson, 2025

Bounce: Balls, Walls, and Bodies in Games and Play, Carlin Wing, 2026

Bounce

Balls, Walls, and Bodies in Games and Play

Carlin Wing

The MIT Press
Cambridge, Massachusetts
London, England

The MIT Press
Massachusetts Institute of Technology
77 Massachusetts Avenue, Cambridge, MA 02139
mitpress.mit.edu

The MIT Press would like to thank the anonymous peer reviewers who provided comments on drafts of this book. The generous work of academic experts is essential for establishing the authority and quality of our publications. We acknowledge with gratitude the contributions of these otherwise uncredited readers.

This book was set in Stone Serif and Stone Sans by Westchester Publishing Services. Printed and bound in the United States of America.

Library of Congress Cataloging-in-Publication Data

Names: Wing, Carlin author
Title: Bounce : balls, walls, and bodies in games and play / Carlin Wing.
Description: Cambridge, Massachusetts : The MIT Press, 2025. | Series:
   Game histories | Includes bibliographical references and index.
Identifiers: LCCN 2025016507 (print) | LCCN 2025016508 (ebook) |
   ISBN 9780262553216 paperback | ISBN 9780262384520 epub |
   ISBN 9780262384537 pdf
Subjects: LCSH: Ball games | Physics | Materials science | Computer
   animation
Classification: LCC GV861 .W56 2025 (print) | LCC GV861 (ebook)
LC record available at https://lccn.loc.gov/2025016507
LC ebook record available at https://lccn.loc.gov/2025016508

EU Authorised Representative: Easy Access System Europe, Mustamäe tee 50, 10621 Tallinn, Estonia | Email: gpsr.requests@easproject.com

For my mother and father,
body and ball,
pen and mitt.

The Infinites *of* Matter

If all the *World* were a *confused heape*,
What was beyond? for this *World* is not great:
We finde it *Limit* hath, and *Bound*,
And like a *Ball* in compasse is made round:
And if that *Matter*, with which the *World's* made,
Be *infinite*, then more *Worlds* may be said;
Then *Infinites* of *Worlds* may we agree,
As well, as *Infinites* of Matters bee.
—Margaret Cavendish, *Poems, and Fancies,* 1653

We are living through a movement from an organic, industrial society to a polymorphous, information system—from all work to all play, a deadly game.
—Donna Haraway, "Cyborg Manifesto," 1985

Contents

Series Foreword   ix
Preface   xi
Acknowledgments   xv

Introduction: Long Past Play   1

I   The Matter of the Ball

1   Regarding Ricochet   37
2   True Bounce   69

II   Virtual Bounce

3   Squash and Stretch   111
4   From Ping to *Pong*   143

III   Bounded Spectacle

5   Bounce Feel   179
6   Pok ta Pok   205

Conclusion: Culture-Inside-Out   239

Notes   251
Bibliography   297
Index   321

Series Foreword

What might histories of games tell us not only about the games themselves but also about the people who play and design them? We think that the most interesting answers to this question will have two characteristics. First, the authors of game histories who tell us the most about games will ask big questions. For example, how do gameplay and design change? In what ways is such change inflected by societal, cultural, and other factors? How do games change when they move from one cultural or historical context to another? These kinds of questions forge connections to other areas of game studies, as well as to history, cultural studies, and technology studies.

The second characteristic we seek in "game-changing" histories is a wide-ranging mix of qualities partially described by terms such as *diversity*, *inclusiveness*, and *irony*. Histories with these qualities deliver interplay of intentions, users, technologies, materials, places, and markets. Asking big questions and answering them in creative and astute ways strikes us as the best way to reach the goal of not an isolated, general history of games but rather of a body of game histories that will connect game studies to scholarship in a wide array of fields. The first step, of course, is producing those histories.

Game Histories is a series of books that we hope will provide a home—or maybe a launch pad—for the growing international research community whose interest in game history rightly exceeds the celebratory and descriptive. In a line, the aim of the series is to help actualize critical historical study of games. Books in this series will exhibit acute attention to historiography and historical methodologies, while the series as a whole will encompass the wide-ranging subject matter we consider crucial for the relevance of historical game studies. We envisage an active series with output that will reshape how electronic and other kinds of games are understood, taught, and researched, as well as broaden the appeal of games for the allied fields such as history of computing, history of science and technology, design history, design culture, material culture studies, cultural and social history, media history, new media studies, and science and technology studies.

The Game Histories series will welcome but not be limited to contributions in the following areas:

- Multidisciplinary methodological and theoretical approaches to the historical study of games.

- Social and cultural histories of play, people, places, and institutions of gaming.

- Epochal and contextual studies of significant periods influential to and formative of games and game history.

- Historical biography of key actors instrumental in game design, development, technology, and industry.

- Games and legal history.

- Global political economy and the games industry (including indie games).

- Histories of technologies pertinent to the study of games.

- Histories of the intersections of games and other media, including such topics as game art, games and cinema, and games and literature.

- Game preservation, exhibition, and documentation, including the place of museums, libraries, and collectors in preparing game history.

- Material histories of game artifacts and ephemera.

Henry Lowood, Stanford University
Raiford Guins, Indiana University Bloomington

Preface

Bounce emerged out of a project called *Hitting Walls*, which began in 2006 as a photographic investigation of portable glass squash courts. The first photograph in *Hitting Walls* was made by opening the aperture of a 4×5-inch camera for the duration of a single long point played between Mohammed Abbas of Egypt and James Wilstrop of England. The players' rapid movement around the court over the course of the exposure left only the faintest traces of them visible on the negative, leaving viewers of the image to take in the tournament audience staring rapt at a seemingly empty court surrounded by the dramatic architecture of New York City's Grand Central Terminal. Wilstrop went on to win the match, only to lose the crown to Abass's compatriot, Ramy Ashour. The tournament's title sponsor, Bear Stearns, would collapse just two years later, initiating the 2008 financial crisis. As part of a desperate and ultimately unsuccessful bid to stave off a massive financial crisis, the US Securities and Exchange Commission facilitated JP Morgan's purchase of Bear Stearns for two dollars per share. Photography is used to record history. And history happens to photographs, shifting their meanings in big and small ways over time. In 2009, I went back to photograph the JP Morgan Tournament of Champions.

Over the years, *Hitting Walls* developed into an ongoing iterative project with works made in a variety of media and forms ranging from large-format photographs to web grabs, experimental videos, sound installations, performances, participatory events, and historical and theoretical writings including *Bounce*. All the works in this project take ballgames as points of departure for approaching interwoven histories. Each operates independent of the others. Taken together, they construct a grammar of motion practices and perform a practice in motion.

Figure P.1
Mohammed Abbas (EGY) vs. James Wilstrop (ENG), Bear Stearns Tournament of Champions, Grand Central Terminal, New York, 2006. 5'x7' color print face-mounted to plexiglass. Photograph by the author.

My initial intention for *Bounce* was to pick up the history of squash and overextend it beyond its nominal enclosure by tracing a genealogy of balls hitting walls stretching back to a prehistory found in games like court tennis and rackets, and forward to a post history found in the first computer and video games that translated bounce into electronic spaces. As the project ricocheted and rolled further and further afield, squash receded almost completely from view. But it remains the game that I played with sheer drive and loved with reckless abandon, as well as the prompt for my first wonderings about what differing conceptions of "the World" different kinds of walls and balls support.

Figure P.2
Ramy Ashour (EGY) vs. Laurens Jan Anjema (NDL), JP Morgan Tournament of Champions, Grand Central Station, New York, 2009. Photograph by the author.

Acknowledgments

As a young reader, I usually skipped over acknowledgments sections, impatient to get to "the beginning" of the book. It was only after I began my own attempts at writing in earnest that I finally understood that this is precisely what acknowledgments are: stories about how books begin and come to be. Today, I take great pleasure in reading what authors offer of the worlds and relationships that have engaged and sustained them over the course of their writing. And I have looked forward to the chance to give my own nod to the people who have helped me find my form and keep my footing over the many years that it has taken to bring *Bounce* into being.

First and foremost, thank you to Raiford Guins and Henry Lowood for inviting me to submit the project to their Game History series at MIT Press in 2016, and then for patiently cheering me on as I spent the next eight years wrangling old and new thoughts and words. They also asked me to join them in editing *EA Sports FIFA: Feeling the Game* (Bloomsbury Press, 2022), an experience that enhanced this book in many ways, including by giving me an opportunity to take a first run at the chapter on *FIFA* that appears here. Thank you to Noah Springer for guiding me through the review and publication process. Thank you also to the anonymous reviewers whose feedback strengthened the project by showing me what was landing well and what needed further work. And to the copyeditors, designers, and everyone at the MIT Press who helped polish and prepare this work for publication.

A host of other editors have supported the project along the way. A lively to and fro with Sina Najafi over "Episodes in the Life of Rubber Bounce: Playing with a Rubber Ball" (*Cabinet*, 2015) helped clarify the history of rubber and elasticity as a key throughline for *Bounce*. Mary McDonald and Jennifer Sterling gave me a chance to explore this further with a chapter on vulcanized play and the history of the Dunlop Rubber company for their edited collection *Sports, Society, and Technology: Bodies, Practices, and Knowledge Production*. Aaron Trammel encouraged me to turn what had been a conference performance into an article for *Games and Culture* on the question of captured play. Jamie Banks and her coeditors, Robert Mejia and Aubrie Adams, of *100 Greatest Video Game Characters* asked me to write about the *Pong* ball as a famous video game character, helping me realize that sometimes a ball can be square! David Shaftel gave me a chance to reflect on the history of the Maravilla Handball court for *Racquet* magazine. I initially wrote "Not without Matter or Substance" for the *Bulletin of the Serving Library,* at the invitation of Sarah Demeuse, and later revised and republished the piece in *ROMchip*. Caitlin Zaloom invited me to write two review essays for *Public Books* that helped me do foundational thinking for *Bounce*, first on Eric Simons' *The Secret Lives of Sports Fans* and second on the phenomenal pleasure of watching women play sports in the context of the FIFA Women's World Cup. Amid the constraints of the first year of COVID-19, Tom Leeser and Charlotte Kent prompted me toward using some freer language in "Swerve and Return," a catalog essay for *Tech/Know/Future: From Slang to Structure*. Writing that text freed me up to play with the first person in *Bounce*.

Curators and collaborators have also made deep contributions to the development of the thought and form of this project. *Bounce* emerged from and is an iteration of *Hitting Walls*, a project that began in 2006 as a photographic investigation of portable glass squash courts and developed into a set of works made in a variety of media and forms that take ballgames as their points of departure. In 2010, Mark Allen and Sasha Archibald at *Machine Project* invited me to present the first of what became a series of ball-making workshops. These workshops helped expand the project far beyond squash and occasioned my first encounter with Amanda Perez and the particular history of handball in Los Angeles. The iteration presented at *Cabinet* connected me to New York's handball traditions. Eddo Stern's invitation to run one for his UCLA students enlivened the finishing of this manuscript. Mariechen Danz's invitation to create a performance and contribute a text to "Unmapping the Renaissance," allowed me to begin my thinking about ricochet. Donald Moss invited me into a series of conversations about Freud's notion that there is, "first and foremost, a body ego," and an exhibition at *Second Story* hosted by Don and Lynne Zeavin, where I was able to allow the sounds of a bouncing ball and a piano playing to overlap and produce infinities. My collaboration with John Dieterich around art, music, sport, play, and improvisation helped feed my thinking and sustained my artistic practice through these many years of writing. My deep thanks to John, and to Keiko Beers, Sylke Meyer, Luke Fishbeck, and all the other co-conspirators who supported and participated in our projects.

It is difficult to say when this project began. Did it begin as a dissertation on bounce? As an art project about squash? When I first stepped on a squash court? When my dad first tossed me a tennis ball? Edgar Morales, Bryan Patterson, Alicia McConnell, Bill Doyle, Satinder Bajwa, Mohammed Ayaz, and Liz Irving all helped shape me into the squash player that I was. As did the many years spent on court, running sprints, and hanging around hotel rooms with Ivy Pochoda, Louisa Hall, Margaret Elias, and Colby Hall. Nicol David and Latasha Kahn offered friendship and sat for interviews early on, when I thought this was a very different kind of project.

For my early artistic and intellectual foundations and for providing key pivots and places to land along the way, I thank Chris Killip, Castle McLaughlin, Jacqueline Hassink, Sharon Harper, Stephen Prina, Alfred Guzzetti, Rob Moss, and all the faculty and staff of what was called at the time the Visual and Environmental Studies Department at Harvard. After serving as my undergrad adviser, Chris hired me twice, first as a teaching fellow and later as visiting faculty. I learned years after the fact that he used to read *Squash News* to keep tabs on what I was up to. Chris passed in 2020. I so wish I could hand him a copy of this book and see him quirk his eyebrow and nod "very good," not giving away for a second that he had already gotten a copy and read the whole thing. Ellen Birrell, Allan Sekula, Michael Asher, and others saw the very beginnings of *Hitting Walls* at California Institute of the Arts (CalArts) and encouraged me to follow my desire to develop its historical and theoretical aspects when the opportunity to do doctoral work presented itself. I carry the spirit of those programs and their people with me. Both Michael and Allan passed away while I was working on my dissertation, and I feel their absences sharply. They remain two of the readers I most often have in mind.

When I realized that some of the *Hitting Walls* projects were veering toward historical writing, and wanted to learn how to do this work in a deeper way, Nick Mirzoeff ushered me into the Media, Culture, and Communication doctoral program at NYU and offered his continual conviction that artists can and should take on tasks of this kind. Deep thanks to him and my other dissertation committee members—cochairs, Arjun Appadurai and Martin Scherzinger. Arjun's written words were guiding my thinking well before we began to work together. I thank him here for our conversations and for his commitment to helping me hold disparate strands of thought together. Martin took me on both as an advisee, as an artistic collaborator, and over many years now as a dear friend. I thank him for his different treatment of these different roles, for his trust in me, and for his ongoing dedication to rigorous thought, aesthetic practice, and deep play. Many other members of the NYU community were instrumental along the way. My other committee members—Erica Robles-Anderson and readers Nicole Starosielski, Caitlin Zaloom, Raiford Guins, and Michael Ralph—stewarded the development of *Bounce* in its first form. Mary Taylor, Gina Young, and Shima Gorgani, helped me navigate the institution. Rod Benson, Mara Mills, and Alex Galloway provided foundations in the classroom, and Lisa Gitelman and Ben Kafka offered friendship and wisdom in the in-between moments. The members of the NYLON and Oikos working groups, along with the baristas and bar-backs at Outpost Café, regularly reminded me that the core of this work is a practice of thinking and being with others.

Since arriving at Scripps, a member of the Claremont Colleges, in 2016, the school has supported this project through faculty research grants, sabbatical semesters, and the freedom to teach classes that feed back into my research and work. During these years, I have had the company and support of an amazing group of artists and scholars. A special thanks to my media studies colleagues at Scripps—T. Kim-Trang Tran, Nancy Macko, and Aly Ogasian—for your friendship, support, good humor, and LA art adventures. Thanks to Amy Crown for keeping Scripps Media Studies in good order, thanks also to my colleagues in Intercollegiate Media Studies, especially Elizabeth Affuso and Ruti Talmor for keeping the train on the tracks, and Eddie Gonzalez and Matthew Pagoaga at the IMS production center for supporting my classes and many *Hitting Walls* projects. The Science, Technology, and Society reading groups sparked ongoing learning and fostered friendships with David Seitz and Jemma Lorenat, among others. The lecture that I created for Scripps' Core I ended up contributing to some of the thinking that I do in chapter 2. Thank you to the Core I crew for your feedback and camraderie through the years of co-teaching that course. Thank you to my students from across all of my years of teaching for your energy and all you have taught me. Jason Nguyen, Nick Imparato, and Francesca Palmera were research assistants at different moments for *Bounce* and related *Hitting Walls* projects. Devon Ma's research paper in *Theories of Interaction* informed some of the thinking that I have since done on athletes as avatars. Juliet Koss and I became friends while sharing a hallway in Baxter and continue our shared project of unicorn spotting from coast to coast. I have shared good food and good thought with Elizabeth Affuso, Kevin Vennemann, Michelle Decker, Jemma Lorenat, David Seitz, Tessie Prakas, Julin Everett, Anne Harley, Martin Vega, and many other colleagues from Claremont. Ming Ma and I have made a tradition of attending on- and offbeat performances in Los Angeles. Ruti Talmor and I have spent countless hours coworking, walking up and down hills, and wending our way through work and life.

Many librarians, archivists, historians, enthusiasts, players, and organizers have supported this research. My thanks go to the staff at Kew Garden Archives and Collections, British National Archives, the British Library, the Royal Society, Lords Cricket Ground Library, the University Club of New York Library, the New York Public Library, the New York City Parks Department Archives, the Strong National Museum of Play Collection and Archives, the William Higinbotham Game Studies Collection, Hampton Court Palace, Société Sportive de Jeu de Paume, the International Tennis Hall of Fame, and the Tournament of Champions who gave me their time and advice. In recent years, the Huntington Library and Getty Research Institute have provided places to focus on writing.

Thank you also to those in the squash world who shared their thoughts and expertise with me. Peggy Beard generously arranged for me to sit in on the United States Tennis Association (USTA) technical committee meeting and visit facilities for testing bounce. I think of that gesture often and wish she was still here so I could convey my thanks. Amanda Perez introduced me to the history of handball in Los Angeles and so much more. Raul Herrera, Aztlan Tenochtitlan, and the other players and supporters welcomed me to the San Fernando City Mesoamerican Ballgame team practices. Art historian and *ulama* scholar Manuel Aguilar-Moreno generously took time to speak with me about the game and continues to include me in the many events that he organizes in Los Angeles about the game and broader Mesoamerican history and culture.

Alex Colston took on the job of providing a structural edit and a line edit for *Bounce*. The coherence and sharpness of the final manuscript owe much to his work and guidance. Many dear ones lent me their eyes along the way. Charlotte Kent, Lisa Haber-Thomson, and Cathy Fairbanks read chapters at crucial moments. Stephanie Boluk and Pat LeMieux invited me to share the manuscript with their class and gave me pivotal feedback right as I headed into the last round of substantive revisions. Hannah Zeavin has spectacularly read every word of this book, multiple times.

I have written, spoken, and otherwise circled these words for years with the people brocaded on my heart. I have spent the better part of two decades walking around cities talking on telephones with Dac Anh Nguyen, Victoria Restler, Liz Glynn, Ruti Talmor, Lisa Haber-Thomson, Hannah Zeavin, and Catherine Wing. Our conversations across distances and over dinner tables combine the everyday and the extraordinary in ways that I would not be without. This project has been repeatedly buoyed by the media-slacker crew—Laine Nooney, Dave Parisi, Jacob Gaboury, Stephanie Boluk, and Patrick LeMieux. Even transplanted to *Discord*, we remain media slackers. My PhD cohort was an incredible group: Kouross Esmaeli, Matthew Hockenberry, Jason LaRiviere, Patrick Davison, and Kavita Kulkarni. I am so lucky to have shared rooms and sat around tables with the five of you. I take continual inspiration from Louisa Hall and Ivy Pochoda, dear friends and former traveling companions in the world of squash who made their turns toward writing well before me. In different moments over the years, Jonas Becker, Christy Spackman, Xiaochang Li, Marissa Dennis, Jessica Feldman, Liz Koslov, Victoria Restler, Michael Braithwaite, Therese Workman, Claire Lehman, Thomas Page McBee, Hi'ilei Hobart, Greg Gagnon, Christina Xu, Seda Gürses, Noah Feehan, Harlo Holmes, Matt Whitt, Jesse Green, Cory Woolman, Margaret Elias Gerety, Kate Levine, Avra Van der Zee, Jérémy Grosman, Becca Franks, Luca Benedetti, David DiGregorio and Mariechen Danz have kept me company in thinking, art making, performing, presenting, writing, and living. While I was writing about the dynamics of bodies in motion, my own body broke down on numerous occasions. In New York, Christine Bratton, Jeremy Quinby, and Maria Holm showed me how to reorient, repattern, and rebuild. In Los Angeles, Katrina Umber, Kasey Lockwood, Akua Two Hawk Maat, Piltzinkoyotl, and Tri Chung have introduced me to new ways of holding and healing.

It has been almost two decades since I first landed in Los Angeles to begin an MFA program at CalArts. From our time as roommates driving up and down the 5 to CalArts to our kids' birthday parties, Liz Glynn has been there every step of the way. Since my return to Los Angeles in 2016, Ivy Pochoda, Eve Ruether, Corey Fogel, Becca Lee, Esme Germaine, Annie Powers, Luke Fishbeck, Sarah Rara, Yi Sheng, Brian Taylor, Jin Maley, Omid Arabian, Mojdeh Zavosh, Sylke Meyer, Kimberli Meyer, Katherine Rand, and Amanda

Perez have all helped make this city into a home. Cathy Fairbanks and David Zuttermeister and I laughed our way into being "climb over the ladder across the back wall" neighbors. They opened up *A Dog's Breakfast* for me to have some focused hours around the corner from home in the final stages of writing. Katrina Umber, Jesse Robinson, and their boys are just a few blocks away. Calder and Silas reminded me that ball piles are for diving into and playing with. For four years, my dear friend Kouross Esmaeli was a colleague at the Claremont Colleges and part of our neighborhood compound in West Adams. He and Khaled Fahmy make my heart happy.

As I was drawing this project toward its conclusion, I gave birth to my son, Elijah Miles Wing. The task of concluding a project like this while caring for an infant is possible only with the support of others. Thank you to my parents. And to Lisa Haber-Thomson, Victoria Restler, Christina Xu, and Catherine Wing, who came to stay at various moments in the tender early times. Samantha Fett, Audra Shapiro, and Morgan Campbell supported my recovery and search for new rhythms. Leslie Rivera's love and care for Elijah allowed me to turn to the work of revision, manuscript preparation, and copyediting. Hannah Zeavin, Geoffrey G. O'Brien, and their son Malachi watched over E while I holed up with late-stage revisions in their basement. Ive Shipman and Freesoul El-Shabazz Thompson wound their way into my and Elijah's lives at just this moment and brought brightness and lessons in compassion, community care, and love that inflected the spirit of this project in its final stages.

To my family: My aunts Ellen Birenbaum and Mary Roman, who let me retreat to their home to write with Easy and Roxy Roman, the best of the gamboling dogs, for company. My brother, Ethan Wing, who had seen me up and down some snowy mountain roads and his son George. My sister, Catherine Wing, again, and Steve Smith and Will Wing-Smith. My brother, Matthew Wing, and Caitlin Lewis, who have been on hand to offer encouragement, perspective, and the location of the best crumb cake, bagels, and babka in New York City. Ernestine Solomon, who taught me how to face the world. I write and live with her in mind and heart. My parents, Rusty Wing and Audrey Strauss, whose love and support is unwavering. You are all, in all of the most important ways, my occasion for being here.

Finally, Elijah. I was waiting. You are here!

Introduction: Long Past Play

It may be said there are things of more importance than striking a ball against a wall. There are things indeed that make more noise and do as little good, such as making war and peace, making speeches and answering them, making verse and blotting them, making money and throwing it away . . . He who takes to playing Fives is twice young. He feels neither the past nor the future "in the instant." Debts, taxes, domestic treason, foreign levy, nothing can touch him further. He has no other wish, no other thought, from the moment the game begins, but that of striking the ball, of placing it, of *making* it.

—William Hazlitt, "The Death of John Cavanagh," *The Examiner*, Feb. 7, 1817

Chasing History

I went in for all kinds of ball play from a young age—wiffleball, baseball, basketball, volleyball, and most intently, squash. In 2007, in my late twenties, I came to the end of a brief career as a touring women's professional squash player. As I began pursuing a master's of fine arts in photography and media at California Institute of the Arts (CalArts) in Los Angeles, I came across a black-and-white photograph in a small book in the school's library. The image clustered five exaggeratedly tall and tilting basketball hoops in Cadman Plaza Park in Brooklyn, New York, a park that I played in as a child. The photograph was of *Higher Goals*, a temporary sculptural installation made by the artist David Hammons as a commission for the Public Art Fund, a nonprofit that funds public art in New York City, in 1986. Hammons studied art first at CalArts, and then at Otis College of Art and Design in Los Angeles. In the 1970s, he moved to Harlem, where he built an art practice grounded in symbolic play and the everyday materials of Black life. For

Figure I.1
David Hammons, *Higher Goals*, 1986. Photograph by Pinkney Herbert and Jennifer Secor.

Higher Goals, Hammons attached basketball hoops to the tops of telephone poles. He collected over 10,000 bottle caps from the streets of New York City to wrap around the poles in geometric patterns; each pole nodded to different African, Islamic, and Florentine design motifs.[1] The work is a one-liner and a political treatise. Towering two to three times higher than a standard basketball pole, the hoops are literally out of reach at first glance. In the face of the many intricate material and immaterial factors that structure access to, and desire for, goals and games of all kinds, it's a sharp laugh at aspiration. At the same time, the work testifies to the artist's belief in the fundamental necessity of absurdist acts and improbable pursuits. Hammons has described it as "an anti-basketball sculpture."[2]

The date of the work caught my attention. In 1986, I was playing tee-ball weekly in Cadman Plaza, alongside a pack of other six-year-olds. Enthralled by the New York Mets' dream run to the World Series, I was determined to be a professional baseball player when I grew up. Hammons built the installation on site over eight weeks, documentation from the Public Art Fund attests, and it was on view for almost a full year from April 10, 1986, to March 27, 1987.[3] Staring at the photograph of the work, I tried to retrieve some long-forgotten memory of encountering these tall hoops. Their unreachable height returned me to my six-year-old self's burning desire to one day play for the Mets, long before I understood and disconsolately accepted professional baseball's strict gender segregation.

In the background, the marble columns and clocktower encased in scaffolding of Brooklyn's Borough Hall touched off my longing for a city that I knew by heart. The bottlecaps studding the poles told me: here was a fellow collector of scattered small things. The work and my past self's proximity left me wishing that I had noticed it then—to be able to locate myself in relation to it, to have participated in its history. In the place of memory, I found a space of speculation—unknowability shaped by known contours. Like the more general task of living with, and in relation to, multithreaded histories, the task of recovering human activity is a continual confrontation with the unknowable, an ongoing gathering and stringing together of a few known things. For many years now, from distant and disparate historical moments I have been stringing together instances of bounce and scenes of ball play. I've thought for a long time about how balls, bodies, and ball courts—like bottle caps, telephone poles, and basketball nets—hold history and open spaces for speculation.

Around when I first happened upon the photograph of *Higher Goals*, Allan Sekula handed me a copy of C. L. R. James's *Beyond a Boundary*. Allan is an American Marxist artist and theorist whose photographic, film, and written works tell intimate and sweeping stories of global labor in the vein of James, a Trinidadian Marxist historian best-known for writing *The Black Jacobins*.

At least, my memory is Allan handed me the book, but in retrospect, it is unlikely that he gave me a physical copy. It's likelier that he visited my studio, where I was making large-format photographs of portable glass squash courts. Looking at these, and listening to me talk about how I was wrangling the resonant aspects of this colonial sport into aesthetic form, he told me about James's work, saying that he thought I might want to read the last book he published, *Beyond a Boundary*. This was a work, he told me, about how one can come to know the world through a ballgame. Allan's gesture of handing James's book to me, figuratively or otherwise, is a common one—one that I now do in my role as a teacher in turn. In this way, knowledge is regularly handed off and passed around—a certain routine, a ritual in educational contexts, and beyond. Recommending a book to a friend, a lover, a child, a parent, a neighbor, a student, a class, a colleague, a group, a fellow, a stranger is a common gesture. When a recommendation lands true, it is a gift.[4]

James begins the preface of *Beyond a Boundary* with the question, "What do they know of cricket who only cricket know?"[5] His autobiographical framework unfurls an understanding of how sport situates the self in the world, especially through the formative power, deep pleasure, and embodied politics that gather around sport as a cultural practice. Written as he struggled, in the wake of his rejection of Trotskyist Marxism, to articulate his transformed worldview, the book reflects James's new recognition of the formative role of his childhood in Trinidad, and cricket in particular, on his politics. Playing, watching, loving, following, and in all ways deeply knowing cricket gave James his first, embodied relationship to politics. His account ranges across the dynamic formations of race, class, colonization, education, politics, and art—and from his childhood to England, where what had been latent and more or less unarticulated, surfaced into explicit political consciousness. The incommensurable contradiction of James's deep love for a colonial game is at the book's heart. He made the game his own. And he was made by the game. This was what I recognized.

This was what playing squash had been for me: fundamental, formative, a source of reliable joy and profound liberation and, at the same time, a source of some of my earliest, initially unarticulated, and embodied understandings of the world. Playing squash, I learned to coordinate eye and hand, arm and leg to racket and ball. I attuned to feedback loops and ricochets, to how to take, make, and give others space during the back-and-forth of a point. I learned the rules of physics and the rules of the game, eventually also learning the way those rules were ensnared in the workings of gender, race, and class in the United States as systemic forms of oppression, in the ongoing presence of colonialism in today's geopolitics, and in the circulation of cultural styles and stereotypes by and between players, organizers, sponsors, and fans from around the world. James and Hammons are two among many artists and thinkers whose works have guided my own perambulations and aided my own working through of the contradiction that we both make and are remade by whatever we take into our hearts.

Before I chased the history of bounce, I spent some twenty years chasing squash balls around four-walled courts all over the world. Hoping for a future tennis partner, my father sent me for lessons at the Heights Casino at age seven. The Casino was a members-only racket club in our upper-middle-class Brooklyn neighborhood in New York City. A coach there quickly converted me from tennis to squash. From that point on, I pursued squash without question; I loved to play. I loved finding the edges of my capacity over and over. I coaxed my parents into letting me compete at age eleven, which meant spending innumerable weekends driving up and down the East Coast for tournaments held at clubs and college campuses.[6] I was good, although never the best. I pursued squash through junior national and international competitions, college, and years on the women's professional tour. At the top of my game, when I lined up a shot, my eyes could find and follow the ball's two yellow dots as they rotated in and out of sight. My body still recalls that experience of time—one expanding across the entire arc of a ball's bounce. I would wait, and then stretch forward in the front-left corner, deciding when to meet the ball and where to send it.

A squash ball is a small thing, literally and figuratively. Squash is a minor sport.[7] Developed in England in the early nineteenth century, it was spread around the world by the British Empire. The game's colonial history is reflected in the current women's and men's world rankings, with top players hailing from England, Egypt, Australia, Malaysia, India, Canada, Hong

Kong, and New Zealand, along with a scattering of European countries, and on rare occasions North America. Squash was the first British game played from its outset with a rubber ball. The sport is named after the ball's elastic, pneumatic, squashable nature. Today's squash balls are similar in size to table tennis balls and most often made of black rubber, although sometimes they are white for better visibility on glass courts and television screens. Made in large batches, the balls are marked with colored dots to indicate their degree of bounce, stamped with the company brand, and boxed up and sold in pro shops and on Amazon. Unlike table tennis's lively, hollow, celluloid sphere, when a squash ball hits the floor, it lands with something of a plop. Until the rubber is warmed up through repeated contact with racket strings and court surfaces, it has almost no bounce. At unpredictable moments in the middle of a point, a ball will break at the seam, revealing tiny beads of hot condensation.

This ball is an industrial object. As such, it's a technology whose carefully calibrated specifications allow (or are supposed to allow) players to count on its behavior. This specified reliability is what I call throughout this work *true bounce*—a bounce that is regular, predictable, and reliable enough for players to count on its behavior. As for many other kinds of balls, the bounce of a squash ball is regulated: the World Squash Federation (WSF) specifications for a squash ball include narrow parameters for weight, diameter, stiffness, seam strength, rebound resilience, and percentage of allowable change under high- and low-temperature conditions. Squash players, like other all athletes, spend thousands of hours training their bodies and timing their gestures around their sport's particular kind of bounce. The national origins of today's top-ranked squash players reflect the way that the game's spread followed the British Empire. Accordingly, the true bounce, on which the squash world relies, is part of a longer story. These small, stamped, hollow, usually round, elastic objects that I came to know so intimately in my early years tell a sweeping story of the relationship between rubber, bounce, and sport. And chasing them around the squash court, year after year, attuned me to the continual collapse and creation of vast infinities that occur with every interaction.

Long after I put my racket down, I have continued to pursue the ball and the bounce, asking questions about how things move and what movement carries—how objects and environments and gestures hold, carry, transform, paper over, and cast off culture and history. In the opening of their edited

Figure I.2
Dunlop Pro squash ball. An official ball must have a diameter of between 1.56 inches to 1.59 inches.

collection *Contemporary Absurdities, Existential Crises, and Visual Art*, Charlotte Kent and Katherine Guinness write, "There are absurdist acts, which offer experiences, provocations, and speculations. They undermine known values and authorities; they point and prod to keep the momentum going. They ask without wanting an answer. They reveal how doing things that are habitual may no longer work. And in these troubled and troubling times, some have called ours, this, an age of absurdity."[8] Chasing a small rubber ball around rooms and across histories seems a gloriously absurd activity. *Bounce*, then is about bounce—about movement and its conditions—about following something as far as it can go and learning to let go and leave alone.

This work tries to depict a certain feeling of time—one that expands and contracts across the arcs of moments. In its early days, when a back injury stopped me from moving in familiar ways, the project itself began to bounce. The inquiry ricocheted from squash to handball, calculated the trajectory from court games to computer games, until I found I was following an array of bouncing balls. This array outlined a path through histories of physical and electronic ballgames—bouncing across the spectrum of play, game, and sport, and into the domains of physics, industrial science, animation, and computing. The research led to *ulama*, an Indigenous ballgame first played over three thousand years ago; to the bounce in Early Modern era court tennis; to the bounce programs of the 1950s and other precursors to *Pong*; and to *FIFA Football*, a sports video game series first released in 1993. This work traces the throughlines between each of these instances of bounce, and make a wish: to have a relationship with history that experiences its gravities and momentums—to make palpable how different images, objects, beings, and environments hold knowing and not knowing together. Organizing this collection of instances creates, as it were, both a historical arena and a collective bounce—one that frames and provisionally contains while projecting possible worlds.

Games of Bounce

Almost all cultures engage in some form of ball play. Playing with balls and other kinds of bounding objects is a basic activity. Through it, humans, as well as other animals, hone what computational neuroscientist Beau Cronin calls "the quotidian spatiotemporal genius of the human brain."[9] Ball play is a way to learn, particularly to learn what kinds of motion to expect from one's own and others' physical bodies. When play turns to game, these activities become narrative encounters of the self with the social. When ballgames become institutionalized as sport, players and spectators alike perform both ideologies and cosmologies with their bodies. In other words, ballgames model worlds. There is a long history of people using bounding objects—and ballgames—for modeling worlds. Across the centuries, these in-between objects, with their rule-bound rituals, act as dynamic material metaphors for political, social, mathematical, and physical principles and systems. In the process, they naturalize theories of social and natural order. What comes of simple games of bounce that serve as models for complex societies? To answer this question, this book turns to an array of bouncing balls that underwrite representation, interaction, and simulation. The answers draw out some of the material histories, cultural contests, and worldviews that are embedded in the instruments, architectures, and gestures of games of bounce.

In what is now called the Americas, first the Olmec, and later the Mixtec and the Maya cultures, developed entire cosmologies around a game played with a rubber ball and rubber bounce. Played for over three thousand years, in many iterations, the ball stands for the sun, and playing the game is understood to keep the universe in balance. As the grandest iterations of the Mesoamerican ballgame reached their apex in the Early Modern era, with thousands of ball courts built across the region, on the other side of the ocean, priests, kings, students, scholars, aristocrats, merchants, innkeepers, and artisans were gathering around the younger game of court tennis. As tennis courts proliferated across what was not yet Europe, William Shakespeare and Michel de Montaigne used the game as a metaphor for war and conversation (violent and amicable exchanges), Galileo and Isaac Newton used it to think through planetary motion, and Jacob Bernoulli picked it up to demonstrate early probability mathematics. While traditions of playing both ballgames persist today, their cultural power has been attenuated by the rise of new forms of ball play.

Over the last 200 years, a host of other ballgames have come to dominate the popular imagination. Huge swaths of airtime; large volumes of ink; and vast investments of economic, social, and scientific resources are given over to the dramas of soccer, basketball, baseball, American football, tennis, golf, rugby, cricket—the list goes on. With the rise of live broadcast television and the invention of digital computing in the twentieth century, ball play was redirected to electronic space. This created new conditions of bounce, changing the terms of both bodies and play. Lists of the most watched television broadcasts of all time inevitably include multiple sporting events—FIFA World Cup finals, the Super Bowl, the opening and closing ceremonies for the Olympics, the ICC Cricket World Cup final, and Muhammad Ali's final fights. Some of these lists are comprised almost exclusively of ballgames.[10] In the United States between 2020 and 2024, NFL games alone accounted for anywhere between 72 and 93 of the top 100 TV broadcasts, with college football, other sports, and news and politics (which are often covered as if they were a form of sport,) taking up most of the remaining slots.[11]

Many of the ballgames offering models of worlds were, in turn, modeled—much more literally—by sports video games. In the world of computing, the most iconic of the early video games, *Pong*, is an extreme abstraction of tennis and table tennis. *Pong* is just one of countless electronic games organized around bounce, and bounce logics prevail in countless games that do not explicitly simulate sports. Indeed, the history of computing is littered with bouncing balls. Bounce programs made their initial appearance when computing machines were first attached to oscilloscope screens, shifting computing machines toward tasks of real-time interaction and simulation. They proliferated from there as the challenge of registering, measuring, simulating, and representing different kinds of collisions became essential to a range of computing contexts: from robotics and molecular modeling to video gaming and computer animation.[12] Today, the field of computer graphics underpins a massive gaming industry and lies at the heart of many routine operations of daily computing. This is part and parcel of the unglamorous but indispensable task of giving our computers visible graphic user interfaces. In this context, modeling bounce brings together the domains of simulation and representation. Both physics engines and animated avatars (from Luxo Jr., Pixar's hopping desk lamp mascot, to your blinking cursor) have bounce as their base.

Put simply, bounce lies at the heart of hundreds of thousands of physical games, digital games, sports, and everyday actions. This book does not aspire to address all, or even most, of these games and actions in detail. Instead, I pair emblematic instances of everyday and extraordinary kinds of bounce. This aims to map a theory of bounce across a *longue durée* by bridging the gap between electronic and nonelectronic games, broadening the scope of game history. Today, the word game often acts as a shorthand for a video game. Even when this is not the case, the extant writing about games, with a few key exceptions, holds itself apart from writing about sport. In contrast, *Bounce* draws games, play, and sport together to reveal a dense entanglement of technical, cultural, and media histories, one in which bounce operates as a fundamental mechanism of mediation. And what bounce primarily mediates, because ballgames and bouncing balls serve as a foundation for both global broadcast spectacle and computing's common sense, is liveness. This, then, is a history of motion practices—a long look at live, lively, and real-time performances as arenas and enactments of culture.

Figure I.3
Ball Ellipsis (Globe Stress Ball, Fun Express Globe Stress Ball, and SAGE 50 Years Earth Stress Ball), 2024. Photograph by the author.

Bounded, Bounding, Bouncing Objects (or Balls)

The object most associated with bounce is a ball. Balls are mostly round, but not always. They can be solid or hollow. Occasionally, they are woven such that air passes right through them. Today, balls are commonly made of rubber—solid rubber spheres, hollow rubber cores wrapped in felt, blown-up rubber bladders secured to leather casings—because of that material's extraordinary elasticity. Rubber's elastic properties give it an uncanny ability to mediate impact and to make that impact measurable. Yet they can also be made from cork, wrapped in cloth, or stuffed with various materials—like wood, reed, gutta percha, foam, celluloid, and other kinds of plastic. Balls can be composed of almost anything at hand, really. Pelé famously practiced with mangoes. Just so long as the object behaves as desired—so long as it flies off a club or bat, spirals in the air, or bounces—it's a ball.

The word "bounce" developed from the Middle English word *bunsen*, meaning "to thump or thwack." In the early sixteenth century, an interjectional appeared. A person might shout "Bums!" "to imitate the report of a gun or other loud sudden noise, and (a little later) to express sudden or violent movement." In Europe, early cannon use was often for the purpose of imposing psychological, rather than physical, damage. At this time, this previously rare verb began to be used more frequently. In its earliest use, to "bounce" meant "to beat," "to knock, bang," or "to make a sudden or violent movement of a bounding nature."[13] "Bounce" began as a word for expressing the experience of shocking sonic and physical impacts. It came to describe the type of movement produced through such impacts—the quality necessary to survive those kinds of interactions, and even thrive in their wake. Today, the conventional meaning of the verb "bounce" is "to move with a sudden bound . . . to bound like a ball; to throw oneself about."[14] This definition of "bounce" can barely be thought of without a ball. Indeed, today, anything that bounces in some sense becomes a kind of ball.

There is a long tradition of balls being used as exemplary bodies in the natural sciences—so much so that it is possible to say that much of Western physics has been a theory of bouncing balls. Historians of science, such as Edith Dudley Sylla and Marco Beretta, describe how ball sports have been used to conceptualize key concepts in physics and mathematics—from planetary motion and optics to probability theory and Markov chains.[15] During the "On the Ball" roundtable moderated by historian of science

D. Graham Burnett, writer Thomas Bindle—acting as the voice of the "last out" baseball autographed by Mets left fielder Cleon Jones, who caught the final pop fly in the 1969 World Series—put it this way: "Balls, we must recall, have long played a privileged role in the history of both matter theory (atomism) and physics (dynamics)—dating back to antiquity. In fact, you could go so far as to say that for most of human history 'science' has been 'ball theory': falling balls, balls rolling down inclined planes, billiard balls, balls of fire. 'Getting the ball to tell its story,' in this sense, amounts to nothing less than *knowledge of nature*—and, thereby, fate."[16] It is not just ballgames, but balls themselves that model worlds.

The Newtonian corpuscular worldview was sidelined by the rise of quantum mechanics, and yet ball and bounce maintained their centrality. Bounce became a key method for gaining knowledge of objects through techniques like echolocation, sonar, and radar. These produced knowledge of the limits, shapes, sounds, and interrelations of things. Brian Hayes, the former editor of *Scientific American*, argues that one result of computing is that simulation has joined deduction and empiricism as methods for producing scientific knowledge.[17] He explains that in the context of computing, bounce is one possible answer to the question of what happens when two objects collide. When computer objects are constructed and set in motion, they will not interfere with each other until the program includes rules of interference (based on rules of physics). As Hayes puts it: "Nothing bounces unless you tell it exactly where and when to bounce."[18] When a computer object bounces, it becomes a new kind of ball.

Whether physical or virtual, balls operate as exemplary test objects for demonstrating and simulating theories of natural order. This is because, in part, they are one of the most basic and common play objects to begin with—central both to simple children's games and spectacular sporting rituals. Tossing a ball against a wall, playing catch, practicing dribbling, and other kinds of informal ball play create an embodied knowledge of the world. Practical ball play coordinates one's body and movements to the conditions of the ball and the play space. More formal ball play weaves together a different coordination of natural order and social order. All sports are aleatoric, meaning that they are structures of planned chance. Planned chance is a reliable kind of uncertainty. When a game begins, we do not know who will win and who will lose. We do know, however, that there will be a result

Figure I.4
Serena Williams mid-serve. "Serena Williams vs. Li Na Full Match | 2013 US Open Semi-final." US Open Tennis Championship YouTube Channel. Screenshot by the author.

at the end of the play. There will be a winner and a loser (or, occasionally, a tie will happen). What a physical ball introduces is a second, more uncertain kind of uncertainty into the fray. A ball dances along the edge of our predictive capacity, always almost, but never fully, under control. And, at least in the Anglophone sporting world, this second kind of chance—the chance of the ball—is especially important to our contemporary understanding of play. Other kinds of contests are raced, run, rowed, and swum; wrestled, fenced, fought, and boxed; timed, weighed, measured, and judged; but ballgames are *played*. Only an athlete who contends with balls (or pucks, or shuttlecocks, or other third-party objects)—and the kinds of distance, mediation, and indirection and misdirection that they introduce—earns the title of "player."

To-and-Fro

The first known textual reference to tennis involves a story about what might be called an anti-ball. It appears in a dialogue written by a thirteenth-century monk. He was a clerk of the holy orders known for being unable to hold things in his memory. One day, a devil offered the clerk a magic stone that would allow him to know everything so long as he held it in his hand. The clerk took the stone in his hand and became a perfect scholar. Soon afterward, he fell ill and died. As tennis historian Heiner Gillmeister relates it:

> A band of demons snatched his soul from his body and rushed it to a horrible valley steaming with sulphorous vapours. Here the devils divided into two teams, each taking a position at one end of the valley, and began to play. "And those standing at the one end hit the poor soul after the fashion of the game at ball, and those at the other end caught it in mid-air with their hands." The devils' claws had the sharpness of iron nails, and so tortured was the poor soul that no martyrdom imaginable could compare.[19]

The parable can be read as a lesson about the relationship between knowing and passing. An object that has to be held continually to serve its purported purpose, like the magic stone, is an anti-ball. It only is what it is so long as it is not passed. The clerk's punishment for having mistaken knowledge as something to be clung to and possessed—rather than passed around and along—was for his own soul to become a ball. His soul was condemned to be batted forever back and forth by the demons' piercing claws.[20] Knowledge is not meant to be—and cannot be—securely held and contained in a single individual—a good caution to all of us collectors. Knowledge exists through passing. It exists through a to-and-fro.

To-and-fro dynamics are present in many theories of self and social formation. Repeatedly hitting a ball against a wall—or with an opponent across a net substituting for a wall—is an organized echo of the *fort/da* game that Sigmund Freud describes his grandson Ernst playing.[21] Ernst's game consisted of repeatedly throwing an object away, announcing it to be "gone." Freud suggests that Ernst initially created the game to enact his mother's absences and thus master his inability to control her comings and goings. From these games, Freud theorizes that our compulsion to repeat is related to our desire to master loss and uncertainty. Later, he reports another iteration: Ernst throws toys that he is displeased with on the floor and tells them to "Go to the war!" The objects now stand in for the father who has gone to the front. This variation also enacts control over a desired going (so the child can have his mother to himself).[22]

Fort ("gone") is a game on its own, and mostly what Ernst plays. Yet when Freud observes Ernst playing with an object attached to a spool of string, which allows him to recall the object at will—to make it *da* ("here")—he believes that he has witnessed the completion of the original game. True mastery of another's comings and goings—or at least of one's feelings about another's comings and goings—must include a gesture that, at least symbolically, makes the person return. In the scene of the game's completion, agency is amplified by way of mediation—the string—in all its material specificity. In the case of court games, the role of the string is played by either the wall or another player. They both recall the ball to themselves. This recall's condition depends on interrelation between the qualities of the ball, the wall, and our own and our opponent's gestures. The pleasure, in part, lies in testing our capacity against the capricious physics of others.

The theory of communicative play developed by British children's psychoanalyst D. W. Winnicott offers us another way to think about this. We build relationships with others through playing with and between objects—balls, walls, bodies, words. This is experienced as simultaneously independent of us and mastered by us.[23] For Winnicott, play is a precarious intermediate space between the inner, subjective, psychic self and the external, objective world: "The thing about playing is always the precariousness of the interplay of personal psychic reality and the experience of control of actual objects. This is the precariousness of magic itself, magic that arises in intimacy, in a relationship that is being found to be reliable."[24] The precarious space of play is produced with, and through, the use of transitional objects. Infants simultaneously experience omnipotent control over these objects, and yet they experience the objects—a developmental achievement, according to Winnicott—as having an external reality of their own. Through play, infants build a system of references and relationships based on continual testing of what, in any given moment, counts as or feels like "me" versus "not me." Crucially, for Winnicott, there can be no communication, and thus no healthy living, outside of play: "Only in playing is communication possible; except direct communication, which belongs to psychopathology or to an extreme of immaturity."[25] He offers us a qualitative judgment in favor of those sorts of communication that occur via rebound and ricochet; lucky and unlucky bounces; the sudden swerve. In Winnicott's model, we find a scaffold through transitional objects. Our early attempts to master our relations to ourselves and others scale up. All culture—art, sport, religion—builds on this first structure of relations between self and world that is formed through playing with objects—physical and symbolic. While in ball sports, balls are flung forth and bounce back via mediating floors and walls, in language, it is words that fly. John Durham Peters describes Socrates, the first theorist of communication, worrying over "writing's ability to throw voices."[26] Communication, then, is a problem of world building. And ball play is a way of considering how we build worlds with others through basic physical and symbolic interactions.

Michel Serres's theory of the quasi-object has similarities to Winnicott's transitional object, while using markedly different language that aims to describe a different scale of relations. Serres uses a ball as his primary example of a quasi-object, writing that a ball "is not an ordinary object, for it is what it is only if a subject holds it." A ball is an object defined by having a subject. It is simultaneously "in the world" as an object, even as it "marks or designates a subject, who without it, would not be a subject."[27] It thus operates as "an astonishing constructor of intersubjectivity."[28] Quasi-objects, which are also quasi-subjects, distribute subjectivity and meaning. They enact collectives through the passing of the *I* and the decentering of self. Serres writes:[29]

> The ball isn't there for the body; the exact contrary is true: the body is the object of the ball; the subject moves around this sun . . . The "we" is less a set of "I's" than the set of the sets of its transmissions. It appears brutally in drunkenness and ecstasy, both annihilations of the principle of individuation. This ecstasy is easily produced by the quasi-object whose body is slave or object. We remember how it turns around the quasi-object, how the body follows the ball and orients it. We remember the Ptolemaic revolution. It shows that we are capable of ecstasy, of difference from our equilibrium, that we can put our center outside of ourselves. The quasi-object is found to have this decentering. From then on, he who holds the quasi-object has the center and governs ecstasy. The speed of passing accelerates him and causes him to exist. Participation is just that and has nothing to do with sharing, at least when it is thought of as a division of parts. Participation is the passing of the "I" by passing. It is the abandon of my individuality or my being in a quasi-object that is there only to be circulated. It is rigorously the transubstantiation of being into relation.[30]

I can almost feel Serres remembering the moment when Emmanuel Petit sent a corner kick to Zinedine Zidane, who in turn directed the ball with his head into the back of the net during the 1998 FIFA Men's World Cup final. Or perhaps he is recalling a different, perhaps gutting moment when the pass did not go as planned. For Serres, a *we* emerges not from an accumulation of *I's*. *We* emerge from passing—or bouncing—the *I* between bodies. There are echoes here of the wretched clerk's soul being batted back and forth between demon ballplayers for all eternity. This picture of the world depicts an existence that is simultaneously enacted and threatened with eradication through the ongoing decentering and circulating of the *I*—through the "transubstantiation of being into relation."[31]

Taken together, Serres's ecstatic decentering of the "I" and batting about of the clerk's soul prompt some questions: Does the matter of the ball (be it rubber or soul) matter? If the matter matters, how does it matter? How do conditions of bounce affect things? And what architectural, environmental, regulatory, and gestural regimes determine these conditions? How, finally, does all this impact rebound and redound to the kinds of *I's* passed around, and the kinds of *We's* created through these passings? From these questions— and from these theories of to-and-fro—an understanding emerges that bounce never belongs to a single object, surface, or environment. Instead, it is a property distributed among things: a common name for those kinds of interactions through which entities emerge with their respective shapes and speeds relatively intact, and with their identities provisionally and temporarily confirmed.[32] That is to say, they survive.

In the following chapters, I show how different kinds of bounce operate as cultural techniques: specifically, as special kinds of interactions for constituting and confirming identity. Through bounce, objects are produced as whole, intact, and yet flexibly recognizable entities. The nouns emerge from the verbs, the objects from the actions, if only for a flash—if only for some apparently brief or extended moment.[33] This capacity for bounce interactions to confirm identity makes bounce a key metaphor for those interactions through which knowledge is produced. When we know the bounce of something, we have a knowledge of identity in terms of observed and observable behavior under a given set of conditions, where interaction is conceptualized as impact.

There are a range of possible outcomes of interactions besides bounce: breaking, fusing, passing by or passing through, orbiting. Consider Wile E. Coyote, Marvin the Martian, and other cartoon characters who regularly accidentally blow themselves up. These are other mechanisms through which identity is created, challenged, changed, denied, and put in dynamic relation. Yet bounce is the mechanism through which identity is, at least apparently, maintained and thus confirmed through the interaction. A consideration of bounce, then, reconfigures a classic question about the relationship between techniques of observation and the subjects and objects that those techniques produce. When the emphasis shifts from the object to the interaction, in and beyond the world of Newtonian mechanics, knowledge appears fundamentally mediated by bounce and its variants. The ways that light, sound, and particles respond to impact have been taken as a central object of study in physics. These objects of study have turned and transformed their attendant interactions—reflection, echo, and ricochet—into techniques of observation and technologies of capture and transmission. In turn, we use ballgames as material metaphors to model worlds, and we use bounce (of light, sound, sonar, radar, radio waves) to produce technical models of bodies and environments and to transmit signals around the world. Following bounce, then, requires moving to-and-fro between bodies and media. These bodies appear in media and act as media. Through their examination, we better understand the various ways that bodies—physical and figurative, corporal and corporate—are produced, maintained, changed, and made to matter through mediating interactions.

Games and Media/Games as Media

On November 6, 2005, a commercial for the new Sony BRAVIA liquid crystal display (LCD) flat-screen television debuted during the broadcast of the Premier League football match between Manchester United and Chelsea. The "Balls" commercial, as it came to be known, featured a slow-motion montage of 250,000 delightfully unruly multicolored bouncy balls, cascading down the steepest street in San Francisco—unruly, or un-ruly, in the sense that they are not part of a rule-bound game, even as they make visible the "rules" of the physical environment governing their bounces, rolls, and ricochets. Jose Gonzalez's spare cover of the song "Heartbeats," by the Swedish electronic duo the Knife, infuses the scene with aching lyrics and a hypnotic rhythm—"To call for hands of above/To lean on,/Wouldn't be good enough,/For me no."[34] In the breathtaking blitz, the luminescent spheres bounce in and out of recognizability—both as rubber balls and as exaggeratedly large color pixels on viewers' television screens. This oscillation points to the screens themselves as the commodity of the ad. The song concludes, and the tagline "Color, like no other" flashes on the screen.

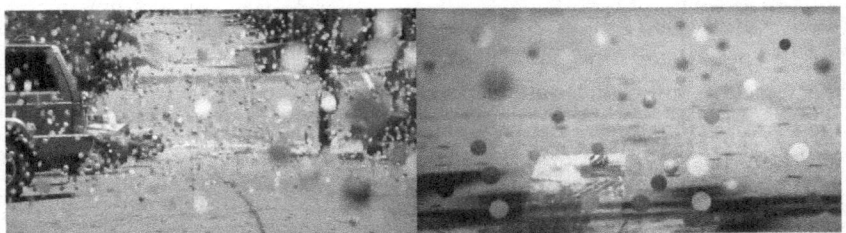

Figure I.5
Scenes from the "Balls" ad created by Fallon for the Sony BRAVIA LCD television screen. Most of the balls in these images are made of transluscent rubber tinted bright colors—blue, green, yellow, orange, red, pink, or purple. In the daylight, and thanks to the video editing, they appear to be glowing. Screenshot by the author.

The image fades briefly to black, only to fade back up to a black flatscreen; it rotates into vision out of the black abyss like the monolith in *2001: A Space Odyssey*. The ad promises the kind of pleasure that viewers may have taken from viewing the deliriously absurd scene of bounding, bouncing balls, let loose on a cityscape on their presumably cathode ray tube television screens. The ad suggests that this pleasure will accompany all their viewing experiences once they make the switch to the Sony BRAVIA LCD television.

To watch LCD television is to be awash in giddy, bouncing colors. Watching the ad on my laptop's LCD screen, I imagine the sentiment of the song's recurring chorus leaking through the television screens that connect Manchester United and Chelsea fans, heart and soul, to their beloved teams—a sudden, rising, subtle sense of the screens mediating their connection, contouring it, pixelating it. Perhaps what felt direct and present a moment before suddenly feels distant—suddenly not quite good enough. The ad's analogy of bouncing balls to bounding pixels, along with its debut during a Premier League match, present bounce as the key connection between the play of ritual ballgames like football (or "soccer," as Americans call it) and the transmission of this play to viewers. It is bounce that entangles ballgames and media broadcasts, mediating experiences of liveness for players and spectators alike. (Chelsea won the match 3–0, by the way.)

What does it mean to say that bounce mediates experience? One way to define mediation is as a set of cultural techniques that generate material and symbolic distinctions and determine what counts. The concept is often used to approach epistemological questions—questions about ways of coming to know the world. Different processes of mediation generate different ideologies. As Arjun Appadurai puts it, "Whatever the ideology of matter and mediation that defines a particular cosmology, it is in and through some such ideology that matter comes to matter."[35] The relationships between matter and mediation—the different ways that matter comes to matter—represent different conditions of communication and knowledge production. They shape ways of being in, and coming to know, the world.

John Durham Peters describes this realm of media studies as focusing on "media as modes of being" rather than on media objects or media institutions.[36] (Here, he draws on Elihu Katz's distinction among three traditions of scholarship on media and communication, which conceive of media as transmitters of information, ideology, and organization, respectively.)[37] The conception of media as modes of being has also been elaborated on by Friedrich Kittler and Bernhard Siegert, among others.[38] Kittler and Siegert both approach media studies as "a way-of-seeing field," emphasizing cultural techniques (or technics): practices such as reading, walking, cooking, viewing, and coding that "generate the very concepts that are then used to conceptualize these operations . . . threshold operations that process the exchanges between nature and culture." The framework of cultural techniques presents "being" as fundamentally mediated. From there, we can ask how identifiable things emerge through the enactment of physical and conceptual distinctions.[39] In other words, this tradition of media studies is fundamentally concerned with bounding practices—practices of making, unmaking, and remaking physical and conceptual boundaries.

The concept of a *boundary* (in the sense of "limit") and *bounding* (in the sense of "a delimiting action") have themselves been the subject of much scholarly work. Anthropologists and sociologists have used the concept to study relational processes, such as group formation and intergroup relationships. James Griesemer, Susan Leigh Star, and Geoffrey Bowker elaborate on the idea of *boundary objects*. These objects are simultaneously plastic enough to be interpreted in diverse ways and immutable enough to maintain their integrity.[40] Historians of science have demonstrated the constitutive nature and political consequences of the bounding practices for the empirical project.[41] Physical and electronic ballgames and bouncing balls are objects; and they are subjects. (In particular, they are the subject of this book.) This mediation between subject and object outlines a linguistic (and conceptual) slip from noun to verb—from boundary to bound to bounce, from the name for a limit to a name for the gesture that creates a limit—rebounding and redounding back to a name for the energetic, leaping movements across a surface that demonstrate that limits exist. This slip is partly literal, partly a pun—a bit of a one-liner. All ballgames have marked boundaries. Balls themselves are, of course, playfully bounding objects. So to write about ballgames and bouncing balls is to write about bounding objects and practices. It is to write about, in a word, mediation.

Games are media. Media historian Lisa Gitelman challenges readers to think about how all media work as the simultaneous subjects and instruments of historical inquiry.[42] Ballgames are regular subjects of both daily and long-form journalism, documentaries and dramas, video games, fantasy sports leagues, and so on. They are attended by a plethora of material media: scorecards, player cards, team uniforms, and other branded paraphernalia. Critical sports studies scholars have offered numerous histories and cultural analyses of diverse kinds of casual, amateur, and professional play and the kinds of fandom and spectatorship that attend them. David Rowe, Sut Jhally, Varda Burstyn, Mary McDonald, and Jennifer Sterling have all analyzed aspects of what Jhally calls the "sports-media complex"—the often disavowed politics behind massive cultural and economic investments in global sport spectacles.[43]

While ballgames have occupied a central place in sport studies, sports video games have not been a primary focus of game studies scholars. There has, however, been a steady increase of attention both to sports video games and to the transformation of video games into e-sports in recent years.[44] This work—led by Mia Consalvo, Abe Stein, Emma Witkowski, T. L. Taylor, Henry Lowood, and Raiford Guins, among others—has laid the ground for understanding the intersection of sport, video games, and media, taking up similar questions to those in sport studies. Yet they also examine how individual performances—and institutional structures—are shaped by conditions of contemporary networked media forms ranging from social media to live streaming.[45]

While it is apparent how games are *subjects* of historical inquiry, it may be less immediately apparent how they serve as *instruments* of historical inquiry. Unlike many kinds of audio, visual, and electronic media, games and sports do not capture or transmit sounds, images, or information in a manner easily organized and stored in archives. Instead, as Marshall McLuhan identified early on, games are "media of interpersonal communication."[46] Most simply, games are media through which various kinds of communication between people take place. But techniques and technologies of communication—be they a biological technique, such as a whisper, or a broadcast technology, such as radio or television—offer models of culture. In his discussion of games, McLuhan returns again and again to the word "model," calling them "dramatic models of our psychological lives," "faithful models of a culture," and "live models of complex social situations."[47] As live dynamic models of culture, games become instruments of historical inquiry. Different games and sports (and the various ways of playing and watching them) capture and transmit organizations of space and time, rules of play, body techniques, collective strategies, patterns of movement, individual and collective values, and various competitive and collaborative ways of interacting with others. Where Serres describes the "*I*" getting passed around, McLuhan's "models" suggest how the "*We*" gets passed down.

Art history, cinema studies, and media studies have all examined and theorized questions of spectatorship. But sport generally fall outside this tradition's scope. In "The Work of Art in the Age of Mechanical Reproduction," Walter Benjamin uses the example of "a group of newspaper boys, leaning on their bikes, discussing the results of a cycle race" to claim common ground between cinematographic and sporting technologies. He argues that, in the cases of both film and sport, "everyone watches the performances displayed as a semi-expert."[48] For Benjamin, the sports fan, in occupying the position of a semi-expert, gains a degree of authorship. The race, match, or game, then, is always the product of multiple authors. It is common property—a collective creation—one generated from within common conditions of relations at a particular historical and technological moment. Sport performs a slip of scale: the spectacular acts performed discretely—yet loudly—by individual athletes belong, in some way, to everyone. In more Foucauldian terms, sport occupies a particular place in relation to panoptic organizations of space. Sport requires sightlines and spectators. Observation is needed, even deeply desired. So in contemporary sport, Bentham's panopticon becomes the condition of possibility for the agile body. This is a positive panopticon—one that openly demonstrates its ideological power. ("Positive" here simply means present to or openly avowed rather than surreptitious and unconsciously adopted.) Affect is reliably produced through sport's aleatoric systems. The sightlines are sanctioned by both players and spectators.

Both player and spectator share investment, acutely tuned in to difference, and thus fascinated with both standardized instruments and repeti-
tious infrastructure. They have a deep knowledge of how a ball bounces
differently off grass compared to clay; how it bounces differently fresh out of
a pressurized can compared to when it has been worn out by a hundred trips
through the ball machine. The higher the level of engagement with play, the
more intensely these contingencies are felt. Professional tennis players may
go through several rackets in a single match because they are so attuned to
the slight loosening of the strings with every impact of the ball. In this arena,
contingency is not liberatory but rather a problem to be mastered. This longing to master both bodies and materials makes sport a modern problem,
which enfolds projects to stabilize environments by controlling all, or perhaps just enough, variables to set up the possibility of a brief transcendent
moment. We expend tremendous effort on setting the conditions of possibility for "flow states": states defined by Mihály Csíkszentmihályi, drawing on
John Macaloon as states "in which people are so involved in an activity that
nothing else seems to matter; the experience itself is so enjoyable that people
will do it even at great cost, for the sheer sake of doing it." [49] Sport is a ritual
practice designed to make something phenomenal (in both the literal and
spectacular sense) more likely to occur.[50]

A Note on Method

Since childhood, I have pursued a mostly solitary kind of play that might
loosely be called a game. I pick up things—small things, discarded things,
things that happen to appear at hand. Like David Hammons, I am a collector. In my earliest years, it was baseball cards, hyperbolically shaped pencil shavings, rocks, seaglass, matchbooks, shells. More recently, pennies,
receipts, candy wrappers, day calendar pages, and other paper refuse to fold
into cranes, another round of pencil shavings, more rocks and shells, photographs of the ceilings I sleep under, photographs of airplanes I spot in the
spaces between buildings, and piles upon piles of discarded playing balls.
When positioned next to or piled on top of each other, these things index
the cultures and histories from which they emerge. *Bounce* is also a collection:
a handful of histories of different ballgames, bouncing balls, and bounding
practices gathered together and presented as a collection of instances of this

Figure I.6
Close up of *Los Angeles Plays*, 2022, an installation made with discarded balls collected from across Los Angeles. Installation and photograph of the installation by the author.

particular kind of interaction. Following an interaction through time and across contexts is a method for doing history that differs from historical methods that emphasize origins, authorship, period, or invention. In different moments, I found chasing bouncing balls into the odd corners of history surprising, disturbing, delightful, and intensely demanding. My hope is for readers of these collected histories of bounce to have a similarly wide range of experiences.

The question of how a ball bounces can be asked about most ballgames. Answering it is a way of attuning to particular histories and the particularity of history. My past life as a professional squash player was the departure point for this project, and is why I first began thinking about bounce in the context of games played on courts. More than any other kind of ballgame, court games are games of bounce. In *Games and Sport in Everyday Life: Dialogues and Narratives of the Self*, R. S. Perinbanayagam (2006) defines court games as games played in confined or bounded spaces that allow the measuring, situating, and placing of the self.[51] For Perinbanayagam, court games orient the self to the world through ongoing iterations of to-and-fro play within given boundaries.

Accordingly, rather than following a chronological structure, the book is organized into three parts—structured as a kind of to-and-fro. Each part contains a pair of chapters. Each chapter introduces a different kind of bounce: *ricochet* in Early Modern era court tennis and *true bounce* in the tennis of today in part I; *squash and stretch* in animation and *ping* and *pong* in computing in part II; *bounce feel* in EA's *FIFA* video game series; and *pok ta pok* in the Mesoamerican game *ulama* in part III. The chapters on the early and contemporary forms of tennis fit snugly into Perinbanayagam's understanding of court games as referring to games such as tennis, squash, and handball. However, over time, I came to understand the term as including a larger set of physical and electronic ballgames played on bounded courts and fields. Part III's focus on *FIFA* and *ulama* employs this more expansive understanding of court games.

Throughout, I draw on archival materials, fieldwork, and interviews to ground the historical work. What most compels me about history is the way that it converges in the instant—there are always so many different histories suddenly, often sharply, at play in any given moment. I use first-person descriptions of my encounters with different games—either by way of a site visit, spectatorship, or sitting down to play to locate the games in particular instances and in the present. Moving back and forth between research and experience, *Bounce* explores how ballgames and bouncing balls carry, hold, and transmit history and at times also suppress, cover over, and cast it off.

Part I. The Matter of the Ball

Part I is the "to" of the book—the serve that starts the rally. It is animated by questions brought to the fore by Serres's concept of the quasi-object: What are the material conditions of bounce? How does the matter of the ball matter? Chapters 1 and 2 introduce the history of bounce through the most classic and popular of the court games: tennis. A dialectical logic moves across the chapters. They discuss the earliest form and the more recent (and more familiar) form of the game, respectively. Together, the chapters in this section establish that bounce is distributed between ball, walls, players, and environments. The material conditions of all of these are at play in shaping subjectivities and collectivities, including the conditions of those distant bodies, architectures, environments, and landscapes that provide the foundations for any given form of play.

Chapter 1, "Ricochet," introduces court tennis as a game of the Early Modern era that puts *ricochet* on display. Court tennis began with the ricochet of balls off slanted awnings onto cobbled medieval streets and developed into a game played in enclosed and codified asymmetrical courts. This chapter uses Antonio Scaino's 1555 publication of the rules of that game to set in motion one of the book's central throughlines—the idea that ballgames model worlds. In its heyday, this enclosed and dialogical form of ball play, organized around a just barely contained kind of ricochet, allowed people in Europe to get a handle on new and contradictory natural, social, and cosmological orders of their day. Court tennis embodied the tension between the local and universal—one that would come to sit beneath the entire empirical project in the seventeenth century.

The model of the world put forward in modern tennis is very different from the earlier version of the game. Chapter 2, "True Bounce," tells the story of the pursuit of *true bounce* in modern tennis. The industrialization of rubber, the rise of color television, and the introduction of ball-tracking technologies all contribute to the production of true bounce. The first half of the chapter situates the introduction of rubber balls in tennis as part of a larger story about the vulcanization and industrialization of rubber. With the rise of ball sports in nineteenth-century England and the pursuit of reliable and standard kinds of pneumatic rubber bounce, true bounce came to serve as the foundation for modern ball sports today, and rubber became a foundational material of modern sport and modern life.

The second half of chapter 2 discusses two rare changes to the rules of tennis, each of which introduced a key element to the construction of true bounce. The first, the color shift from white to optic yellow felt, made tennis balls more visible to television viewers. The second, the incorporation of Hawk-Eye ball-tracking technology into tournament play through computer-generated, animated representations produced animated and apparently objective representations of exactly where the ball bounced. These two visualization technologies set up the transition to part II, which introduces the histories of bounce in animation and computing.

Part II. Virtual Bounce

Whereas part I of *Bounce* addresses the matter of the ball and the material conditions of bounce, part II turns to virtual bounce. This middle part of *Bounce* is the "-and-" of the book: the pivot between part I's "to-" and part III's "-fro." The chapters in part II shift from part I's sharp focus on the kinds of bounce that appear in a specific game to offer a broad examination of the centrality of bouncing balls to the fields of animation and computing. The question that animates both chapters is: How is it that computing and animation both call on the history of the physical ball, and yet the virtual ball is discernible only as a simulation through the mechanics of bounce? If we know from Serres that physical balls are already somewhat strange quasi-objects, the move into virtual space doubles and intensifies that strangeness. Balls are quasi-objects in virtual space by way of quasi-actions, or interactions: bouncing movements, whether through analog or digital electronics or animated sequences. In both cases, the ball loses much of its materiality—its heft—yet it retains its role of communicating both the physics and the character of animated worlds and computational spaces. However, the way that bounce appears and operates in these two domains is importantly distinct. Where animators move from ball to bounce—from object to action—computer programmers move from bounce to ball—from action to object. The relationship between the chapters in this part follows a chiastic logic, reflecting the deep entanglement of the fields of animation and computer graphics today, while bringing forward their distinct histories.

Chapter 3, "Squash and Stretch," introduces the history of bouncing balls in animation. With an emphasis on the Disney Studios technique of *squash and stretch*, I build on Paul Wells' argument that sport and animation are two foundational motion practices of modernity. Historically, the ball is the ur-object for animators. One of the first things beginning animators learn how to do is to set a ball in motion: to make a bouncing ball. A range of techniques have been used to create the heaps of bouncing balls that have appeared in the history of animation. Squash and stretch—a technique developed by Disney animators of squashing and stretching the drawn boundaries of objects to represent them in motion—is particularly important because of its prevalence and because, when elaborated, it becomes a way to communicate the personality of a given object or character—as lively, saggy, sleepy, sloppy, or otherwise. This chapter presents animating bouncing balls as a basic technique that allows animators to set bodies in motion and give animated characters their character.

Chapter 4, "From Ping to *Pong*," traces the history of the two conceptions of bounce central to the history of computing: the conception of signals as projectiles, *ping*, and the conception of the world as a rule-bound game space for testing different models—or simulations—of bounce and other kinds of interactions, as demonstrated by the iconic video game *Pong*. In computer (or video) games, the contrast with physical ballgames is starkly apparent. In physical ball sports, a great amount of effort goes into creating standard and fixed spaces of play, so that the uncertain bounce of the ball is not completely beyond a player's capacity. In animated and computational spaces, the effort goes toward predictably producing the unpredictability of the ball in the first place. There are many different technical ways to program bounce. But whatever the approach, before a virtual ball can appear and play with a player's predictive capacity, it must first be purposefully produced by a bounce program. The history of sending signals and programming bounce reveals that in computing, bounce acts as a key method for producing differential knowledge, facilitating real-time interaction, and simulating movement.

How bounce operates in animation and computing is necessary to understand how it appears in both physical and electronic games today. Together, chapters 3 and 4 show how the transformation of the physical ball into virtual bounce entails innovations and advances in animation and computing. Via different kinds of virtual bounce, a burgeoning social logic of collective participation in judgment and experience emerges. This emergence tracks across computational and cinematic developments and in turn creates new possibilities for both collective and individual experiences of gaming and spectatorship in the various kinds of contemporary global play.

Part III. Bounded Spectacle

From part II's pass through animation and computing, the final part of the book—the "fro"—is a return: a discussion of two specific ballgames that explores the possibilities and conditions of collective and individual experiences of global play. Chapters 4 and 5 tell the stories of two very different mythic games and examine the various body techniques and bounded spectacles that accompany them. Games and sports are rule-bound frameworks—bounding logics—that support a wide range of cultural performances to begin with. When a game becomes mythic (in Roland Barthes's sense of myths as masks that sustain the social order and reinforce class distinction), it participates in additional cultural frameworks beyond those embedded in the gameplay rules. The pleasure and spectacle of mythic games are produced in the interaction among the players, the spectators, and the sets of standards and specifications for player and fan behavior, game equipment, game architecture, gameplay broadcasting, and so forth. When a game becomes spectacular, it becomes bound by the techniques and specifications of its spectacle.

Building on the history of *Pong* and its predecessors, chapter 5, "Bounce Feel," considers what Steve Swink calls the "game feel "of the interactive programmed bounce in the *FIFA* video game series from Electronic Arts (EA). For over twenty years, *FIFA* (renamed *EA Sports FC* in 2024) has been the most popular and profitable sports video game on the market. The chapter takes up my experience as a woman, and a former athlete, sitting down to play *FIFA*, and braids this with the history of the development of the feel and representation of the game's interactions. Along with the bounce of the in-game ball, I examine how the bounce of light off the motion-capture suits donned by women's national team members helped to facilitate—in 2016, twenty-three years into the video game's existence—their belated inclusion in the game series. Following Iris Marion Young's phenomenology of feminine body comportment in "Throwing Like a Girl," and Legacy Russell's description of glitched bodies as those that pose a threat to a social order, the chapter tracks a set of technical and cultural glitches in the game's feel and hegemonic common sense.

In contrast to the global reach of soccer and *FIFA*, chapter 6, "Pok ta Pok," again picks up the theme of rubber bounce to tell the story of the elastic bounce of one of the world's oldest ballgames: the Mesoamerican ballgame, often called *ulama*. After the Spanish conquest, the game survived multiple attempts to suppress its play. Today, *ulama* has been revived by players and activists who organize in person and over Instagram and YouTube to share ancient techniques of ball-making, learn the game, and recruit new players. While I use the name *ulama*, the Nahuatl name used by the Associación De Juego de Pelota Mesoamericao (AJUPEME), throughout much of the chapter, I chose to use the onomatopoeic Mayan name for the game, "*pok ta pok*," for the chapter title because it describes the game's kind of bounce and carries the sound of the game's to-and-fro. Like to-and-fro, *pok tak pok* evokes the rhythm of play—the embodiment of hitting a heavy, solid rubber ball back and forth, bouncing between players' bodies and court surfaces. The chapter concludes with a short swerve into my encounter with the oldest standing handball court in Los Angeles and the ways that this encounter laid the ground for my connection with the city's *ulama* players.

Both chapters in part III relate a historical account: *FIFA*'s history is just a handful of decades long, but it is based on a game that is centuries old. *Ulama*'s history stretches across three thousand years. Both chapters draw more heavily on ethnographic methods and phenomenological description than the prior ones. In its simulation of the world's most popular professional sport, EA's pursuit of cinematic realism is an open and universalizing bid for total virtual control of representation. By contrast, the present-day revival of *ulama* through a network of delegations in Mexico, Belize, and the United States is an attempt to preserve the game and decolonize sport set within and against the logics of a closed world. Taken as a pair, the chapters in part III of *Bounce* illustrate new kinds of enclosure and togetherness—new kinds of bounding—that have emerged with cinematic and informatic transformations and entanglements of physical and virtual worlds.

With some sense of these transformations and entanglements still hovering, *Bounce* bounds on to its conclusion, "Culture-Inside-Out," which brings together bouncing babies, body techniques, Freud's body ego, Pierre Bourdieu's habitus, and other shiny bits and pieces for a final proposition that, while sport is of course part of culture, it is also, as importantly, culture-inside-out.

Against History

Not everyone has balls to play with. Balls are challenging objects to write and talk about—especially when pluralized, and especially for a woman.[52] Ball play is gendered, raced, classed, and structured by histories of colonialism. Gender is a particularly strong structuring force for ballgames because historically, these games have been predominantly played by men. Gender segregation in sport both reinforces and is reinforced by the construction of the gender binary more broadly. The construction of "fair play" on the grounds of gender segregation is what makes trans athletes easily available for vicious targeting in the current culture war in the United States, and beyond. The emergence of modern sport has long been told as a story of men and boys at play: a world grounded in motion practices and object relations made by and for men. To give just one example, the documentary *Bounce: How the Ball Taught the World to Play* does not feature a woman touching a ball until an animated—rather than live-action—avatar of soccer star Mia Hamm appears after the film's sixty-minute mark.[53] The world—and the balls that teach the world about it—is one strikingly devoid of women. This world simultaneously makes a spectacle and specter out of gender.

Still, I enjoy watching men at play.[54] As a former woman athlete, this sometimes feels like a guilty pleasure. In squash, as in many other sports that favor body types more commonly found in men, the best women players can beat almost all the men in the world. Yet they often cannot beat their ranked equivalents, the top men players. When top women athletes do manage to beat everyone else in their category—and perhaps threaten to beat their male equivalents—they become suspect. Writing about the case of the runner Caster Semenya, Jennifer Doyle describes the predicament of the female athlete who beats others in her category by such a large margin that she "takes sublime flight from" or "runs out of" the category of gender.[55] The problem, as Doyle puts it, comes from the inability of female athletes to serve as a measure of human capacity.[56] When they exceed their category, they do not become superhuman. Rather, they become a category of human that they are not supposed to be. This misogyny was loudly declared in the world of electronic gaming over the course of #GamerGate. That instance is just one of many such examples that demonstrate that the gendered claim to territories of play does not disappear when physical distinctions are no longer available to excuse them. The decades-long exclusion of women from the *FIFA* game series does the same.

Modern sport relies on normative categories, such as gender and age, in order to make contests fair. This model has played a critical role in facilitating girls' and women's participation in sports in greater numbers today than ever before, but it also has consequences. Those who repeatedly suffer these consequences are athletes whose capacities and bodies are nonconforming. Their existence gives the lie to the premise that these categories adequately represent the range in diversity of human bodies and selves. Sport crashes the vast range of bodies and selves into normative categories, even as it selects for exceptions and outliers. Perhaps this is why Doyle declares that the "athletic figure is queer . . . elemental, fleshy, and intersubjective."[57] Unlike sport, the academic world does not employ categories that explicitly acknowledge the physicality and sociality of the work at hand. Nevertheless, Sarah Ahmed opens *Queer Phenomenology: Orientations, Objects, Others* with a meditation on Edmund Husserl at his writing desk. His family is out of sight behind him in the garden beyond. This makes a similar point about the many presumptions built into the role of an academic. Enfolded within these presumptions is the idea that those inhabiting nonconforming bodies, identities, and positions can only ever partially, awkwardly, and often with injury occupy society.

For many years, I watched men play with longing. As a thirteen-year-old girl, I realized that I was never going to be able to beat my longtime childhood friend and classmate Peter Kelly again. Part of the pleasure of watching sports lies in this kind of longing—a condition of spectatorship. Such longing is not restricted to a particular identity position. It does not need to stay stable. It can seek out new objects. It is a type of longing that somehow stays—for the most part and on most days—on the right side of envy. In the context of watching a musician perform, Barthes described it as an erotic relation, one that opens a "new scheme of evaluation" and develops "beyond the subject all the value which is hidden behind 'I like' or 'I don't like.'"[58] Sports offer us a chance to extend ourselves—in more dimensions than two—in a manner that feels asymptotic. We imagine approaching capacity. Through sport, game, and play, we feel out our relationships to other bodies in motion, pass subjectivity off and around, and project quasi-selves to and fro. These are acts of orientation where, at once, we bound ourselves and become bounding beings.

The act of writing history, likewise, is an emblematic interaction. Writing orients us toward some contact with thinking together about the many ways that it is possible to configure relationships between objects and worlds. The world of vast global communications systems works on principles of pings and pongs, relays and responses. There are ever more channeled virtuosities and greater efforts to turn every environment into a predictable ground. What this book offers, finally, is a first step toward a material history of to-and-fro: a set of stories about different ways of scaling bodies and selves to worlds. Through bounce and its many variants—ricochet, rebound, spring, ping, *Pong*, coefficients of restitution, squash-and-stretch, and other kinds of elastic motion—these bounding stories return readers to the conditions of possibility and unpredictability of extraordinary individual and collective gestures.

Figure I.7
Ball Portrait (Globe Beach Ball), 2024. Photograph by the author.

The Matter of the Ball

1 Regarding Ricochet

To play tennis, more than any other game, or be emotionally involved in watching this kind of game, is to engage the self in the significance of ratios, economies, and measurements in the social and the interactional life, to recognize boundaries, limits, and parameters. The boundaries within which a single agent has to drive and place the token of the self, which is the ball, with the help of the extension of his or her body, which is the racquet, is the exemplification of the measured processes by which the self is to be presented.

—R. S. Perinbanayagam (2006), *Games and Sport in Everyday Life*

Through the Netting

Court tennis is not a common sport today. Only forty-seven courts remain in the world, hidden behind the tall walls of private clubs and historic castles and estates. Until the late nineteenth century, the game was simply called "tennis" in English, only acquiring modifiers once it was eclipsed by the upstart game of *lawn tennis*. Court tennis was "real tennis" to some, a plaintively defiant adjective, and "royal tennis" to others, referring to the aristocratic court's enthusiasm for the game. In the United States, the name "court tennis" sufficed to underline its enclosed space of play, emphasizing its distinguishing feature.[1]

I first learned about this older version of tennis via YouTube. While I was searching for footage of fives, another now-obscure racket sport, a court tennis match popped up in the sidebar. I was watching what looked like a bizarre game of doubles tennis on an indoor, oddly asymmetric, walled court. Indeed, the courts in this version of tennis are double asymmetric (meaning that bisecting them along either their length or width results in two nonidentical areas). This was not the tennis I knew. I bounced from

Figure 1.1
Through the netting, a view of US Open Court Tennis Championships, R&T Club, New York (January 2014). Photograph by the author.

video clips to club and association websites to reading definitive and enthusiast histories. Finally, I sought the sites of play themselves and found my way to the Racquet & Tennis (R&T) Club on Park Avenue—home to the only court tennis court in New York—for the 2014 US Open Championships.

The R&T Club is a private, men's-only social and athletic club established in 1876 and rehomed to its current Italian Renaissance-styled clubhouse in the heart of Midtown in 1918. The club allows women to enter some areas for specific social and spectator events, such as the US Open. This kind of space was not unfamiliar to me. It had the stiff yet worn feel of clubs where I played squash as a teenager; with forest green, maroon, and navy carpets and dress codes requiring players to wear all whites. These were places where my grandmother refused to watch me compete because, for much of her life, clubs like this had not allowed Jews to join.[2]

I stepped off the elevator, and ran into a man whom I had once crushed on, unrequitedly, back when we were teenagers playing squash tournaments up and down the East Coast. In the decade or so since we had last seen each other, he had become a value equity manager and acquired the kind of weight that often comes with age and abundance. As he told me a bit about the players who were currently on court, I peered through the netted window into the vast enclosure.

The court appeared as a strange mash-up—part tennis court, part football field, part medieval town square, part inside-out pinball machine. One huge, smooth sidewall ran from end-to-end and floor-to-ceiling, interrupted only by a buttress jutting out of the far corner on one end. The court's other three walls were lined with pitched roofs, on which the ball would bounce around until it rolled off and dropped into play. Under these "penthouses," as the sloping rooftops are called, there were long, horizontal, netted windows called "galleries," through which I was looking. These spaces are like pockets in billiards: when the ball is hit into the netted gallery, the attached bells ring, and the point ends. I tried to decipher how the chase lines, marking the floor at intervals, prompted the teams to switch sides at—to my eyes—seemingly random moments. Watching without knowing what I was seeing, I listened to how the court's many angled surfaces contained and returned both ball and sound—the ball booming against the slopes and thwacking off rackets and court walls, amid muffled echoes of the players' and scorers' voices, and ringing bells.

As players sent the ball flying here and there, every additional wall and angle multiplied the possible trajectories and momentums of balls and bodies even as they contained them. The tennis ball looked normal: it was the usual size, boasting the familiar neon yellow felt. But it did not behave like the balls that I had tossed back and forth with my dad in gas station parking lots when I was growing up, and watched Andre Agassi and Martina Navratilova slam them at their opponents. Decidedly less elastic—and less lively—this ball rolled, skidded, and died in the corners of the court. Even when sent sky high by the angle of a player's racket, it did not rebound dramatically off the floor, penthouse roof, or wall. The game seemed simultaneously vital and vestigial, a testament to what cultural forms do and do not hold constant. I wanted to understand what this game had been historically, and what it is now.

The pair of chapters in this section—chapter 1, on court tennis, and chapter 2, on the modern tennis of today—tell a series of stories about physical bounce and how changes to the materiality of balls, the architecture and boundaries of the court, and the formal rules of play change the kind of bounce at the center of the game. In this chapter, I argue that court tennis is a game organized around the containment and display of *ricochet*. Defined as a series of two or more interconnected impacts that compound the energy of those impacts, the word "ricochet" implies unpredictability. Echoes of an initial interaction, a bounce that ricochets is one that threatens to exceed human capacity and containment. This stands in stark contrast to the bounce of modern tennis, which (as I will show in chapter 2) aspires to be a *true bounce*, one that is standardized, reliable, and predictable. Today, the object most associated with bounce is a ball—that exemplary quasi-object and constructor of intersubjectivity that Michel Serres describes.[3] I show how changes to the tennis ball were fundamental to changes in bounce. However, part of my argument across these two chapters is that the close association of bounce with the ball, to the exclusion of the surrounding architecture and environment, is something that happened with the late nineteenth century's introduction of rubber and pressurized air, which together made the balls dramatically more elastic, shifting more focus toward these flying objects and away from the role of the surface and external environment. In this chapter, the argument is more distributed across the material and social elements (the balls, the court, and the rules) that condition the game's bounce.

What is there to make of the change from ricochet to true bounce? As R. S. Perinbanayagam argues, all court games are about orienting the self to the world through interactions with uncertain objects and surfaces. What this first section of *Bounce* aims to achieve, then, is to show how the differences between these two iterations of tennis make for different activities that orient the self to the world. They model different ways of knowing the world, particularly the attendant changes in conditions of scientific inquiry, empirical judgment, knowledge production, and techniques of observation. The model of the world that court tennis offers is one of restricted play—one that both contains and amplifies chance and uncertainty within bounded architecture, putting them on display. The world in question is one in which the field of play is asymmetrical—surroundings play a crucial role here—and the honed skills include responding to the unpredictable. Virtuousity, in this context, is the capacity to contain this unpredictability, and become able to play with it. This is strikingly different from the model of the world offered by modern tennis. In modern tennis, play is still restricted, but the irregular walls have been removed. The field of play is symmetrical, and all game elements are made as standard as possible. The way that court tennis produced a certain field of play—with its lucky bounces, hazardous positions, bad hops, complex glances, angular rebounds, and achingly slow rolls—has been reduced dramatically. Instead, in modern tennis, a good court supports true bounce. The remaining difference—the difference between both the players themselves and the bounces that they produce—stands squarely in the spotlight.

Modeling the Universe in a Closed Court

In the Early Modern era, a "game of ball" became wildly popular across Europe.[4] From the fifteenth through the seventeenth centuries, as European powers began their "colonial encirclement of the world," the game bridged the continent's own diverse social, political, and linguistic boundaries.[5] Most of the continent's major cities, towns, royal palaces, noble estates, and educational and theological institutions boasted courts. Priests and kings, students and scholars, aristocrats and merchants, innkeepers and artisans all played, watched, gambled, and gathered around the game. As early as 1427, the *Journal d'un bourgeois de Paris*, which recorded the history of ordinary people, listed a woman named Margot as the best tennis player of the day. Religious and secular authorities alternately endorsed and banned play of the game. Court tennis was the first European game to have its own guild, established in Paris in the late 1400s, and the first to have a published book of rules, penned by the Italian cleric Antonio Scaino in 1555. By the close of the sixteenth century, there were over 280 *jeu de paume* courts in Paris alone, surpassing the number of churches.[6] A person visiting London could find tennis courts at many sites around the city, including Essex House, Whitehall, and Ironmongers' Hall.

What started as a street game became an elite pastime. In Italy, at the turn of the fifteenth century, the courtly class engaged in a frenzy of competitive court building with "the Medici in Florence, the Sforza in Milan, the Este in Ferrara and the Gonzaga in Mantova . . . ordering the construction of tennis courts on their sumptuous estates."[7] Holy Roman Emperor Charles V of Spain had multiple courts at several of his palaces. In 1528, Henry VIII built a closed-play establishment at Hampton Court to complement the open-play attraction that Cardinal Thomas Wolsey had initially built to lure the king out to the estate for visits. Meanwhile, the great thinkers of the day—Galileo, Isaac Newton, Jacob Bernoulli, William Shakespeare, Thomas Hobbes, and Michel de Montaigne, among others—picked up the game as "an-object-to-think-with": making metaphors out of the instruments, architectures, and gestures of play.[8] The rise of court tennis heralded the beginning of the end of the medieval tournament, signaling the emergence of ball sports as a dominant and transnational form of competitive play in the Western world.[9]

Figure 1.2
A casual game at the *jeu de paume* court at Château de Fontainbleu, a longtime summer residence for the French monarchy (July 2014). Photograph by the author.

What is there to make of the fact that the primary form of institutionalized play of the Early Modern era in Europe involved people throwing themselves—bodies, souls, and bank accounts—into playing and gambling on an elaborate game of ball? The game coalesced into a common cultural form at a time when grounding relationships—between heaven and Earth, God and man, man and matter, and sovereign and subject—were up in the air. In the wars between European powers, developments in ballistics—and the regular use of cannons—forced bodies and buildings to endure, and adapt to, new levels of sonic and physical impact.[10] The Spanish crown, then bent on accumulating metal to make gold and silver coins, sponsored ventures by Ferdinand Magellan and Christopher Columbus to confirm the Hellenistic theory of a spherical Earth, initiating centuries of conquest and colonization. The rediscovery of Lucretius's poem *On the Nature of Things* signaled a "swerve" toward a new atomistic metaphysics.[11] Nicolaus Copernicus's heretical theory of heliocentrism slowly rippled through intellectual circles, undermining the authorities of both the Catholic Church and the new Protestant orders. During this historical stretch, the movement

Figure 1.3
Giovanni Battista Bracelli, from *Bizzarie di Varie Figure*, c.1624. Courtesy Rosenwald Collection, National Gallery of Art, Washington.

and manipulation of spherical objects of whatever size constituted a pressing matter.

This enclosed and dialogical form of ball play allowed people to get a handle on new and contradictory natural, social, and cosmological orders of the day.[12] As European conceptions of the world underwent dramatic shifts, court tennis offered a way for players and spectators to mediate relationships through an embodied practice of ricochet in a measured and restricted space.[13] In its heyday, court tennis modeled new empirical and probabilistic ways of knowing. All ball sports are aleatoric structures (i.e., structures of planned chance), organized to a greater or lesser degree around some kind of bounce. Located between the games of pure mathematical chance and the messy complexity of the world, ballgames make good material metaphors for modeling practices. The first European game to have a written code and to develop standards for its playing space, court tennis embodied the tension between the local and the universal that would come to lie beneath the entire project of empiricism in the seventeenth century.

As the empiricists called for a science based on practice—and a shift in the methods of truth production toward one of observing measurable expressive displays—sport combined the dramatic expressiveness of the theater with the rule-bound framework of games, operating as a "theater of proof," to use Bruno Latour's term, to reliably produce winners and losers.[14] The bounded nature of sport in general, and court tennis more particularly, also created similarities to this new kind of science. Court tennis operated as a "closed world," a term first used by Sherman Hawkins to describe major dramatic spaces in Shakespearean plays and adopted by Paul Edwards to refer to "a radically bounded scene of conflict, an inescapably self-referential space where every thought, word, and action is ultimately directed back toward a central struggle."[15] Within the physical and conceptual architectures of the game, invested players and spectators could attempt to master unpredictability to produce reliable results in a safely contained and measured space. From here, they refigured and refined the boundaries between self and other, and self and world, by developing new methods and capacities for predicting, strategizing, measuring, and mastering an unpredictable and complex kind of bounce.

A Game of Ricochet

Court tennis is part of a genealogy of ball-and-wall games first played in the streets, courtyards, and cloisters of medieval Europe. Gathered together, the names and variations call out a constellation of features—the ball, the palm, the fist, the racket, the cord, the square, the court—across which bounce is distributed.[16] The earliest versions of the game—offshoots of football—are thought to have been played in France in the thirteenth century.[17] By the Early Modern era, the common French game was *jeu de paume* (game of the palm), also called *court-paume* (court-palm) in scattered places. *Jeu de carre* (game of the square) was a variation played on a slightly larger, more symmetrical court. Prior to this, in northern France, sometime around 1300, long before the game was enclosed and codified, people played an open-air version called *jeu de bonde*, or "the game of bounce."[18] Heiner Gillmeister, one of the game's foremost historians, translates *jeu de bonde* alternately as "the bouncing game" or "the bouncing ballgame."[19] Gillmeister draws a connection between the word *la bonde* and *boundadou*, the French Provençal word for the service slab—the surface that a player sends the ball flying off of to start a point—which in turn corresponds to *steute* in the Saterlandic version of the game of ball, both of which Gillmeister translates as "the bounce."[20] From our vantage point today, the ball appears as the exemplary object of bounce—the object that in itself implies the type of movement in question. This view is shaped by a deep association with balls and the extraordinary elasticity of rubber bounce, but this association did not begin in the Western world until the nineteenth century.

In this earlier era, tennis balls did not have much bounce on their own. Most had leather casings, sewn together in various patterns, and were stuffed with materials ranging from human and animal hair to cloth and lumps of lead wrapped in stocking wool.[21] Gillmeister's etymology shows how *la bonde* first referred to the bounded surface area where the ball was hit to begin play. Bounce, therefore, was a notion tied to, and produced by, the limits of the court; which suggests that initially, responsibility for the type of movement produced from this principal interaction was attributed to the bounded surface and only later moved to the bouncing ball.

The first meaning of the word "bound," derived from the Old French word *bonde*, is "limit." Both the French word and its English derivation began as a noun referring to "a landmark indicating the limit of an estate or territory," and more broadly to "a limit or boundary, that to which anything extends in space."[22] The second meaning of "bound" as a "leap" derives from the Old French *bondir,* meaning "resound, rebound, or echo." Whether we look to echoic mechanisms or to marking territory, certain actions, such as springing upward and leaping forward, depend first on a set of boundaries or limits—an environment that conditions the movement. A type of movement, "bound" not only demonstrates but produces limits by making them visible or audible—in a word, *observable*. Through this process, objects are identified and identities constructed.[23] While etymological dictionaries locate the emergence of this second meaning of "bound" and *bonde* in the fifteenth and sixteenth centuries, the game *jeu de bonde* demonstrates how the French word had moved in this direction much earlier. *Jeu de bonde,* then, is a site where the distribution of bounce is made apparent through play. It reminds us of the role of the surface ("the bounce") in mediated interactions: any interaction—between foot and floor, hand and racket, racket and ball, or ball and wall—creates and temporarily holds open a space of uncertainty whose closure completes the interaction.

From the early street play of *jeu de bonde* to the elaborate enclosed game of court tennis that I encountered at the R&T Club, this family of games shares what is called the "chase rule."[24] This rule constitutes the most striking difference between the early forms of the game and the tennis of today. Unlike the more familiar modern tennis, in court tennis, the player on the service

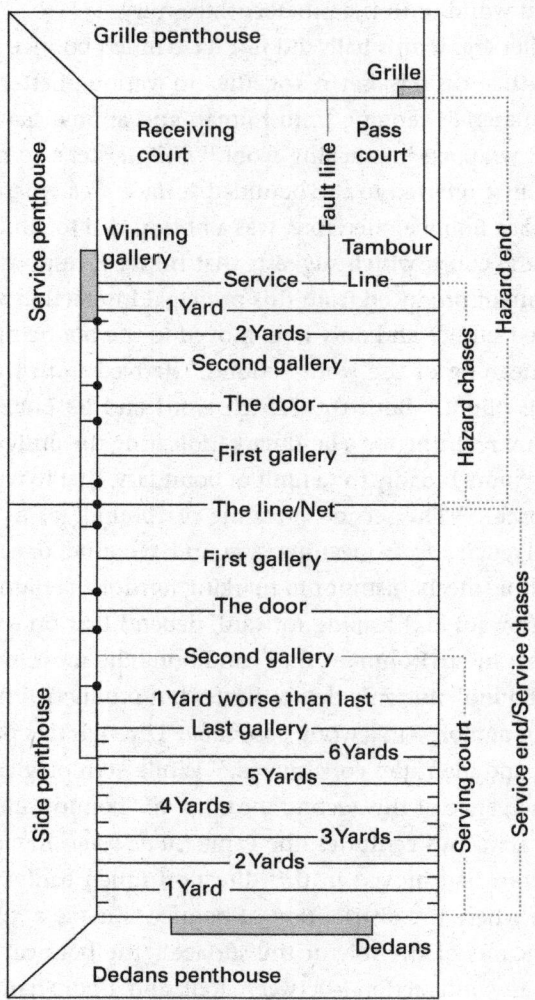

Figure 1.4
A diagram of a court tennis court with the chase lines labeled. Created by Atethnekos at English Wikipedia.

end continues serving and the player on the hazard end receives indefinitely. The word "hazard" most likely derives from a colloquial Arabic word for "chance" (*al-zahr*, meaning "the die"), which aptly describes the disadvantages of being on the receiving end of the ball. The hazard end was difficult to defend because of the multiple chance elements that the player on that side had to contend with, ranging from the opaque choice of shot by the server to the physical structure of the architecture, which included the angled buttress of the tambour, guaranteed to send the ball flying off in impossible directions. The only mechanism for players to change ends is the chase rule, which makes ricochet not simply an element, but also a mechanism of play. The chase rule dictates that when a ball bounces for a second time—meaning that the rally in question is over—the location of that second bounce and the chase line that it is closest to are marked.[25] Chase lines are predrawn numbered lines that run horizontally across the floor of the court. (This is what those lines that reminded me of American football turned out to be for.)[26] If a player fails to hit a ball before it bounces a second time, they chase down the ball. The place where the ball is caught—or in later iterations, the place where the ball lands on its second bounce—is marked, and the closest chase line to this location becomes "the chase."[27] When a chase is set, no points are awarded to either side. Instead, the chase is noted and then set aside until game point is reached or until a second chase is set. At this point, the players change sides and play the "chase point." During this special kind of point, the chase line in question becomes a second net of sorts. The player attacking the chase line must hit every shot so the second bounce of the ball lands closer to the back wall than the chase line. The player defending the chase must return any ball that seems as if it will pass the chase line on its second bounce. The job of both players is to act and react based on where they project the ball will land on its second bounce.

All game action occurs (either by swinging or by choosing not to swing) based on a guess about the ball's future position. The walls play an important role here. If a ball bounces once on the floor and then rebounds off the back wall such that its second bounce on the floor lands in front of the chase line, the attacking player loses the point. Just as the walls of the court can remind us of the mediation that is part of any interaction, the importance of knowing when to refrain from hitting a shot can remind us that certain kinds of inaction are in fact actions. This extends to the physicality of hitting the ball. Because the solid ball is so unpredictable and the racket is an imprecise instrument, the first fundamental for the player is not to swing with great force, but instead to hold their body still and strong to turn into a perfectly angled wall that the ball can rebound off of.

Because the chase is the only mechanism for changing ends, and because being on the service side is a much more advantageous position, players are always anticipating the location of not just the first but also the second bounce of the ball. They play every point with an eye toward possible chases that will reconfigure the boundaries of play and the positions of the players in a future rally. Players on the receiving end try to create chases to initiate a change of sides, while players on the serving end try to avoid them. Good players know that they might have to defend the set line by any shot they hit, so they try to hit the ball with the aim of setting the terms of a future point. In a historical moment that saw European powers fighting wars over each other's territory as they began expansionist colonial projects that would encompass much of the world, the rule of the chase asked players to play between the two meanings of the word "bound"—"boundary" and "bounce." By determining and playing the second bounce of the ball in one rally to their advantage—to reconfigure the boundaries of play and the positions of the players in a following rally—they could reset the terms of play. Court tennis requires players to measure bounce with their bodies, and the chase, in particular, requires them to demonstrate the capacity to control and predict a complex chain of bounces. It is this requirement to think multiple steps—literally multiple bounces—into the future that first made court tennis a game of ricochet.

Just as the secondary meaning of *bonde* emerged after the play of *jeu de bonde*, the initial rebound of *ricochet*, in the context of ball sports, preceded the related use of the word to describe the physical movements of an object in both French and English. English takes the word "ricochet" directly from French. It appears first in the Middle French phrase *fable du ricochet*, an endlessly repeated thing, especially an endless exchange of questions and answers. A little later, *chanson du ricochet* was a form of song that took this endless repetitive back-and-forth as its model.[28] "Ricochet" appears as an independent word in the early seventeenth century to describe the action of a stone skipping across water, and it later came to mean a series of interconnected events. In the late 1600s, the Marquis de Vauban, a famed military engineer and strategist for Louis XIV, adapted this interconnected series into a military strategy called "ricochet firing," a tactic initially used for sieges in France's wars with its immediate neighbors, and later in its colonies. The technique consists of "firing a projectile such that it is made to glance or skip along a surface with a rebound or series of rebounds."[29] Cannons were often fired at low angles and less than maximal power. The damage was done not through the brute force of the shot punching through a wall, but via a skipping, bouncing cannonball that destroyed the structure from the inside out.

Vauban may not have had court tennis in mind when he developed his military strategy of ricochet, but he would have been quite familiar with the game, given its popularity with the French aristocracy and royalty. Indeed, several of the castles that he designed included tennis courts. In the game's context, the enclosed architecture induced a skipping and skidding bounce but was never itself threatened by it. In the realm of warfare, the game was discursively present: its agonistic structure and back-and-forth lobbing of objects, offering an apt metaphor for violent conflict. In Shakespeare's *Henry V*, the Dauphin gives the English king a set of tennis balls, to which the king replies:

> We are glad the Dauphin is so pleasant with us.
> His present and your pains we thank you for.
> When we have matched our rackets to these balls,
> We will in France, by God's grace, play a set
> Shall strike his father's crown into the hazard.
> Tell him he hath made a match with such a wrangler
> That all the courts of France will be disturbed
> With chases.[30]

The creation of closed courts, such as the ones that would have been found at Vauban's castles, further added to the unpredictability of the bounce. Gillmeister writes that after "closed play" was invented, "balls came ricocheting off the walls from all sides, which severely tested the reflexes of the players."[31] Every additional wall and angle multiplied the number of possible trajectories and possible momentums of both balls and bodies. The advent of the fully enclosed court turned a game of bounce (*jeu de bonde*) into a fully realized display of ricochet.

All the World's a Court

In his *Treatise on the Game of Ball*, Antonio Scaino suggests that those who are so inclined can "think of the court, closed in on all sides by walls and barriers, as nothing more nor less than this troublesome world."[32] Scaino's text, published in 1555 by the House of Gabriel Giolito de Ferrari and Brothers, is the earliest known codification of a modern sport in Europe. As a set of rules, it articulates a common form for the game (in the sense of a sharable standard), thus shifting the emphasis from a common kind of play (in the sense of locally popular and nonelite). But the *Treatise* is more than a set of rules. It is a philosophical argument that claims the game of ball as a true art, "beneficial for body and mind, especially of benefit in the purification of the spirits through which the soul performs all its functions, even that of understanding."[33] At the time, Scaino was a young cleric and budding Aristotelian philosopher living under the patronage of the prince of Ferrara. He rallies these traditions to argue that the game of ball is both a standard, against which all other games should be measured, and a model of the world, wherein participation offers players and observers an opportunity to understand the natural, social, and cosmological orders of the world. For Scaino, all the world is not a stage: it is a court.

The *Treatise* is one example of how early modern sport was used to model the natural world. This topic was taken up at a conference on tennis and the Scientific Revolution, organized by Marco Beretta and Alessandro Tosi at the Museo Galileo in 2013. In Beretta's paper "Training Tennis Players through Natural Philosophy," he devotes special attention to Scaino. He quotes a passage of the *Treatise*, in which Scaino, to explain the behavior of a ball hitting off various surfaces, turns to echo:

> All natural things (so the Philosophers say) having a defined limit and fixed course . . . it must be known that those things forced by some obstruction to turn aside from the original path travel toward the place to which they have been diverted only so far as their force & movement permit. In this manner Echo is born, for the sound created by the percussion of the air born in some low cavern or other such place sounds back, as it returns, the same as when produced. Therefore, when the Ball flying through the air strikes a wall it is forced to rebound, moving as far in some other direction as is allowed by the force imparted thereto.[34]

Scaino's description comes from Aristotle. Beretta's interest in this example lies in the existence of another commentary, one that Scaino wrote forty years after the *Treatise*, on Aristotle's *De Anima*. There, Scaino makes the same argument, but in the reverse direction, explaining the phenomenon of the echo by comparing it to the bouncing of a ball.[35] He could assume that knowledge of the ball's various motions was common enough to make

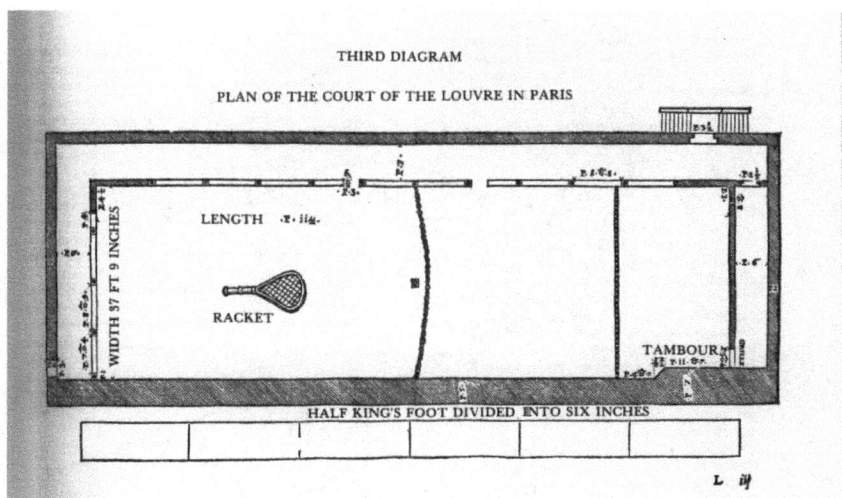

THIRD DIAGRAM

PLAN OF THE COURT OF THE LOUVRE IN PARIS

LENGTH

WIDTH 57 FT 9 INCHES

RACKET

TAMBOUR

HALF KING'S FOOT DIVIDED INTO SIX INCHES

L

Figure 1.5
Plan of the *jeu de paume* court at the Louvre in Paris, from Scaino's *Treatise on the Game of Ball*, 1555.

it the reference by which other motions could be explained. Where the ball appears first as an object to be explained, through comparison to echo or to ballistic trajectories, a few decades later the ball is used to do the explaining. Using the ball to explain echo thereby calls up the experience of those who had played or watched the game of court tennis. Beretta cites the nineteenth-century authority on rare books, Jacques-Charles Brunet, who argues that although only a limited number of copies were printed, "the influence of this small volume was not at all limited to a restricted circle of readers."[36] Scaino's discussion provided an at-hand metaphor that his readers could share with others, enabling widespread understanding of emerging ideas.

Tennis became an explanatory object, and a graspable metaphor, for many new scientific concepts during the Renaissance. In its shape and motion, the ball could stand in for a particle, a projectile, or a planet. In Pisa, in 1632, "Galileo chose an example that he knew would be familiar to all of his readers—the curved trajectory of a ball hit by an expert player—to explain a scientific concept in his second dialogue—the combined rotational and translational motion of a mobile."[37] In Graz, Johannes Kepler used the ball to explain the laws of planetary motion. Newton and René Descartes both

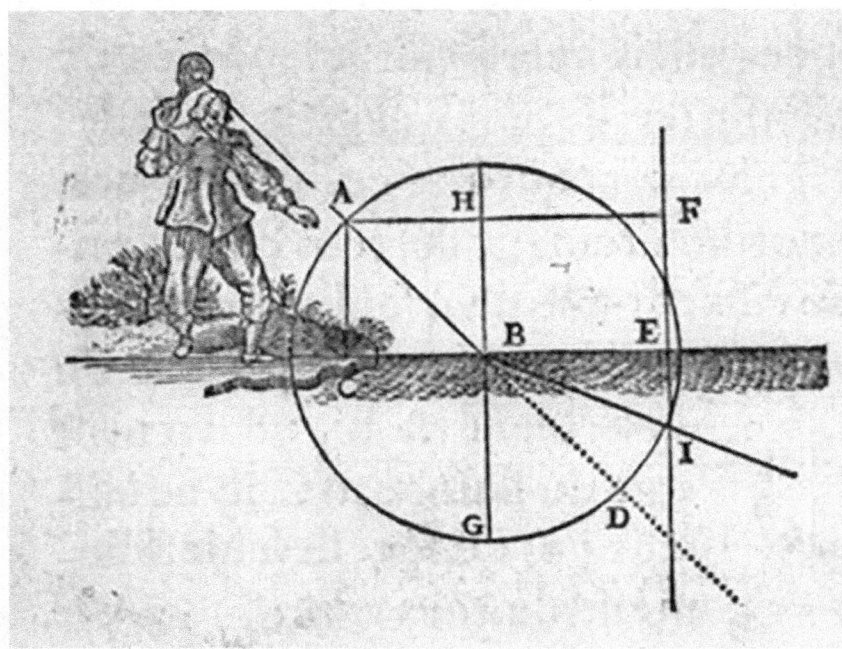

Figure 1.6
Diagram of refraction from René Descartes's "Dioptrique," 1637.

picked it up to help explain the behavior of light. In their introductory essay, Beretta and Tosi write: "The practitioners of early modern science could not remain indifferent to the authoritative influence of Galileo, Descartes and Newton, and the motion of the tennis ball, like the complicated rules of the game itself, would continue to be used to illustrate the most disparate ideas, from elasticity theory to iatromechanics."[38] Looking at contemporary physics, math, and computer programming textbooks and articles demonstrates that the tradition of using a ball, in particular a tennis ball, as an exemplary body and test object continues today (see chapter 4, on computer bounce, for more).

What gets pushed to the side in this picture, at least to some degree, are the walls. Beretta and other historians of science, perhaps because of a contemporary bias toward the object (in this case, the ball), tend to leave the walls of the court out of their accounts. But, of course, walls play a key role in the production of echo. When discussing the tennis of the Early Modern era, the walls matter because, as we have seen, it was a game played in, and against, architecture. Here, we see how explanations of the natural order get entangled with explanations and workings of the social order. The enclosed nature of the game was central, and yet inexplicit, to the model of the world that it offered.

Scaino details a number of games in the *Treatise*, but he holds up the favorite game of his patron, "game of the ball and cord" ("cord" being the early version of the net), as the most noble, attributing this perfection specifically to the controlled conditions of the enclosed court:

> The nobility . . . consists principally in the perfection & excellence the things contain within themselves. First on account of place, for it is played in a court surrounded on all sides by walls & limited in space . . . we may affirm that *the cord game is more perfect than those played in the open, because the cord game is played in a restricted space*, others in indefinite space. Therefore, as eye & judgement are surer in seeing the Ball & gauging the quality of the court in the cord game, it follows that this game is less subject to chance than the others, & more regular, skilful & so more perfect. Then, too, it is more perfect also in the way it is played, for the cord game is played in a more restricted way than the others & is therefore more perfect.[39]

The virtue of the enclosed version of the game, for Scaino, was how it brought unpredictability into the folds of predictability, providing an opportunity to master chance elements by restricting play to a reliable space. Even though the walls produce unexpected bounces, players can more reliably see and respond to this ricocheting ball because an enclosed court removes the uncontrollable elements of the outdoors: the ground, the light, and the weather. Scaino's argument equates containment with perfection on the authority of Aristotle, but the logic of enclosure would have been familiar to him from his clerical training. The religious cloisters played an important role in the game's initial development across the Continent, and there is a sympathy between the two logics of enclosure. Both value a removal from the uncertainty of the world, employing bounded practices as techniques in the pursuit and production of knowledge. Within the bounds of monastery life, the chaos of the secular world was limited, but it still existed through the slippages of human frailty. Within the closed world of the court, ricochet did not run rampant the way that it might outdoors but was amplified and put on display. Players and spectators could experience, in an embodied way, how things that felt unknown and unpredictable were right at the edge of the human capacity to fully and accurately predict. There was a promise that if you just practiced a little harder and watched a little more closely, here was a system that, on a given day, you could master. One way to understand the difference between court tennis and modern tennis is that the latter shed its walls but retained the former's impulse to enfold unpredictablility as it moved further and further toward the desire and display of a predictable bounce.

Of course, not just anyone could master court tennis. It was a truly difficult game. Players had to direct their bodily movements toward an erratic, unlively ball. In one instant, this would require a split-second reaction, and in another instant, a period of patient positioning as it rolled slowly down off a penthouse. For the young Scaino, the game's difficulty made the physical capacities and inner characters of the players visible to any viewer who knew how to observe properly:

> This game is of so excellent a nature that, although many practice it, few nevertheless become perfect therein; for it requires only men who are dexterous, agile, graceful, noble-minded & quick at making, without loss of time, sudden decisions that help the hand & mind of the excellent Ball player in whose bearing & habits a judicious beholder should with ease be able to judge of his value in the profession of arms, in wrestling, running, throwing the Pallone & in divers other gentlemanly pursuits.[40]

Here, Scaino imagines the game as a mirror that reflects already existing qualities: a sorting device to identify who has what qualities and in what quantities. But the game did not simply reflect the players' innate qualities; it taught them what qualities to seek and helped them develop themselves in socially preferred directions. In Renaissance Italy and across the Continent, the restriction and the bounding, of both space and self, were highly valued.[41] Drawing on Norbert Elias's work in the *Civilizing Process* and *Courtly Culture*, sociologist Robert Lake argues that the game's popularity with the aristocracy was combined with its elaborate code of etiquette, and this developed and demonstrated key qualities of the new courtly culture: self-restraint and foresight.[42] (These qualities are still prominent in modern tennis, which R. S. Perinbanayagam, following Erving Goffman, calls a game involving "the performance of self in tight situations.")[43] Lake is primarily concerned with the way that the game served as a symbol of prestige and a vehicle for social mobility among nobility. He argues that the geographic dispersal of court tennis allowed the European aristocracy to struggle against each other for

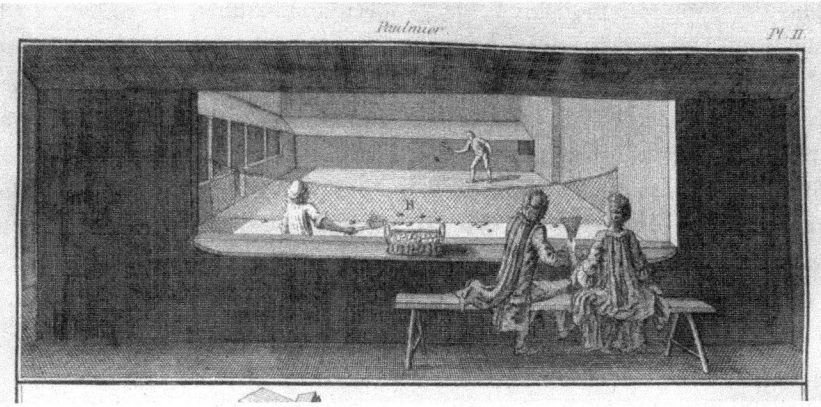

Figure 1.7
A subsection of Plate 2, *Art du Paumier-Racquetier et de la Paume*, M. de Garsault, 1767.

status, evaluating "shared sensibilities," such as self-restraint and foresight, that solidified their burgeoning class identity—different from, and yet still bound to, Christian values.[44] The game can do this work of creating shared sensibilities only because, as Scaino continues, the player is revealed not only to themselves and to their opponent, but most important to any observer: "Anyone of good judgement and knowledge can read from the face, gestures & words of such a player (as though he were looking at the image of himself in a very clear mirror) all the inner sentiments of his mind."[45] For Scaino, the game brought to the surface (*bonde*) a player's true self. As a player responds to the ball bouncing off the enclosed court's surfaces of walls and floors, their soul surfaces, in turn, for the "judicious beholder" to assess for character and for likelihood of victory.

This judicious beholder was aided by the enclosed court's restricted viewing conditions which, in Scaino's mind, helped foster the skill of proper observation. At that time, the only way to watch a game played in a four-walled room was to cut windows into the walls or to watch from above. Viewing was restricted to small numbers of people, and the perspectives were controlled and framed by the architecture.[46] Scaino argues that in conditions of containment, the "eye & judgement are surer in seeing the Ball & gauging the quality of the court."[47] This is key to the game's perceived perfection. According to Lake, "Evidence suggests that most viewing galleries had space for only a few dozen spectators, so the competition to spectate, let alone play the game, was probably fierce."[48] Spectating required one to have good standing. The game's ability to contain chance and grant understanding to those of "good judgement" leads Scaino to make his most striking claim: that the game of ball "should be held to be chief & sovereign of all other games as it is also, of all other games, the standard, measure & foundation."[49] For Scaino, a game at its best offers access to (for the knowledgeable observer) the true nature of objects, people, God, and the universe, and as the best possible game, it becomes a method for taking the proper measure of all things.

Even Odds and Credible Counting

First and foremost, what gets measured in court tennis is chance. The game's contained display of ricochet made the court into a reflective model of the world—the game's rule structure, its "game design" in contemporary parlance, grants this model its force. The chase rule, which was shared across many variations of the game, created a structure for playing between bounce and bound that honed prediction of the second bounce. Scaino's published treatise, meanwhile, gives its very first chapters over to a lengthy articulation of the rules for betting on the game. Betting was assumed to accompany play. It linked playing to watching and made who won points, games, and sets—and the order in which they won them—matter both emotionally and materially. Throughout the *Treatise*, Scaino refers to the rules for betting (along with the rules for scoring) as *questo artificio* (this device).

A similar term is used by Clifford Geertz to describe betting in his famous analysis of the Balinese cockfight, where he argues that "the center bet is a means, *a device*, for creating 'interesting,' 'deep' matches."[50] Historically, gambling on games is, to use a term from Stephanie Boluk and Patrick Lemieux, one of the most common metagames. In cockfighting, court tennis, and in so many other games that cultures gather to bet on, a tremendous amount of effort goes into attempting to create contests with 1:1 odds. Called "even odds," they are the riskiest odds to bet on. Betting intertwines social and economic risk, which for Geertz is why it is the device that makes contests operate as simulations of the social matrix of Balinese society. He defines these ongoing simulations as "deep play,"[51] a phrase that he borrows from Jeremy Bentham, who uses it to refer to actions that humans take that are irrational and against their own economic interests. In Geertz's "deep play," money and status are risked together: the more money staked on a given contest, the deeper the play. Geertz follows Max Weber to argue that "the imposition of meaning on life is the major end and primary condition of human existence."[52] When you lay yourself on the line, you increase the meaningfulness of an event. The possibility of experiencing "what a human feels like when on top of the world and when at the bottom of a pit" compensates for the economic loss that may result.

Betting on court tennis has taken countless different forms. In the first published rules, Scaino describes three kinds of victory: simple, double, and triple (also known as "furious") victory.[53] When points (which Scaino also calls "chases") were won by both sides, in no particular order, and the winning party eventually won the game by either two or three points, a simple victory was granted. The victor, in this case, was entitled to the amount that was originally bet because he had "prove[d] himself by one to be good & capable of defending his honour against the strength of the adversary."[54] When one player or side won the game in four straight points, they were rewarded for their demonstration of dominance with twice the amount of the original bet—a double victory. But the greatest reward was reserved for a very particular kind of game. For this reward to even be possible, a player had to come very close to losing. If a player lost the first three points—in danger of being shut out and losing double the original bet—but then stormed back, winning the next five points in a row, they were entitled to triple the amount of the original bet. Scaino calls this a "furious" victory. This triple reward is "a very great incentive & very sharp spur for inciting the minds of the players to so beautiful and glorious a task."[55]

By tying both moral and economic values to winning, these rules of betting pushed the play. The three possible victories depended on the sequence and the outcome of the play. Any initial bet was even money, but depending on how the game progressed, the wager held the dramatic possibility of becoming 2:1 or 3:1. By varying the value of sequences of play rather than just recognizing outcomes as a binary of winning and losing, these rules create a powerful feedback loop: the closer a player comes to losing, the more they stand to receive if they win. The betting spurred on different intensities of play, and different dramatic magnitudes of achieved excellence, in turn, resulted in different degrees of victory. By betting as a player, you staked yourself. By betting on another player, you took on economic risk that your own body could not control.[56] These modes of invested play and invested spectatorship, created the conditions for deep play for all involved.

While these elaborate betting rules are not necessary to commence a competitive game of court tennis today, the handicapping system is still employed with some regularity. The handicapping rules of court tennis work explicitly to create contests with 1:1 odds. Handicapping makes a contest "fair," where "fair" is understood to mean "equal" in the sense of all contestants having the same chances of winning and losing. Today, the primary strategy for creating equal contests is to put players into categories and then to proscribe players in one category from competing against players in another category: Men are separated from women. Amateurs are separated from professionals. Athletes of certain weights, ages, heights, or abilities are shunted into discrete matches. Rather than being presorted into fixed—and, thus, fundamentally troubled—categories, a system of handicaps levels the playing field by assigning advantages to players of clearly unequal skills and capacities.[57]

A handicap aspires to make play fair by offsetting differences between players.[58] Starting from the presumption that players are unequal, handicaps estimate this inequality beforehand and respond with a rule or other structure that supports potential equality despite that differential. What makes a "good" game is a contest where both players stand an equal chance of winning. Handicaps, thus, make games of skill more like games of chance. The large amount of effort that goes into ensuring that contests are equal can easily be read and felt as ethics. And as a model that contemporary sport might use to create new forms of competition that do not rest on gender and ablest binaries. In court tennis, the handicap system produces the greatest possible uncertainty about the outcome of a contest because equal contests are the most exciting for players to play and for spectators to bet on.

While the betting rules enabled both players and spectators to invest in play—and hone their observations—the scoring system enabled players and spectators to trust (as much as possible) that the winner was the better player. Much of court tennis' scoring system is still used in modern tennis today. A player wins points if the other player makes a certain kind of mistake, such as hitting the ball out of court or into the net.[59] The major distinction between scoring in court and modern tennis is that, in court tennis, when a player misses a ball that lands in play, instead of a point being awarded, a chase is set. Since the chase is the mechanism for players to swap between the service and hazard sides of the court, rallies in court tennis were not just contests for points but contests for the more advantageous position.

From the early 1400s through today, most games of tennis have been scored 0, 15, 30, 40, game.[60] A score of 40–40 is called "deuce," and at that point, a player needed to win two consecutive points to win the game. This last rule is key. In most sports, any margin of victory is acceptable.[61] Winning is winning, no matter how small the margin. But ball–wall sports have to account for the mediation of the surfaces to offset the high probability of the lucky (which is always also the unlucky) bounce. Good players learn how to call on this sort of luck—aiming their shots for the places on the court that are most likely to produce it.[62] In Scaino's championing of the scoring system, he explains the inclusion of a margin as an attempt to offset any unfair or unexpected behavior on the part of the environment.[63] A player must master both the opponent and the environment, but they are not expected to be wholly responsible for the ways in which the environment itself can "cheat," so to speak. By winning two consecutive points, a player provided credible evidence of their superior skill. This logic of margins appears twice: first, in the rule that games be won by a margin of (at least) two points, and second, in the rule that a set must be won by a margin of (at least) two games. This scoring system, perhaps inadvertently, made it theoretically possible to have an infinite game or an infinite set. In the right conditions, any given game—and any given match—could hypothetically go on forever.

Having to win a game by two points—and a set by two games—controlled for the environmental variable and ensured that the result of any particular game could be trusted. A game might be theoretically infinite, but practically, there was always a conclusion. When a set concluded, the result was trustworthy because the outcome was a result of *replication*: "the set of technologies which transforms what counts as belief into what counts as knowledge."[64] In *Leviathan and the Air Pump*, Steven Shapin and Simon Schaffer identify replication as the key mechanism for transforming belief into knowledge in the seventeenth century, marking a new shift to empirical truth production.[65] They revisit debates between Thomas Hobbes and Robert Boyle over Boyle's air pump and Boyle's broader commitment to repeated systematic experimentation as a mode of producing incontrovertible fact. For Boyle, an experimental hypothesis was proved or disproved through repeated tests.[66] With this method, he had found an aleatoric framework for knowledge production in which failure was as productive as success in validating the functionality of the framework. He saw his air pump, and other pneumatic experiments, as having made "a matter of experimental fact" that a vacuum exists, that air has weight and spring. This was later reformulated into Boyle's law, which states that the pressure of a gas—its degree of elasticity—exists in reciprocal proportion to the degree of expansion, and undergirds all pneumatic technologies, including modern tennis balls.[67]

Hobbes, a philosopher and avid court tennis player, debated Boyle over the validity of his experimental pneumatics—and over the broader project of empirical science—which advocated for knowledge based on observations made with the hands, eyes, and other instruments of measure conducted by credible people and communicated via printed material containing the written word. For Hobbes, experiments were fundamentally unreliable ways of producing truth. An experimental result, like a result in court tennis, might be trustworthy on a given day, in a given place, but that didn't mean that it could carry the weight of unassailable truth in the manner of natural law. Hobbes's most famous student, King Charles II, "made a sport of placing bets" on the outcome of the Royal Society's pneumatic experiments, a practice that was common from the start of the society in 1660.[68]

As Shapin and Schaffer reexamine Boyle's debates with Hobbes, they repeatedly deploy the metaphor of a game to describe the project of empirical science, drawing on Ludwig Wittgenstein's notion of a language game, which theorizes that the rules of a language are analogous to those of a game.[69] Shapin and Schaffer compare empirical science to a game to emphasize how Boyle, and the other early empiricists, used bounding practices to remove the scientific fact from the realm of both political and metaphysical accountability. Boyle's claim that his air pump created a vacuum, a space that contains nothing and thus ontologically does not exist, posed both a physical and political problem for Hobbes, who believed that "there was only one cause of the movement of a material body: the movement of a contiguous body."[70] Shapin and Schaffer explain that "Boyle identified Hobbes' criticisms as denials of interpretations, not of facts [and that] . . . by assimilating their disagreements to conflicts over interpretations, Boyle here suggested that he and Hobbes were playing the same game and that Hobbes played it badly.[71] In naming Hobbes's criticisms "denials of interpretations," Shapin and Shaffer suggest that Boyle purposely missed the point. Hobbes was arguing that knowledge and truth should not share the same structure as games. They should not offer the kinds of results that can be bet on.

Unlike natural laws, both laboratory experiments and ballgames require specialized architecture. For court tennis to be recognized across the Continent—and, therefore, for it to be capable of facilitating competition between members of Europe's courtly class—it needed to introduce some degree of standardization by having not only the same set of rules, but also the same ball, court, and rackets. A standard is a method for creating common ground, an agreed-upon measure. In sport, standardized spaces and activities create the possibility of fine-grained knowledge through repeated practice, gaining a capacity to predict the behavior of the ball, the court, or a regular opponent. The goal of standardization in sports is to eliminate home court advantage in order, in theory, to create a meritocracy. Of course, home court advantage continues to exist in every sport, no matter how much effort is put into standardization. This is a matter of building materials, environment (altitude, temperature, etc.), crowd, and other elements. In sports such as soccer or baseball, a range of tolerance is written into the rules. The rules of the game recognize the environment and the local architecture as factors but try to eliminate them by turning them from variables into constants.[72]

In their turn to the linguistic with Wittgenstein's language games, Shapin and Schaffer inadvertently suggest how the structure of the era's bounding game might have prepared the ground for the historical empirical shift in knowledge production. The problem of local versus universal, one visible in court tennis, was beneath the entire empirical project of the seventeenth century. The empiricists called for a science based on practice, a shift in the method of truth production to one based on repeatedly observing measurable expressive displays. The problem with this shift, as argued by Hobbes at the time—rearticulated more recently by Shapin and Schaffer, and also by Bruno Latour—was that the experimental method is not actually, epistemologically universalizable: a fact is always a fact under the conditions of its production—often those of the laboratory.[73] Laboratories, as Latour describes them, are "closed and protected places": "theaters of proof."[74] However, unlike theatrical productions, laboratory experiments have to produce measurable and credible results.

The demand for measurable and credible results is what makes empirical science like sport. Even though any given game's result depends on the conditions of the place and time of play, sports operate as theaters of proof that reliably produce winners and losers, validating their framework in the process. Through Boyle's law, and other such laws, knowledge is formulated as statements about the relationships that things will have under specific local conditions. Like the closed court of court tennis, experiments contain and display uncertainty, setting up conditions for results to be reliably produced, mirroring what Boyle argued. Countering Shapin's and Shaffer's argument, J. J. MacIntosh, Peter Anstey, and Jan-Erik Jones argue that Boyle did not claim that experimental facts were universal: his embrace of bounded knowledge was a full rejection of the kind of abstract universalization that Hobbes saw as a necessary condition of truth, and thus accountable to metaphysics and politics.[75]

Court tennis was never fully standardized. No two courts were exactly alike, nor were they expected to be.[76] So the net or cord, penthouses, galleries, dedans, and the grille are more properly described as the elements common to most court tennis courts. The number, scale, and placement of these architectural elements varied. The singularity of every court meant play was always local to place. Home court advantage had material meaning.[77] Still, the courts were similar enough that players could travel around, performing on different courts across Europe, and agree that they were playing the same game. Once it became fully institutionalized, court tennis was paradoxically a deeply local and broadly common cultural form. This paradox was a key source of the game's value. Court tennis demonstrated that a play environment could be standard enough. The game modeled a method for obtaining local results understood to be universally true within its bounded context.[78] In both sport and science, standard enough conditions—within playful bounds—opened new possibilities for prediction.

Creating Equal Chances with the Art of Conjecture

Bernoulli, the father of probability mathematics, picked up the empirical method of repeated experiments and applied it to court tennis to demonstrate a new method for knowing the world—a mathematical method, an "art of conjecture." Games are special objects in this art because they are spaces where decision-making is performed within a rule-bound framework. Meaning to form an opinion based on incomplete information, the term "conjecture" relates to the future, etymologically, by the root "-ject," meaning "to throw": a gesture of propelling, with force, through the air. In his history of probability as a concept, Ian Hacking describes how, during the Early Modern era, "games were seen to serve as models of all sorts of decisions

Figure 1.8
Looking into the royal tennis court at Lord's Cricket Ground, which was under repair at the time (August 2014). Photograph by the author.

under uncertainty."[79] Hacking dates this emergence to the publication of the Port Royal Logic in 1662—one year after Louis XIV, the Sun King, ascended to the throne. Used to articulate early ideas about probability, most of the games—ones that Blaise Pascal and Christiaan Huygens used to build their theories—were games of pure chance, such as dice and cards.[80]

In contrast, Bernoulli gives special attention to another game, which mixed chance with skill, played on the public court of his hometown of Basel.[81] Court tennis was part of a class of activities that combined mathematical chance with manly honor. These included some card games, chess (court tennis was described as "chess in motion"), and cockfighting. These were not games of pure chance, like dice. Neither were they games of pure (or purer) honor, associated with the development of explicit war skills such as fencing and archery. Bernoulli first put forward theses on court tennis in 1686, returning to the subject with *Letter to a Friend on Sets in Court Tennis*, which was included with his *Art of Conjecturing* when it was posthumously published by his nephew in 1713. In the *Letter to a Friend*, Bernoulli used a game of skill instead of a game of pure chance to demonstrate how probabilistic knowledge—and mathematical and scientific reasoning—could be applied to practical and ethical life. Court tennis was found to be "very fit and worthy to increase the precision of your thought."[82] Physically enclosed and rule-bound, court tennis provided a framework where one match played by two opponents on a given day was essentially equivalent to a match played by them on another day.

Bernoulli was interested in calculating the expectations of winning at any point in a game (rather than predicting the outcome of match)—a search for a method to calculate probabilities as knowledge. The method that he developed required extensive observation that had a different nature than Scaino's. His thesis about the handicapping system demonstrated how to calculate the number of points that should be given to a weaker opponent to make a contest equal (similar to a handicap in golf). His approach paid attention to accumulation. He proposed counting the strokes won and lost by each player (rather than the number of games) to "measure the strengths of players with numbers." This paved the way for his groundbreaking theorem for how to calculate degrees of certainty.[83] This theorem, commonly known as "the weak law of large numbers," states that in situations where the ratios of cases (how often an outcome will be A versus B) are not known *a priori*, one can find that ratio *a posteriori*, assuming that one is willing to collect sufficiently large amounts of data.[84]

Bernoulli writes that *a priori* knowledge "is rendered absolutely impossible by the action of a thousand hidden causes. But this does not prevent us from knowing the players' relative abilities almost as certainly *a posteriori*, from observation of the outcome many times repeated."[85] His proof of the first limit theorem reflects his interest in a certainty produced through knowledge of ratios: numerical representations of the likelihood of a future outcome created by accumulating past outcomes of analogous events. In the case of court tennis, he still presumed that the world was more like a sport than it is. A game of court tennis is complex, unpredictable, and challenging, but radically simpler activities of daily life, in which rules are not explicitly stated and agreed upon up front, are resolutely goal-directed in ways that daily life is not. The abstract image of the game that Bernoulli used to make his argument was simpler still.

Bernoulli is concerned with equality of the contest, a chance for each player to begin with an equal expectation of winning. It is crucially not about equality of outcomes. In *Classical Probability in the Enlightenment*, Lorraine Daston argues that the understanding of probability in this era was based on a notion from aleatory contracts: quantified expectations of results to plan for chance. These contracts calculated an "equality of expectation as a precondition for a fair game."[86] Bernoulli scholar and translator Edith Sylla also emphasizes this point.[87] Equality of expectation, however, was also already embedded in the game itself. The betting structure favored and motivated equal contests by doubly penalizing those who lost without winning a single point or game, and by triply rewarding those who almost lost in this shameful manner but then made a remarkable comeback. Rather than leveling the field—the game in fact depends on the court's many irregular surfaces—the elaborate handicapping systems leveled the players by offering a way for all the players, regardless of skill, to begin a match with equal hopes of winning. This made for the best kinds of contests: uncertain outcomes that are, in turn, the fairest contest possible and the best to bet on.

The notion of probability developed in Europe, Hacking contends, conflated ethical and mathematical senses of the term. Something similar happened with ethical and mathematical conceptions of equality in the equalized chances of games and sporting: equality of expectation overlaid ethical overtones onto the mathematical concept. With an eye on fairness, the sports analogies of probability theories insinuated a larger logic of competition, and this conflated mathematical and ethical notions of equality. Like Scaino, Bernoulli took the game of court tennis as a model of the world, but rather than a matter of personal character, social standing, and the physical laws of the universe, he created mathematical models of a messy world. Like Boyle (in opposition to Hobbes), he was committed to extensive empirical observation, but rather than produce experimental facts in a laboratory, he contended that sufficiently extensive empirical observation, paired with calculation, demonstrated future probabilities of interactions in the world. Chapter 2, on modern tennis, will further explore the connections between fairness, truth claims, the pervasive accumulation of data, and the logic of predictive algorithms for which Bernoulli and others laid the foundations.

Conclusion: Mirrors and Models

At the height of its popularity, natural philosophers and mathematicians pressed the game of ball into service for their points. At large, these games were stages for status struggles and class formations by the courtly class. Historians and sociologists, who focus either on the scientific or the social, miss how the codified and contained framework for games modeled knowledge and truth production in a manner that cut against the aristocratic order of the day. These changes would go on to have deep social and political consequences. Sports are physical, symbolic systems that we *do* with our bodies. When we play ballgames, we perform the physical, social, and political orders of our day. In this chapter, I have argued that we need to look not just at our linguistic systems but also at our actual games—the attendant social physics of bounce—to understand how social formations are configured, thereby structuring the boundaries between different kinds of knowledge. Our physical theories of the world have social and political consequences, and vice versa. The bounding practices that we employ either tie our physical, social, and political theories together—as with Scaino and Hobbes—or hold them apart, à la Boyle and Bernoulli. In either case, they enact relationships between physical and ethical comportment. Excluding politics is not an apolitical position. In this insistence that knowledge and facts are fundamentally contextual, there is strong politics.

At the end of *Leviathan and the Air Pump*, Shapin and Schaffer come down against Boyle and the empirical mode of knowledge production, ending the book with the sentence: "Hobbes was right." The results of experiments should not be understood as true. Yet, while Hobbes observed that there was more at stake—the universe itself—his conclusion was negative and limited. An avid court tennis player, Hobbes was a participant in rule-bound games that involved manipulating objects, but he did not think that these practices produced knowledge that could, or should, be called "truth." In his belief that something true is universal, he misses the implied politics of Boyle's position: knowledge is not universal; it must always be understood to be true in a given context. While this was not Boyle's explicit argument, his understanding of the contingency of context offers a foundation for real pluralism.

Amid the shift to the new empirical method of truth production, court tennis held sway in seventeenth-century Europe before falling out of favor for two hundred years. Through the game's many angled walls—its intricate rules of scoring, betting, and handicapping—players and spectators grappled with controlling, predicting, measuring, and understanding the behavior of irregular and unequal bodies in an apparently universally predictive space. Tennis regained its popularity in the nineteenth century, when a simpler iteration, lawn tennis (as the modern version of tennis was initially known), was created. By this time, the scientific method of systematic experimentation and the mathematics of probability had been greatly elaborated.

Some of the modesty of Boyle was lost. No longer a game of ricochet, the new form of the game embodied these changes. In the next chapter, I will track the movement and development from the *ricochet* of court tennis—a physical and unpredictable technology of interaction—to the *true bounce* of modern tennis—a predictable technology of interaction founded on reliably spectacular rubber bounce, televisual tracking, and virtual accuracy. This new kind of bounce was the product of imperial science and technology, and was transported around the world—first, by the agents of the British Empire and, later, by global television broadcasts. Looking back to court tennis's display of ricochet, enclosed architecture, and invested play and spectatorship, we can wonder anew at the continued dominance of ballgames today and ask: How do different conditions of bounce—and its attendant configurations of rules—underwrite our world-modeling practices today?

Figure 1.9
A basket of ball-making materials at Societe Sportive in Paris (July 2014). Photograph by the author.

2 True Bounce

i sold my soul like a tennis shoe and i derived no profit from the sale of my soul.
people who post frequently on boards appear to know that they are factory equipment and tennis shoes, and sometimes trade sends and email about how their contributions are not appreciated by management . . . many cyber-communities are
businesses that rely upon the commodification of human interaction. they market
their businesses by appeal to hysterical identification and fetishism no more or less
than the corporations that brought us the two hundred dollar athletic shoe.

—humdog, "Pandora's Vox: On Community in Cyberspace," 1994

A Ball in Hand

Some of my earliest memories are of tossing a tennis ball back and forth with
my father. There were always a few lying around. Anytime we took a car trip,
he would tuck one in his jacket pocket in case we found a few minutes to play
catch at a road stop or in a parking lot. Every Christmas, one of his closest
friends, Pete Putzel, would gift him a can of new tennis balls. He would give
Pete an egg carton of golf balls in return. In later years, he developed a habit
of handing off the tennis balls that had lost their bounce to my aunts' dogs,
Easy and Roxy, for a second act of being chased and chewed to pieces. Whenever I catch sight of one of the many faded neon balls that bedazzle the Los
Angeles cityscape—discarded at the edges of public courts, dog parks, school
playgrounds, sidewalk parkways, street gutters, and occasionally highway
entry and exit ramps—it calls up the childhood comfort of always having a
tennis ball close at hand.

In 2018, I began to collect them.[1] Not only tennis balls—I also rescued
soccer balls, basketballs, baseballs, wiffleballs, volleyballs, kids' balls adorned

with Teenage Mutant Ninja Turtles, Mickey and Minnie Mouse, and Sponge-Bob, stress relief balls sporting Telemundo and other corporate logos, the small colored plastic balls sold in packs at the 99 cent stores—picking up these punctums in Los Angeles's concrete city scenes. At some point, I plucked the tennis balls from the larger pile and placed them on the top shelf of a blue library cart next to my desk. The brands stamped across this pile include Wilson, Penn ATP, Penn Coach, Penn 2, Dunlop 4, Dunlop Stage 1 Green (a ball for beginners). The sun-faded Wilson US Open 4 was made on the same production line as the balls that Serena Williams, Roger Federer, Iga Świątek, and Rafael Nadal slammed back and forth across the net at the USTA Billie Jean King National Tennis Center during the US Open. Then there are the ones aimed at canine creatures: Boots and Barkley, Kong, Wild Ones (made from ultra-pet-friendly materials), and Martha Stewart Pets. They are ideal brandable, chasable, chewable media. They carry smells and textures. The felt can be gnawed on and gnawed through, the rubber casing punctured. A few are smaller or larger, presumably made for the mouths of smaller and larger dogs, but most appear to be the standard 2.57–2.70 inches. Almost all of them are intact spheres. A handful have sharp slits or jagged rips, exposing the rubber surface that their covers are glued onto. The wool and nylon covers range in color from white and gray to pink, blue, and green. There is even a black-and-white soccer ball print—a decent joke. But most are some shade of the familiar yellow-green.

The tennis balls that decorate the Los Angeles streets index the way that the game has filtered into everyday life. Courts can be found at public parks, high school and college campuses, private clubs, purpose-built stadiums, and training centers around the world. Modern tennis, first introduced as lawn tennis in the late nineteenth century, derives from the game that took France, England, and Italy by storm in the Early Modern era and shares some elements with that earlier game, including the size of the court and the scoring of games and sets. But it also has crucial differences.[2] The tennis of twenty-first-century television, which Williams, Federer, Ons Jabeur, Carlos Alcaraz, and so many others have taken to its zenith, is a widely played game with a professional circuit that creates global celebrities. Modern tennis commands economic and cultural capital in manners unfathomable in the heyday of court tennis. It is no longer a game of *ricochet*. Instead, it is a game organized around *true bounce*. This chapter tracks the transformation of bounce in tennis over the long twentieth century. The bounce

Figure 2.1
Tennis balls from the *Los Angeles Plays* ball collection. Photograph by the author.

that lives at the heart of today's tennis has been made and modified many times over as it has been adapted and translated across contexts and media. I look at three key changes that have all worked toward the creation of a truer bounce—a bounce that is more regular, reliable, predictable, exchangeable, and, more recently, trackable—and I show how these developments are intertwined with larger-scale changes in material and technological production and world-modeling practices.

The first change to the bounce that I look at is fundamental to the materiality of balls themselves, and took place in the late nineteenth century in Europe and the United States, when the balls used to play court tennis— leather and cloth spheres stuffed full of mixed materials—gave way to hollow, pressurized rubber balls. I argue that these pneumatic rubber objects mark a more general rise of an era of *vulcanized play*: a transformation in sport, games, and play that occurs in step with the introduction of industrialized and plantationized rubber and the accompanying reshaping of landscapes and imaginaries around the world. In this era, rubber became a foundational material for modern sport and more, for modern life. Today, in the world of sport, there is an endless array of rubber objects. Rubber is used to make sneaker soles, bicycle and car tires, swimming goggles, diving masks, exercise balls, gymnasium flooring, gymnastic tumbling mats, yoga mats, running tracks, and sometimes tennis court surfaces. Even some kinds of artificial grass are made with rubber backing and crumb rubber (chopped-up tires) that spray onto player's bodies and leads to worse injuries than those that occur on grass.[3] And, of course, the vast majority of the solid, hollow, and inflated balls made for being hit with feet, heads, hips, hands, rackets, bats, clubs, sticks, and paddles are made of rubber. Rubber balls bounce higher, faster, longer, and altogether more dramatically than balls made of cork cores, wrapped strips of cloth, or inflated pigs' bladders. Their bounce is extraordinary, even magical. While bounce does not require rubber, the history of rubber plays a special role in the history of bounce.

In the case of tennis, the creation of rubber tennis balls allowed for fundamental transformations in the game. The key to the tennis ball's new bounce was rubber's bounding properties: both its material elasticity and—once pneumatic, hollow balls were introduced—its ability to contain air and serve as a boundary between two differently pressurized environments. Pneumatic rubber bounce uses Robert Boyle's central insight about the relationship between the volume and pressure of gases to radically transform Thomas Hobbes's favorite game. The hollow rubber cores of these standardized, industrialized pneumatic objects—filled with pressurized air or nitrogen to increase their bounce, covered with wool or nylon felt, and emblazoned with the names of their manufacturers—act as miniature, closed worlds. Their thin rubber casings allow them, for a time, to maintain an internal atmospheric pressure separate from the external environment.[4] This technology of true bounce pushes the role of the surface—and the surrounding environment—further and further out of focus. So much so that, while the official rules of the game include precise specifications for the ball's weight, size, rebound, and forward and return deformation and color, they have never included specifications for what material a court surface can be made of. To date, there are over 160 different types of tennis surfaces.[5]

The second change to bounce that I examine is the moment when tennis balls were "(re)Made for TV."[6] When tennis became a mass media sport, spectating placed new pressure on the ball's bounce and reconfigured what it has meant for the ball to have a true bounce. A key component of this was the way that color television made the white ball hard to distinguish from the light green grass and white lines. In response, in 1972, the International Tennis Federation (ITF) announced a rare change to the official rules of the game, introducing the now-familiar neon yellow-green felt as the cover for the tennis ball. With the amplification of the bounce of light off the ball, assisting television viewers' eyes, tennis balls became media objects writ large—visual symbols of the game's enmeshment in the technical and social ecologies of live broadcast television as a global sporting spectacle. Once viewers are in the loop—once tennis balls need to not only bounce reliably for players but also be reliably visible to television viewers—an amateur ethos derived from the game ethic of muscular Christianity had to negotiate with other, more professionalized regimes of value emerging from corporate media and the viewers themselves.

The third and final change that I look at is yet another rare change to the ITF's official rules that responded to the representation of the bounce of the ball on television. In 2005, Hawk-Eye Innovation's ball-tracking and visualization technology was approved for use in tournament officiating. This technology uses footage from a host of cameras, set up around the court, to create animated visual estimations of where a ball has bounced, which is then presented as an instant replay that offers irrefutable, objective truth. Two years after it was incorporated into live broadcasts, the ITF approved the technology for use in officiating. This was one year after a now-infamous match between Serena Williams and Jennifer Capriati at the 2004 US Open, during which Williams had multiple balls called out, which Hawk-Eye's animations had shown the TV announcers and viewers were clearly in. The change ensures more accurate calls within the bounded frame of the game, while obscuring its own degrees of uncertainty. In doing so, this development presents the enactment of justice as something best solved technologically rather than socially.

Each of the changes to the bounce discussed in this chapter—the introduction of rubber tennis balls, the shift in the color of the ball for the benefit of the television viewers' eye, and the introduction of tracking technology to live broadcast and game officiating—pushes bounce to be more and more reliable, predictable, visible, and trackable. And yet tennis balls begin to leak and lose their bounce the minute that they are released from their pressurized cans. In this sense, they cannot even stay true to themselves. Hollow rubber balls are fundamentally unstable objects, accounting for the remarkable number of discarded balls that I see scattered around this city. The reliable loss of their initial liveliness leads them to quickly wear out their welcome, and they are left to litter the streets. This dynamic offers a route for thinking through the different kinds of pressure—from atmospheric to economic—that go into the pursuit of true bounce. These seemingly simple objects and interactions connect our everyday gestures to global histories, helping us to understand the material and cultural conditions of existence.

Rubber's Re/liabilities

The sand of the desert is sodden red,—
Red with the wreck of a square that broke;—
The Gatling's jammed and the Colonel dead,
And the regiment blind with dust and smoke.
The river of death has brimmed his banks,
And England's far, and Honour a name,
But the voice of a schoolboy rallies the ranks:
'Play up! play up! and play the game!'
—Henry Newbolt, "Vitaï Lampada," 1892

In 1874, British army officer and member of Wales's landed gentry Major Walter Clopton Wingfield patented "A New and Improved Portable Court for Playing the Ancient Game of Tennis."[7] Having served as a member of the 1st Dragoon Guards in India and China, Wingfield became a member of the Royal Body Guard and resided in London at the time of the patent. He gave his new game two names: *Sphairistike*—a Greek name meaning the art of playing ball, presented in Greek lettering—or lawn tennis. Sold in a box, it was "not much larger than a double gun case,"[8] containing everything required to play the game: a net, poles, court markers, rackets, a drastically simplified rulebook, and rubber balls imported from Germany. Since building a brick court for court tennis was a great expense, Wingfield's focus was on portability and accessibility. In the patent, he emphasizes how "the above-described portable court can be erected in a few minutes on a lawn, on ice, or in any suitable sized space either in or out of doors."[9] While the rules did not dictate specific criteria for the surface, grass was clearly preferred, as was made clear by the inclusion of the name "Lawn Tennis" alongside *Sphairistike* (spelled in Greek lettering) on the cover of the initial box sets. England's

Figure 2.2
A Jeffries Royal Lawn Tennis set manufactured shortly after the appearance of the first *Sphairistiké* lawn tennis sets, with rackets, net, and line posts. Courtesy of the International Tennis Hall of Fame.

many immaculately kept cricket and croquet lawns quickly became popular sites for the game.

The new rubber balls made portability possible. By 1876, the rules stated clear specifications for their material, size, and weight: "The balls shall be hollow, made of india-rubber; they shall be 2½ inches in diameter, and 1½ ounces in weight. Balls covered with white cloth may be used in fine weather."[10] Court tennis balls had been made of leather or sturdy cloth casing—sewn together and stuffed full of some mix of cork, cloth, or animal or human hair wound with twine or animal intestines (usually from a cow or sheep, despite the name "catgut")—and had little bounce. Rubber balls, on the other hand, were "bouncing bombs even on turf," as historian of tennis Heiner Gillmeister put it.[11] Surfaces no longer had to do all the work. He elaborated, "The solid walls encompassing the ancient tennis court . . . the shelter of its roof . . . [and] also its hard tiled floor which alone had prompted a reaction from the stuffed tennis-balls of old" were suddenly unnecessary.[12] The liveliness of the new balls allowed Wingfield to eliminate the court's walls. He also got rid of the asymmetry and the chase rules of the earlier courts, making the location of the ball's second bounce irrelevant.

Tennis was no longer a physically contained game of ricochet. Rubber balls released the game from its enclosure, allowing tennis to become a game of true bounce. The material that secured this release was first introduced to the West through European colonization in the sixteenth century, but it was in the nineteenth and early twentieth centuries that the production of rubber through the industrialization and plantationization of the *Hevea brasiliensis* tree became a major European imperial project. Equipped with elastic bounce, tennis was now able to travel the world as a "hard cultural form," capable of doing the work of empire that Wingfield knew so well.[13]

Many different roots and routes can and have been used to create elastic bounce across history. Beginning three thousand years before the industrialization of rubber, techniques for working with elastic material were developed in Anáhuac (now called the Americas)—refined first by the Olmec (Nahuatl for "rubber people") and later by the Mixtec, Aztec, Maya, Tlaxcala, and other pre-Columbian civilizations. The Indigenous roots of the technology are visible in the name *caoutchouc*—which wound its way into French, German, Spanish, and Russian during the first wave of European colonization—and became the "classical" scientific name for rubber. The term has been variously translated as "juice of the tree," "tears of the tree," and "he who casts the evil eye."[14] The English word "rubber" arrived via a different route. In its earlier uses, the word had two meanings: first, to refer to a tool for (or a person engaged in) rubbing; second, to refer to a set of games (usually three out of five) deciding a contest's outcome, especially to the final game within that set. The deciding game is the "rub," the mediating event distinguishing two players.[15]

The first person known to apply the word to the material was Joseph Priestley. In the 1770s, he used the compound noun "India rubber" to describe a small cube of the material imported from India. In London, an artists' supply store sold the commodity for rubbing away pencil marks. Emphasizing its special capacity to mediate interactions between objects and surfaces—all but rubbing away the material's origins, extraction, and production process—"rubber" stuck. Over the course of the nineteenth century, the word and its material origins became homologous with its commodities: "rubbers" referred to waterproof boots, mountaineering shoes, erasers, condoms, tires, and clothing "made wholly or partly of rubber."[16] In later usage, the word often has taken on erotic or fetishistic overtones and is hence used euphemistically for sadomasochistic desires or practices. When we say "rubber" today, we are almost always referring to a highly refined material that is vulcanized, variably elastic, water-resistant (if not waterproof), and as likely to be made from synthetic as natural elastomers.

Tennis's adoption of a rubber ball was part of the larger rise of rubber in the Western world. Schoolboys, soldiers, sportsmen, botanists, bureaucrats, explorers, inventors, industrialists, and indentured laborers of the British Empire transformed this elastic substance into one of the essential materials of the Industrial Revolution, laying the foundation for the landscape of twentieth-century global sport.[17] This is the initiation of an era of *vulcanized play*, a phrase that simultaneously names the foundational role of industrialized rubber in so many kinds of everyday, institutionalized, and spectacular play; and highlights how Indigenous histories central to this foundation have been rubbed away.[18] Enfolding more than tennis and modern sport—from soccer to cycling and beyond—industrialized rubber is foundational to modern movement and mediated interaction broadly.

Taking up the material problem of rubber, British and American inventors did not look to Indigenous histories of the technology. Instead, they understood themselves as inventors of a new technology and rightful owners of intellectual property. After years of failed attempts and multiple bankruptcies, American chemist Charles Goodyear discovered anew sulfur's capacity to cure rubber of its undesirable properties—its odor and instability chief among them, but it was English merchant Thomas Hancock who won the race to the patent office. Claiming the British patent in 1843, Hancock succeeded just months before Goodyear's US claim in 1844. Hancock named the process of adding sulfur to cured latex at high temperatures "vulcanization," suggested to him by his friend William Brockedon—painter, inventor, and author of books such as *Italy, Classical, Historical, and Picturesque*. Brockedon was following a common practice of the time: drawing on Greek and Latin roots to generate names for new technologies. (Wingfield's choice of *Sphairistike* for tennis is another example of this practice.) Hancock invoked Vulcan, the Roman god of fire and technology—to describe the process of using heat, mastication, and additives to cure rubber of its ills, making it a viable and desirable material—and linguistically erased the technology's origins in Mesoamerican society. Rubbing away that history, Hancock's classical humanist gesture instead gestures to ancient Rome as the foundation of all modern knowledge.

Hancock did not have any special investment in sport. Sports equipment was simply a kind of thing that could be made better with new industrial methods and materials. In his 1857 autobiographical account *Personal Narrative of the Origin and Progress of Caoutchouc or India-Rubber Manufacture in England*, Hancock breaks down an exhaustive list of rubber products into categories: Airproof Goods; Mechanical Purposes; Portable Boats; Nautical Articles; Surgical and Hospital Articles; Domestic Articles; and Sporting Articles. Various kinds of balls appear therein, including an Inflated Globe (for Letters) in the Nautical Articles, a Playing Ball, alongside Chest Expanders (Flat and Round), a Soap Bag, a Sponge Bag, Bathing Caps (for Ladies and Gents), a Nipple, and a Nipple with Shield in the category of Domestic Articles, and two Foot Balls, one Solid Rubber and one Air Proof, featured among the Sporting Articles. This last category makes visible two competing definitions of sport. On the one hand, sport continued to be a direct contest with nature—one that required the assistance of shooting and fishing hats, gun covers, bows and arrows, stockings for horses and greyhounds, and so on. On the other hand, footballs, cricket gloves, and leg protectors evinced a newer sporting life.

When Hancock's book was first published, India-rubber was just one of the subcategories under the British patent office's "Class 132: Toys, Games, and Exercises." Catalogued under the umbrella category of "balls," India-rubber kept company with other materials such as wood, metal, plastic, ivory, glass, and (after 1866) celluloid. Over the next fifty years, rubber transformed into an industrially, commercially, and culturally viable material in Europe and the United States, which radically refigured the demand and desire for it. Rubber went from an occasionally desired to a fundamentally necessary material. By 1909, toys and exercises had dropped out of the British patent office's purview. Now classed under "Games," gutta-percha and Indiarubber became a single category, opposed to a second category of all "other balls not composed mainly of india-rubber, gutta-percha, and the like."[19] Rubber had become the presumed material of ball play.

In the first half of the nineteenth century, the history of rubber technologies had been linguistically dislocated. Their material dislocation commenced

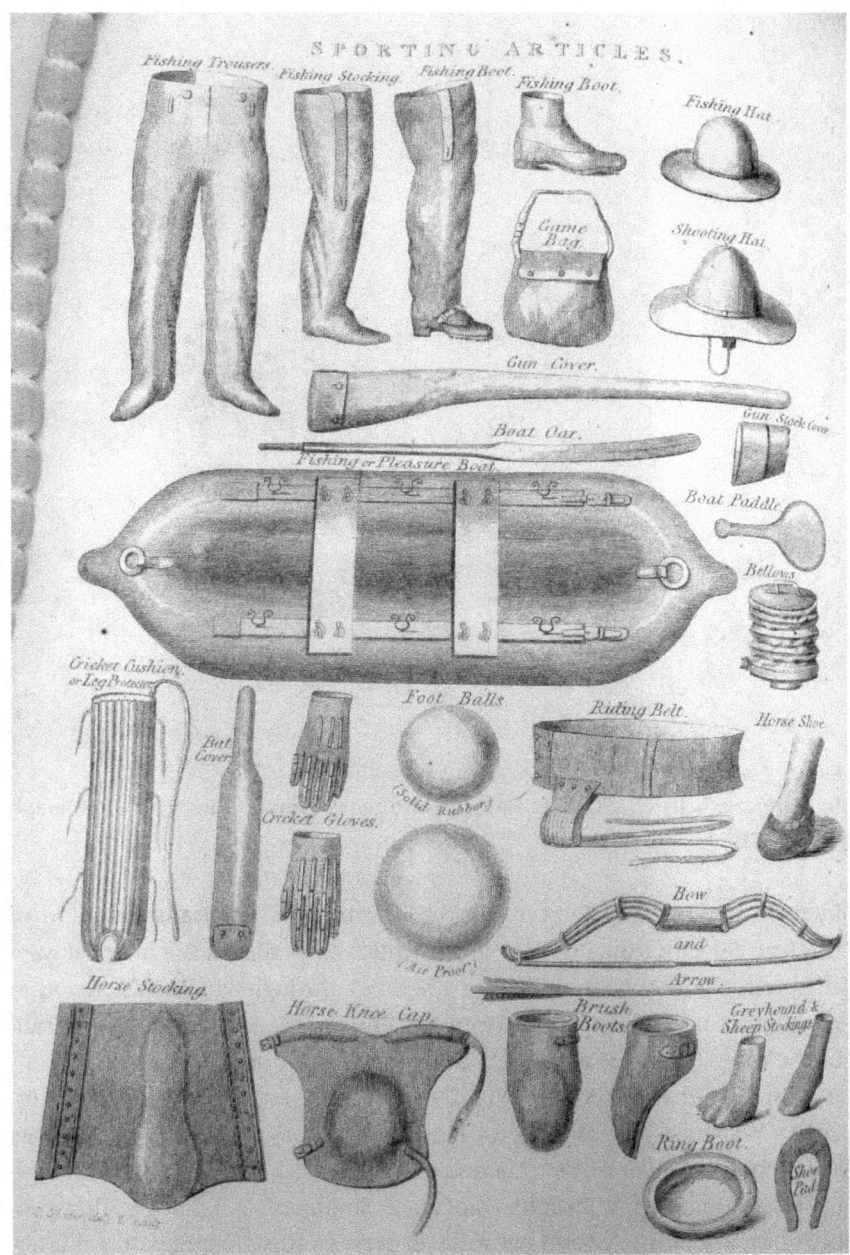

Figure 2.3

A page of "Sporting Articles" featuring both an inflatable and a solid rubber ball. From Thomas Hancock's *Personal Narrative of the Origin and Progress of the Caoutchouc or India-Rubber Manufacture in England* (1857).

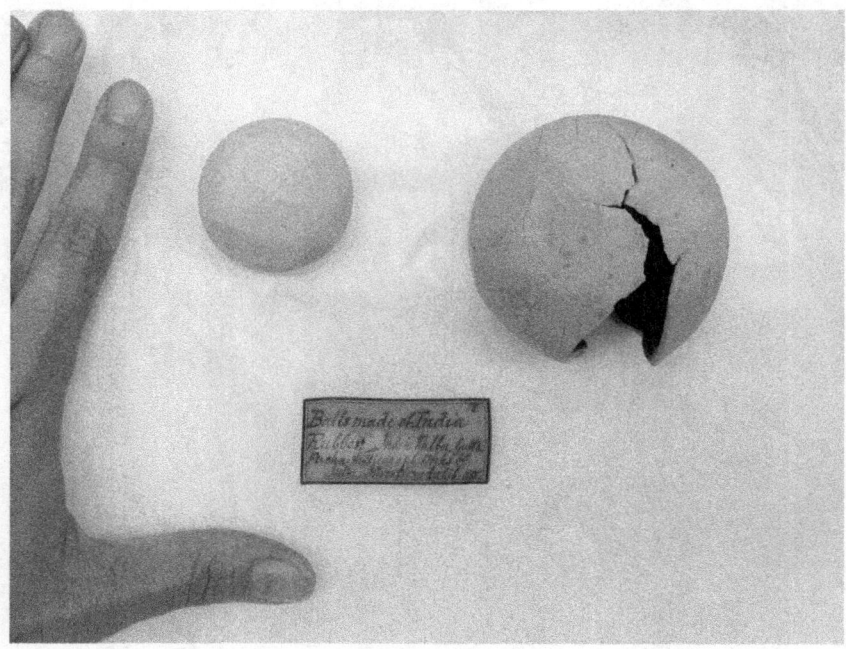

Figure 2.4
India-rubber balls c.1890, Kew Gardens Economic Botany Collection. Photograph
by the author.

in the second half of the century through the physical transplantation of
rubber trees. Once the physical and chemical properties of the material were
adequately (although not finally) stabilized through vulcanization, other
instabilities became apparent. From the British Empire's point of view, rubber's clearest weakness was that the raw material lay outside their direct control. In 1876, two years after Wingfield introduced his new game of tennis,
Henry Wickham left England for Brazil with a commission from Clements
Markham on behalf of Kew Gardens: to bring back seeds of the *Hevea* tree
to start plantations in Britain's colonies in Southeast Asia. Prior to this, rubber's primary property form consisted of great swaths of land, with wild rubber trees growing on scattered parts of it. Rubber barons employed armies
of *caucheros* (rubber collectors) to scour the land and tap the trees. Plantations offered a more condensed, controlled, and efficient property form, with
trees planted close to each other in neat rows. Labor was overseen via brutal
exploitation tactics that secured an ongoing supply of commercial product.[20]

Trees take time to grow. Wickham stole seeds in 1876, but it took decades for the British rubber plantations to fully take root.[21] With the slow but inexorable growth of British plantations promising to undercut existing models in South America and Africa, systems of extraordinary oppression and genocidal violence appeared at the primary sites of wild rubber extraction. The best-known atrocities occurred in the Congo Free State, a territory under the personal control of Leopold II of Belgium from 1885 to 1908, and in the Putumayo region under the control of Julio César Arana, the Peruvian head of the British-registered, Anglo-Peruvian Amazon Rubber Company from 1879 to 1912. Due to murder, starvation, exhaustion, and disease, around ten million died in the Congo region during this period of rubber extraction—on par with the total number of causalities in World War I.[22] The visibility to the Western world of these two sites of sustained murder was due to two reports written by Roger Casement, a diplomat sent as the British consul to visit, document, and report on each site.[23] It is impossible to give an accurate number of deaths, let alone a complete list of names of those who died due to the ruthless pursuit of rubber.

The greatest driver of the demand for rubber was the invention of pneumatic tires—and their adoption by the nascent automobile industry. Scottish veterinarian John Boyd Dunlop's pneumatic tire was initially made not for cars, but for bicycles.[24] Building on his experience inventing and producing rubber appliances in his veterinary work, Dunlop made a test tire for his son's tricycle in 1888. Borrowing the method for inflating and tying off football bladders, he used his son's football pump to inflate his new "air tyre."[25] Soon after, a cyclist sporting Dunlop's tires ran away with a local cycling race. This caught the eye of Dublin-born financier and cycling enthusiast Harvey du Cros, who bought Dunlop's patent and incorporated the first iteration of what would become Dunlop Rubber. The company would keep the inventor's name and image in the public eye long after he himself had been removed from play.[26] The du Cros family built Dunlop into a sprawling multinational rubber corporation, becoming a leading sports manufacturer known for their tennis balls and tennis rackets.[27] It is not an exaggeration to say that sport propelled the history of Dunlop.[28]

In the 1890s, Dunlop was purchasing rubber from manufacturers relying on "wild" rubber collected from trees and vines by indentured, enslaved, or otherwise coerced workers in Brazil, an independent nation under the control of a military dictatorship led by Deodoro da Fonseca, the Putamayo

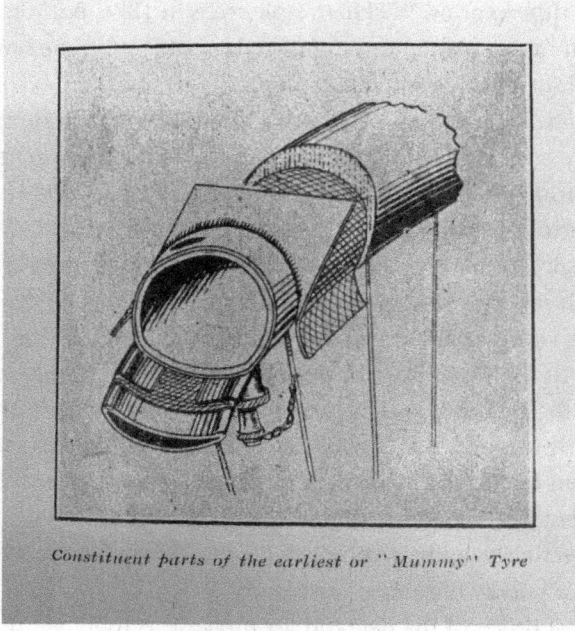

Constituent parts of the earliest or "Mummy" Tyre

Figure 2.5
This drawing of the "Mummy" Tyre was included in Dunlop's *The History of the Pneumatic Tyre,* which was completed and published posthumously by Dunlop's daughter, Jean McClintock in 1925.

region of Peru, and the Congo Free State. Under du Cros family leadership, Dunlop became an "all-up" company, controlling production of raw rubber to make products, from "cord to the finished tyre: manufactured worldwide and sold worldwide."[29] The company established its first rubber plantations. in Malaya (now known as Malaysia) and formed the Virginia Rubber Company (renamed Dunlop Plantations in 1915).

By its peak in 1920, Dunlop had plantations and factories around the world, with especially large holdings in India and the Straits Settlements— colonial territories that are now part of Malaysia. By World War I, the company was the single largest landowner in the Straits Settlements and remained so through the 1970s. The official *Story of Dunlop,* published in 1958, proclaims: "By this means Dunlop not only secured technical control of its main raw material but made an attempt to protect itself against violent price fluctuations in the world market,"[30] Not included in the story are incidents such as the police massacre of striking workers as they marched in protest of the prior

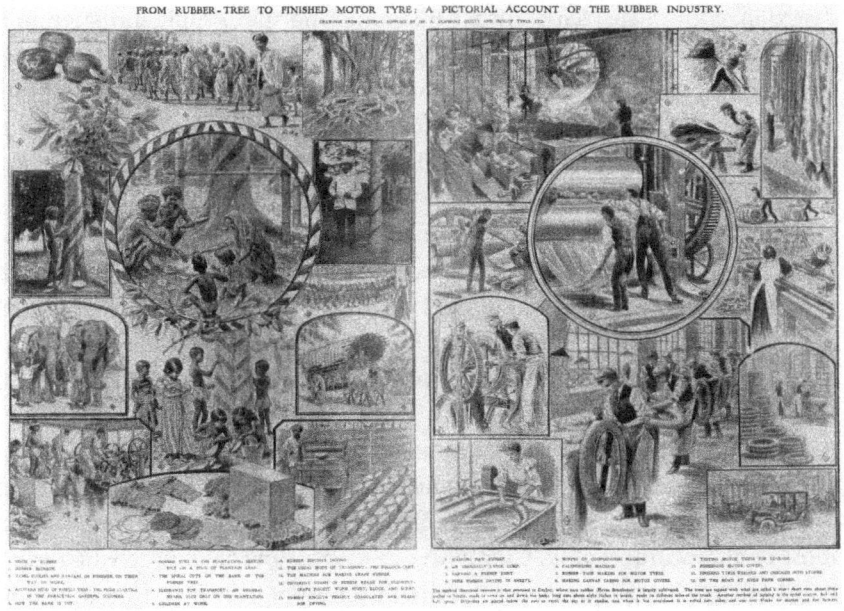

Figure 2.6
A two-page pictorial account of production from tree to tire based on material provided by a Mr. A. Oliphant Devitt and Dunlop Tires Ltd. Courtesy Chronicle/Alamy.

arrest by colonial police forces of two of the strike leaders.[31] The company's violence against the land and the workers' bodies at the sites of production was repressed, only to return, refigured as "violent price fluctuations," from which the corporate body needed protection.[32]

The du Cros family's deep investment in sport reflects the ways the modern notion of sport in turn of the century England was undergirded by new imperial industrial materials, technologies, and practices. Through vigorous physical activity, especially through sport and military and civil service, such imperial practices were disciplined by an ethos of muscular Christianity, which demonstrated Christian ideals of honor and character development. This ethos adapted an argument made by Renaissance humanists in Italy, such as Antonio Scaino, who drew, in turn, on a classical Greek and Roman idea about the connection between physical and moral character. In *The Games Ethic and Imperialism*, J. A. Mangan describes the "profound purpose" of gameplay, seeing to "the inculcation of manliness"[33] and instilling in the public schoolboy "the basic tools of imperial command: courage, endurance, assertion, control and self-control." By promoting "not simply initiative and self-reliance but also loyalty and obedience," Mangan elaborates, "it was therefore a useful instrument of colonial purpose. At one and the same time it helped create the confidence to lead and the compulsion to follow."[34] In his 1959 article "They Taught the World to Play," Charles Tennyson describes Victorian England as "the world's games-master." Tennyson declares football, cricket, lawn tennis, and golf "the great games of the world."[35] Victorian public schools—with their football and rugby fields, handball, rackets, and squash courts—produced soldiers and civil servants who spread these games around the globe. When a British public schoolboy from Harrow, Eton, or Rugby went on to become a commanding officer in Her Majesty's Armed Forces, he took his games with him.

As a game ethic grounded in honor and sacrifice became rearticulated as fair play and team loyalty, it also came to include something more modern: a new emphasis on codified rules, regular training, timed intervals, and, most of all, recordkeeping. Gillmeister describes how "at a time when capitalist entrepreneurs strove for the manufacture of ever better products in even less time, sportsmen began to measure, not the output of the machine, but the efficiency of the human body, by comparing their performances with earlier ones, with a view to improving them. Such experimenting is borne out by the appearance of records: the keeping of times and the results of matches."[36] To keep an accurate record, the instruments and architectures of play had to be constant and consistent for the comparison across matches to be valid. In the case of tennis, the court itself became symmetrical, making comparison of two opponents easier. In addition, the guidelines for betting on the game—ones that Scaino took as his starting point for "the game of ball"—were nowhere to be found in Wingfield's new rules of lawn tennis.

In the case of Dunlop, the game ethic is redeployed in a new setting: that of the multinational corporation. "The battle of Waterloo was won on the playing fields of Eton," a saying attributed, without evidence, to the duke of Wellington encapsulates the view that games train boys to be good soldiers. Harvey du Cros's third son—and eventual managing director and deputy chair of Dunlop Rubber—Arthur du Cros offered his own twist to this famous idiom: "To-day 'the playing fields of Eton' and all they stand for are being out-rivalled by the playing fields of industry and the Nation, a valuable and significant feature of our modern commercial life."[37] The du Cros family used

Figure 2.7
The frieze at the entrance to Lord's Cricket Ground in London, with the refrain from Henry Newbolt's "Vitaï Lampada," "PLAY UP, PLAY UP, AND PLAY THE GAME," etched behind a line of sportsmen and women. Photograph by the author.

these "playing fields of industry" to promote the same game ethic fostered in the British public schools within their company. As Dunlop Rubber expanded rapidly, they created and acquired ever more subsidiaries in Europe, Africa, Asia, and the Americas, becoming a private rubber empire. Meanwhile, those very games—all the ones that Tennyson names, except for cricket—used balls made with rubber bladders or cores. With this series of playing fields, this new model deployed an ethos of fair play—one that aspired to produce an ever more regular, reliable, predictable, and efficient world. Yet this world— and its model—relied on a dramatically unequal political economy to produce the instruments of better, truer bounce.

Testing for Truth

For any sport played with a round ball, the bounce must be true, and for the bounce to be true, a ball must be regular, round, balanced, and evenly surfaced. *True bounce* pushes this further: it is the pursuit of precision, reliability, portability, and substitutability. Valued above other concerns, true bounce became possible with techniques of industrial production to make rubber balls. The notion of true and false bounce first appears in early twentieth-century American and British discussions of sport. Both ball and surface were understood as potentially jeopardizing the trueness of the bounce. Writing in 1903, Eustace Miles comments that both "the Courts and the balls have grown more and more true and uniform, more and more reliable."[38] A *New York Times* article from August 21, 1914, headlined "No Cause for Worry: European War Will Have No Effect on American Sporting Implements," describes the qualitative difference between English and American tennis balls as "a trifle in resiliency or 'true bounce' that only the most expert player can detect."[39] Another article in the *London Times* from the same year details a dispute between the British and American Davis Cup tennis teams. The Americans had worn cleats (which the British call "steel points"), tearing up Wimbledon's painstakingly prepared grass and affecting the bounce: "Steel points . . . produce false bounds, and the more false bounds the worse for the British Isles players, who allow the ball to bounce more often than the Americans do."[40] The writer describes the differences between the players as differences of style. The Americans, he writes, are understandably anxious about "their foothold, which depends upon wearing what they are accustomed to; and their foothold is as much to them as the true bounce is to us."[41] Both anxieties are about not being able to predict the result of an interaction— between ball and surface or body and surface. In the latter example, the ball is true but the surface, marred with holes, makes the bounce false. A smooth, level surface is needed to create "true bounce," presented as a British cultural value that lives in tension with an American value of "level footing."

Supplying true bounce—a bounce that can be approximately the same across variations in location, surface type, atmospheric pressure, and other factors—became an important part of what sporting goods and tire manufacturers like Dunlop did. Likewise, this quality became something that governing bodies would oversee—whether through the rules and specifications of play or through the regulations of industrial supply chains, from plantation and processing plant to factory and distribution center. Organizing bodies usually garner less attention than the spectacle of individual players' performances, but these institutions produce the equipment and build the venues—the fixed ground upon which contemporary athletes train and practice. This is a paradigmatic example of institutionalized play. Since the 1950s, for example, one way that Dunlop has achieved consistency is using "robot players": machines that continuously strike balls with a racket. They "rebound from concrete surfaces back again on to the machine which strikes them again. In this way the playing quality of the melton [fabric exterior] is measured in a very short time and special balls can be designed for use in parts of the world where different climatic conditions would affect

Figure 2.8
Championships Vinnie Rogers Dunlop three-ball can, one of the earliest examples of a vacuum-packed tennis ball can, c. 1927. Courtesy of the International Tennis Hall of Fame.

their performance."[42] Object, surface, and player participate in the production of bounce, but larger considerations loom too. Companies account for variations in climatic conditions—differences in the pressure of the atmosphere—by creating balls with different amounts of pressure, ensuring a consistently true bounce for players across various atmospheric conditions. When balls became industrially designed objects—with carefully calibrated specifications—players could count on their behavior to a different degree. Rubber-soled shoes also changed the way that tennis players and other athletes move across courts, tracks, and other surfaces, allowing them to stop, start, and change direction faster. While rubber carpet has fallen out of favor for surfacing tennis courts, today's synthetic grass and hard-court surfaces rely on plastics. The greater the reliability of objects and environments, the more fine-tuned the virtuosities of the players, who become highly sensitive to minute differences and produce extraordinary demonstrations of the limits of human capacity.

True bounce has been achieved over the last century through regular testing based on specifications issued by the ITF. Created in 1913 by twelve national tennis associations and headquartered in London, the ITF oversees all rules for tennis, wheelchair tennis, and beach tennis. Similar organizations and systems exist for most ball sports. The ITF tests everything: balls, surfaces, rackets, shoes, and player performance. For degrees of difference that are considered tolerable or "true enough," specifications are given in ranges rather than as exact numbers. Manufacturers follow these guidelines in their production of the balls—often adopting even tighter specifications on their end—and then submit samples to the ITF for approval. National organizing bodies add more layers of approval and testing. Any company may submit balls to the ITF for testing, and if they pass, the balls from that production line are sanctioned by the ITF for use in specific kinds of play.

The United States Tennis Association (USTA) is the organization in charge of measuring and testing for these smaller and smaller differences to ensure that tennis balls and court surfaces produce the bounce that players and viewers expect.[43] I visited the USTA ball testing lab in Westchester, New York, in October 2014. Located off a nondescript hallway in the USTA's headquarters, the lab was the domain of Suresh Ponnusamy, an engineer and former professional tennis player who was charged with conducting ball and court surface testing for the association. Brimming with various testing machines—alongside boxes and cans of balls from manufacturers such

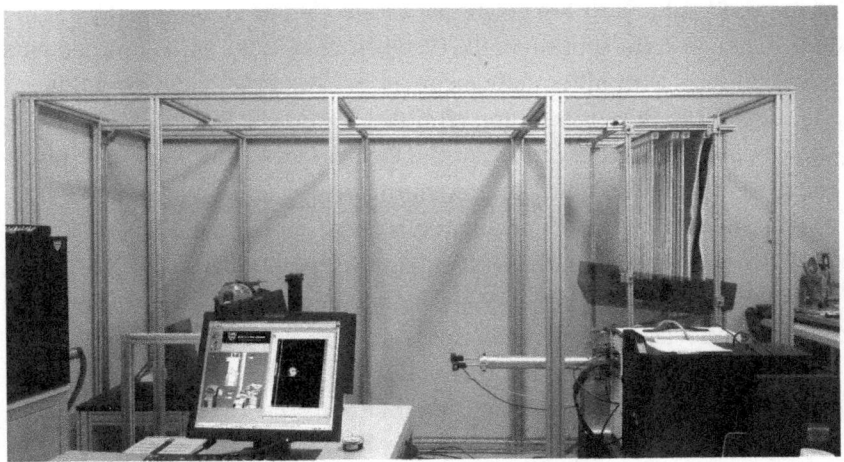

Figure 2.9
USTA ball-testing lab, 2014 site visit. Photograph by the author.

as Penn, Dunlop, Gamma, and Wilson—the lab consisted of a small, car-peted room that opened onto a slightly larger one with laminate flooring. In the second room, two of the newer machines were connected to a desk-top computer running USTA-branded software to record and visualize the results, testing bounce height, ball mass, ball deformation, and other ele-ments. Ponnusamy also used his laptop to run tests and log results from smaller machines in the front room. Ball manufacturers send samples of all the balls that they want to bear the "Approved by USTA" logo on their prod-uct to this lab. The existence of samples makes apparent that these balls are not, or are not just, singular objects. Any given Wilson US Open 4, Penn 2, or Dunlop Pro is the output of a specific design material through its production line. Once approved, these balls bounce into USTA-sanctioned junior, profes-sional, and senior tournaments.[44]

One of its most high-profile tasks, the lab tests the Wilson US Open 4 balls for the consistency and reliability necessary for the US Open.[45] The US Open Tennis Championship is the fourth and final Grand Slam tournament of the year on the professional tennis tour. (The four Grand Slams are the most prestigious events in tennis, and they offer the most prize money and rank-ing points.) First played on grass courts in 1881 at the Newport Casino as a club invitational, since 1978, the Open has been played on the hardcourts of the USTA Billie Jean King National Tennis Center in Queens, New York. Wilson has been the sponsor of the US Open since 1978. Other tournaments have contracts with other manufacturers. For example, Slazenger has been the official supplier of balls for Wimbledon since 1902. While Wilson is also the official supplier of balls to the French Open, they make a different ball designed for clay courts in that tournament.

Like many pressurized pneumatic objects, tennis balls are physically unable to stay constant to themselves. This has been the case since the invention of pressurized balls in the 1920s. Tennis balls begin to "die" the moment that they are released from their pressurized cans, slowly equalizing with the external atmospheric pressure. True bounce requires staving off this inevitability to maintain the equivalence of one ball and others. For big tournaments, samples of already sanctioned balls are retested by national organizations to ensure a particular batch's quality. Any batch can degrade due to changes at the production facility or in sourcing the rubber. Their bounce can change depending on the rubber plantation, Ponnusamy explained, because the rubber quality differs from site to site despite the intense degree of monoculture in rubber plantations.[46] The Wilson US Open 4 tennis balls in the USTA facility represent a standard, one standing in for the ideal state of any Wilson US Open 4 ball fresh out of a can. Because all balls are a little leaky, they are changed regularly. Over 90,000 tennis balls are used during a single US Open tournament.

The USTA lab is an ideal space against which all outside results are measured, but court surfaces cannot be brought into the lab. Therefore, Ponnusamy also does extensive testing on the court surfaces at the Billie Jean King National Tennis Center. In the lab, he uses a machine with a high-speed camera to test the balls on a "standard" surface of granite. Tennis players do not play on granite, but it is a frictionless surface, so it produces a baseline against which results from outside the lab can be tested. The goal is to get every surface—every part of the surface, lines included—to bounce evenly. There are endless possible ways that this can be foiled, though. If a top surface is bouncing too fast, a new coat, with more silicon mixed in, is laid down to slow it. If the surface is too slow, it needs to be played on to speed it up. When it rains hard one day and then it's very hot the next day, vapor gets trapped under the surface, causing the court to bubble and creating an effect where the ball will not bounce at all. To release the vapor, the solution is to drive a nail into the surface. Over time, testers realized that the biggest variable was coming from the ball itself, getting warmer and cooler, rather than from the surface temperature changing. Rubber responds both to impact and temperature: as it heats up, it's bouncier. To solve this problem, they created a constant for the ball temperature by taking the temperature of the ball before each test.

The working over and shoring up of both ball and surface reflect an investment in a kind of play that is the opposite of mastering ricochet. No longer is the goal to put an unpredictable, surprising bounce on display and marvel at the ability of a player to master it. Rather, effort is plied across the board to create a constant bounce, aimed at reducing environmental forms of chance at all costs—or at great cost—so the players seem to emerge as the only sites of difference. On the manufacturer side, batches of balls are tested for regularity and substitutability, checking their temperatures, testing surfaces for consistency, and other evaluations. While the ball approval processes produce a degree of reliable bounce that the everyday player enjoys, the USTA's level of effort and expense is not designed for the everyday game. This level of attention and concern for true bounce is reserved for professional tournaments—and not just for the players, but also for television audiences.

"Made for TV"

Since the founding of the ITF one hundred and twenty years ago, changes to official tennis rules have been few and far between. Two of the most consequential changes were motivated by television, and they both reconfigured and expanded the conditions of the game's bounce. The first is evinced by a somewhat unusual display hanging on the left wall of the second room of the USTA lab. Set apart from the many machines and screens, there are six framed, slightly askew, unequally sized panels of colored felt—five variations on optic yellow set next to a single white panel. At first glance, they seem a casual wink to the history of abstract painting, yet these panels have a utilitarian purpose. They are color test swatches, shades of felt that meet the strikingly vague rules that tennis balls be "white or yellow." First approved by the ITF in 1972, the same year as the initial release of the video game *Pong*, the optic yellow cover was added to the previously approved white cover in response to the introduction of color television. With the shift from white to fluorescent yellow, tennis balls became "Made for TV." In this, they joined a host of everyday objects that, with the rise of color television in the

Figure 2.10
Optic yellow panels on the wall of a USTA bounce test facility, 2014. Photograph by the author.

1970s, were designed (or redesigned) to be seen on television screens. Slowly, these objects populated homes, schools, streets, shops, and the few remaining screenless spaces with television-specific aesthetics.

Tennis's relationship with television began long before the shift in ball color. Sports in general, and ball sports in particular, became central to television beginning in the late 1930s, about a decade after the invention of the medium.[47] What sports offered television was their aleatoric liveness. During a live broadcast of a sports event, no one knows what will happen when the broadcast starts, but they know that by the end, there will be a result—at least in most cases. Even with the advent of streaming—a time when television was remade into an on-demand medium—sports remain one of the few kinds of programming that regularly compel people to gather to watch a broadcast (or stream) at a specific time on a specific day. Live sports have kept live broadcast television going to such an extent that it is possible to say that modern sport—with its liveness, aleatoric structure, and cyclical seasons—is made for television.

Tennis served as a key test object for some of the earliest live outdoor broadcasts on British television. In 1937, Wimbledon was the second live

Figure 2.11
A still from the 1937 Wimbledon broadcast, including highlights of the match between British player Bunny Austin and Belgian player Andre Lacroix. The ball is barely discernable in the footage. The film footage can be viewed on the British Pathé YouTube channel. Screenshot by the author.

outdoor event broadcast by the BBC, less than two months after the first, the coronation of George VI. The BBC's weekly magazine, *The Listener*, explains the broadcast's great success, despite the minor detail that the ball was mostly invisible: "As in the news films, it has seldom been possible to watch the progress of the ball itself. But the strokes and the movements about the court have all been so clearly visible that the absence of the ball has hardly seemed to trouble the viewer after his eyes and his spectator's reactions have become accustomed in a minute or so to the strangeness of it all."[48] The camera was positioned to give viewers a picture of the entire court at once, rather than flipping rapidly between one side of the court and then the other, which helped viewers to adjust their eyes and reactions to reconstruct the path of the ball in real time.[49] The strokes and movements about the court were good enough indicators of where the ball had been and currently headed. Through the frame of a court (tennis, basketball, volleyball), field (soccer, baseball, American football), or rink (hockey) within which the movement took place, the implied movement of an invisible object was made available to the viewer. Television's first job was to reproduce and double the preexisting frame provided by the game's marked boundaries. In doing so, the frame of the television screens—and the viewers at home—became the frame of the court and the game in turn.

This feedback loop between television audience and televised athlete is central to twentieth-century sport.[50] Vilem Flusser's *Into the Universe of Technical Images* describes the looping interaction generated by technical images using the example of a Brazilian scientist succumbing to the spell of watching a Brazilian football club play a German club in Tokyo: "At first, he thinks he has caught his enthusiasm from the enthusiasm of the Brazilian players. Under critical analysis, however, he confirms that these players were enthusiastic because they knew he and those like him were watching them. They were not playing as a function of the match but as a function of the image's transmission."[51] Flusser names here the conditions of possibility for the match and its effects, which are impossible without the transmission. Both the enthusiasm of the players and the spell of the fans are effects conjured by transmission.

While television reshaped the conditions of possibility for modern sport, when color television was introduced to the general public, sport was a vehicle for selling the new medium in turn. Color television was initially invented in 1928—almost a decade before the first live black-and-white broadcasts of the coronation and Wimbledon in England—but due to the two world wars, it was not fully developed and adopted across the majority of television broadcasters and viewers until the 1970s. As Sue Murray lays out in *Bright Signals: A History of Color Television*, in the United States, "regular color programming began in 1954, with an emphasis on sports and spectaculars."[52] David Attenborough—then controller of BBC2, now best known for his voiceovers of nature documentaries—recalls being tasked with taping the BBC's first live color broadcast in 1967 and—as those before him had done in 1937—turning to Wimbledon:

> I was controller of BBC2 in 1967 and had the job of introducing colour. We had been asking the government over and over again and they wouldn't allow us, until suddenly they said, "Yes, OK, you can have it, and what's more you're going to have it in nine months' time," or whatever it was. "And it suddenly dawned on me that the one thing we did have was outside broadcast units. I thought, "Blimey, couldn't we deploy them?" And then I thought of Wimbledon. I mean, it is a wonderful plot: you've got drama, you've got everything. And it's a national event, it's got everything going for it.[53]

Murray shows how sports shored up claims about the dimensionality of color—and "color television's heightened relationship to veracity."[54] "Certain sports programming provided a very utilitarian argument for color—legibility," she writes, "It allowed viewers to better locate a ball, identify teams, distinguish where a body ends and the field begins, and so on."[55] In theory, color television addressed the invisibility of the tennis ball. In practice, however, color did not necessarily deliver this legibility.

Improvements to black-and-white picture quality had made the tennis ball a little less invisible over the years. So much so, that when Attenborough taped coverage of Wimbledon in 1967, it became clear to him that color television actually *reduced* the legibility for viewers. The white ball—stained slightly green from grass—was even harder to distinguish from the white lines of the court, white tape of the net, and green of the court.[56] Attenborough pushed the ITF to explore new colors. In 1972, they approved a fluorescent yellow felt to cover the balls. They named the color "optic yellow" based on research showing that this color would be more visible to television viewers.[57] It was no longer good enough to let viewers imply the movement of an invisible ball. The ball needed to bounce true not only for players, but also for television viewers. Once legibility matters, then true bounce becomes a matter not just of how a ball bounces off a surface, but also how light bounces off the surface of the ball as an image of the ball's movement is delivered to millions of eyes glued to glowing screens.

Seeing Is Believing

> In all court games, every time a player serves or returns service, makes a drive or a drop, or an overhead smash, he or she is touching the world of objects around him or her, contacting them and connecting his or her self to them in delicately measured moves. An inch, a part of an inch, out of the lines of the various boundaries—horizontal or vertical—and the player knows that he or she has stepped out of bounds, has let his or her self be a little looser than he or she should have, has allowed it to wander a bit more than was good for him or her.
>
> —R. S. Perinbanayagam, *Games and Sport in Everyday Life*

The adoption of optic yellow addressed the way that light bounced off the ball not only from the perspective of the players and stadium viewers, but also to the cameras and onto the television viewers' eyes. The next substantive change to the rules of tennis addressed the question of precisely where

Figure 2.12
Diagram of the camera setup for line-tracking system in tennis. Image courtesy of StudioSayers.

the ball bounced. Was it in or out? Did it catch the tape of the net? Did the player's foot cross the line while serving (a foot fault)? In ball sports, the question of how to judge—and who shall judge—whether a ball is in or out of bounds is a matter of justice, revealing core values. Like the rest of society, sport is currently navigating the role that technologies can, and should, play in adjudicating matters of justice. ITF and USTA officials had barely begun exploring the possibility of an instant replay system to game officiating when the now-infamous 2004 US Open semifinal match between Serena Williams and Jennifer Capriati took place.

During the match, television networks were using Hawk-Eye's Shot Spot technology. Hawk-Eye Innovations is a British company, founded in 2001, by artificial intelligence (AI) researcher Paul Hawkins to analyze cricket matches. Presumably named after the expert marksman and fighter in the Marvel Universe, known for his strong moral compass, which at times leads him to disregards direct orders (bad calls) that he has been given, Hawk-Eye makes optical tracking, vision processing, video review, and graphics technologies for sports. Ball-tracking technologies like Hawk-Eye are called "track estimators" because they use multiple computer-linked television cameras, positioned at different angles, to trace the ball's trajectory. The computers process the video in real time and then combine a minimum of six separate views to produce an animated three-dimensional (3D) representation of where the ball landed. Named Shot Spot, these animated representations were first incorporated into the live broadcast of the US Open in 2004, meaning that viewers were privy not only to live-action replays—a long-standing feature of sports broadcasting—but also to animated visualizations that showed a reconstruction of the path of the ball, marking where it had landed on the court. None of these visuals were available to the chair umpire or lines people on the court.

In the first of a stunning series of blown calls, a shot by Williams was called out. CBS's coverage showed Williams's hand on her hip as she looked across the net, puzzled. Then the broadcast cut to a view from a camera at ground level—a fuzzy gesture of a crowd in the background, icons for USOpen.org sitting at the bottom left, and a wavy USA at the top right. A flat, yellow, animated ball, devoid of any brand logo, entered from the left side of the screen and traveled across the flat, green court toward the white line—positioned in the center of the image—accompanied by its shadow. (While there is no brand name on the ball, newer iterations of this kind of replay feature corporate sponsors in the background: IBM, JPMorgan, and Citizens Bank for the US Open, and the Rolex Official Review for Wimbledon.) The animated ball bounced, with an audible "thwack," just catching the outside of the line. The ball's shadow split in two, with one shadow staying behind to testify to the location of the ball's impact, as the other accompanied the ball offscreen. The camera angle swung up, taking an authoritative, bird's-eye-view of the court and the remaining shadow of the ball. The announcers, John McEnroe and Mary Carillo, responded in disbelief: "Ooof . . ." "Are you sure? Eh well, it looked close," and the camera cut back to the action.

The second instance was more blatant. After the ball was called out, Williams walked over to talk to the umpire. Before the replay was initiated, McEnroe and Carillo were already clear: "That was way in." Williams was clear, too: "That ball was SO in. What the heck is this?" Williams *could* believe her eyes—and couldn't believe that the umpire had failed to see what she

Figure 2.13
Williams's reaction to her ball being called out in the 2004 match against Capriati, and the Shot Spot representation of where the ball landed. Screenshot by the author.

(and the rest of the viewing audience) plainly could. This time, a replay of the live-action footage was played first, pausing to reverse and replay the ball's bounce in slow motion. Again, from ground level, the animated replay followed, now centered on the sideline, with the net in the background. As the ball bounced up and into the camera's lens, the remaining shadow was left well inside the line.

By the third time, Williams just rolled her eyes and shook her head. McEnroe and Carillo were suspicious from the start—ready to call it from the instant replay: "That looked like it was in." The next animated close-up of the bouncing ball, again shedding its shadow on the back line, elicited a "WOW." And on the fourth call, they responded to the live call itself, echoing each other: "Oh no." "Oh . . ." "Please!" "Give me a break." "This is getting ridiculous." "This is getting ridiculous." Disgusted by the parade of blown calls against Williams, McEnroe said from the broadcast booth, "Hawk-Eye please."[58]

Pinpointing ball, boundary, and body, Hawk-Eye's multicamera tracking and visualization technology offered a new kind of instant replay.[59] Christopher Hanson argues that instant replays destabilize conceptions of "flow" in television, requiring a reassessment of "the lived experience of television spectatorship," because the replay fractures viewer's temporal experience.[60] In this understanding, Shot Spot's visualizations operate as a "parenthetical phrase," as Barbara Morris puts it, temporarily interrupting the live-action footage the way a parenthetical interrupts a sentence.[61] Like instant and slow-motion replays, Shot Spot redirects the attention to a just-now-passed moment that warrants another look. These parenthetical fragments permit commentators and viewers to judge for themselves if a decision on the field was right or wrong. By slotting Shot Spot's reconstructions into live broadcasts in the same way that live-action replays are interposed, some viewers may understand them as live-action replays rather than animated reconstructions.

TV replays repeat crucial moments of a game—stretching, repeating, and reversing time—and they have always posed a challenge to the hard-won, embodied knowledge that umpires accrue through years of experience making decisions in the instant of play. In *Bad Call: Technology's Attack on Referees and Umpires and How to Fix It*, Harry Collins, Robert Evans, and Christopher Higgins argue that Hawk-Eye's Shot Spot has tipped the scales. Shot Spot reconstructions appear to present a degree of objective truth that has made television viewers and commentators at least equivalent, if not privileged, actors compared to others who watch, play, officiate, and administer the game. Previously, the ontological authority of umpires—their power to make reality—was presumed the better, fairer, privileged way of knowing. This shift in epistemological privilege—from umpires to the crowd, commentators, and television viewers—meant that the television viewer was now in a "better position to make a judgement than the umpire or referee."[62]

So what is the problem here? Why are these new forms of instant replay not an exemplary case of technology delivering a level of justice that unaided humans are unable to deliver themselves? Within the closed world of the court, they do, to a degree. And yet they also warrant a closer look. In his theorization of replay, building on Henri Bergson, Christopher Hanson argues "The instigator of replay—be it a television producer or a TiVo user— . . . exercises a degree of discernment through the processes of selection exercised to privilege certain moments and sequences over others . . . the replay's *discernment* exposes the specificities of the medium—that is, the constitutive elements which define it and differentiate it from other forms but also its obsessions, preoccupations, and trauma."[63] Following and building on Hanson, we can ask: What are the specificities of Hawk-Eye's Shot Spot? And what can they tell us about this new medium's preoccupations and traumas? Given the deep entanglement of sport and television, the new technology is not the only medium at play here. Television's and tennis's obsessions, preoccupations, and traumas are also exposed.

One central obsession, or at least preoccupation, shared by sport and television in this context is with objective truth. Colins et al. argue that a crucial problem is that Shot Spot's animated *estimations* are presented as accurate objective truth when

> exact accuracy is . . . a shibboleth; this is obvious in games such as tennis, football, or rugby where the ball is filled with air. Such balls are "squashy," and one cannot know the position of their outside envelope to within a millimeter because such accuracy does not exist. . . . What happens in the case of the high technologies such as track estimators is that the actual playing area and ball are replaced by a computer-generated virtual playing area and ball, but this is never made clear to the public; rather the virtual is presented as actual. What matters is not exactly where the edge of the ball is but where it appears to be to the human eye, just as has always been the case in sports.[64]

What appears to be objective truth—and what some viewers may mistake as "real" footage!—is an approximation, in fact. Crucially, the 3D animations do not visually represent Hawk-Eye's own margins of error or the uncertainty of the ball itself. This is a choice. And in making that choice, Hawk-Eye creates a kind of "false transparency," Colins et al. argue, wherein justice appears to be done even if it is not.[65] "Judging what is happening on a sports field is a matter not of ever-increasing accuracy but of reconciling what television viewers see and what match officials see . . . it is a matter of justice," they argue.[66] To their point, that justice is reached through relational processes, I would add that these processes are always subject to histories of power. Colins et al. are primarily concerned with the impact of the use of track estimators across many sports on the authority of umpires. They neither address the relationship between justice in a sport and the sport's broader cultural contexts nor what might be driving the particular desire in tennis for the appearance of unassailable objectivity.

Another central preoccupation of tennis and television that Hawk-Eye surfaces is with boundaries and frames. Ballgames are almost all formally concerned with boundary play. In both the early and modern forms of tennis, the core obsession is the interplay of bound and boundary and the question of where the ball bounces is a pivotal mechanism in both iterations. In the modern game, the location of the ball's bounce is neither used to swap sides on an uneven court nor to reconfigure the boundaries of that court. Instead, where the ball bounces—inside, on, or outside the line—determines if the ball is in or out. Without walls to return the ball into play, the lines are the court's marked boundaries and the best possible ball becomes the one that just barely grazes the outside of the line. To "paint the lines" with the ball is modern tennis's virtuosic form of boundary play. Homing in on the game's preoccupation with the relationship between bounce and boundary, Hawk-Eye's animations show viewers what it would be like *if* the ball had indeed painted the line. Viewers are invited to take their time gazing at the shadow that the ball leaves behind at its point of impact. It is represented as the truth of where the ball landed. The roles of estimation, guesswork, and good (as well as bad) faith that were previously central to all play with the game's boundaries have been pushed into the background.

As C. L. R. James lays out so powerfully in *Beyond a Boundary*, his account of coming into political awareness through cricket, sport is never simply or fully cordoned off from the world. A game developed by the elite class in England—first spread along the routes of British colonization—tennis's sites of play reflect the sport's history. Dominated by white men, this upper-class and upper-middle-class sport's official institutions worked for a long time to preserve the game's cultural and class boundaries. In the United States, minorities were officially excluded from the private clubs—home to most tennis play through the mid-1950s—and implicitly excluded for far longer. All the tournaments necessary to qualify for the USTA National Championships, the precursor to the US Open, were held at these clubs.

But this history is not the only history of play. As David Berry notes in *A People's History of Tennis*, the sport was played as enthusiastically by women from the start. Wimbledon was the site of the first action in the suffragettes' firebombing campaign, aimed at getting women the right to vote in Britain.[67] Likewise, the exclusion of people of color from private clubs in the United States led to the development of a parallel history—beginning in the 1880s—of Black tennis clubs in the country.[68]

Serena Williams's career tracks the convergence of these parallel histories. Whenever Williams stepped on court, at least in the United States, her litany of accomplishments was set against tennis's "sharp white background."[69] In 2015, in advance of breaking her fourteen-year boycott of the tournament at Indian Wells, Williams published a reflection on her experience of being booed by fans who believed accusations that her father fixed the outcomes of matches between her and her sister. It opens, "As a black tennis player, I looked different. I sounded different. I dressed differently. But when I stepped onto the court, I could compete with anyone."[70] Williams goes on, "The false allegations that our matches were fixed hurt, cut and ripped into us

Figure 2.14
Williams after yet another of her shots is called out in her 2004 match against Capriati. Screenshot by the author.

deeply. The undercurrent of racism was painful, confusing and unfair. In a game I loved with all my heart, at one of my most cherished tournaments, I suddenly felt unwelcome, alone and afraid."[71] Throughout her career, Williams refused to contain her Blackness. As Samantha Sheppard puts it, in *Sporting Blackness*, "She achieved her stellar and unparalleled athletic career without conforming to white sporting conventions. Rather, she wins blackly, via virtuosic Black body in all of its cornrowed, cat suited, and Crip-walking glory."[72]

The poet Claudia Rankine has said that her best-selling book of lyric poetry, *Citizen: An American Lyric,* begins with reflections on watching Serena Williams's disbelief, accumulated across countless moments over the course of her career, finally overflow in her 2009 US Open semifinal match against Kim Clijsters into an out-of-control rage at a linesperson who called a foot fault against her:[73]

> And insane is what you think, one Sunday afternoon, drinking an Arnold Palmer, watching the 2009 Women's US Open semifinal, when brought to full attention by the suddenly explosive behavior of Serena Williams. Serena in HD before your

eyes becomes overcome by a rage you recognize and have been taught to hold at a distance for your own good. Serena's behavior, on this particular Sunday afternoon, suggest that all the injustice she has played through all the years of her illustrious career flashes before her and she decides finally to respond to all of it with a string of invectives. Nothing, not even the repetition of negations ("no, no, no") she employed in a similar situation years before as a younger player at the 2004 US Open, prepares you for this. Oh my God, she's gone crazy, you say to no one . . .

. . . And as Serena turns to lineswoman and says, "I swear to God I'm fucking going to take this fucking ball and shove it down your fucking throat, you hear that? I swear to God!" As offensive as her outburst is . . . It is difficult not to applaud her for existing in that moment, for fighting crazily against the so-called wrongness of her body's positioning at the service line. [74]

In his obituary of fives (handball) champion and house-painter John Cavanagh, essayist William Hazlitt declares, "He who takes to playing fives is twice young. He feels neither the past nor future 'in the instant'."[75] And yet, instants—replayed or otherwise—both make up histories and are made up of histories. Rankine's work calls for an extension of the imagination beyond the tight frame of a given point, a given call, in order to invite in the many ways that any given moment—any instant—is bounded by and comes bounding out of history. As in this instance, "[Serena] says in 2009, belatedly, the words that should have been said to the umpire in 2004, the words that might have snapped Alves back into focus, a focus that would have acknowledged what actually was happening on the court."[76] Without an expansive understanding of the individual and cultural histories called forward in this instant, the temporal displacement of Serena's reaction renders both it and her "insane."[77]

In Rankine's 2015 *New York Times Magazine* profile of Williams, she reports being moved by Williams responding to a question about her status as the second most marketable women's sports star, after the less talented but whiter and blonder Maria Sharapova, by nodding to the African American players who came before her: "I'm just opening the door. Zina Garrison, Althea Gibson, Arthur Ashe and Venus opened so many doors for me. I'm just opening the next door for the next person."[78] Reflecting on this response, Rankine writes, "A crucial component of white privilege is the idea that your accomplishments can be, have been, achieved on your own . . . Serena reminded me that in addition to being a phenomenon, she has come out of a long line of African-Americans who battled for the right to be excellent in such a space that attached its value to its whiteness and worked overtime to keep it segregated."[79] Once extricated from the individualist ideology of white privilege, accomplishments are almost always shown to be collective.

This returns me to Capriati's role in that 2004 match. A tennis match is a game played between people: an opponent is always also a collaborator. And I wonder: What would have happened if Capriati acted from this underlying shared agreement, refusing one or more of the points awarded to her? A competitive professional player, playing a match broadcast around the world, she is supposed to mind her business, leaving the judging to the judges. Those are the cultural norms for an athlete playing a professional sport. Yet, as a competitive professional player, her eyes would have perceived the truth of at least some of those balls' landings. In nonprofessional contexts, etiquette dictates that players call balls as they see them—that they follow what they see as true—regardless of whether the call is to their own advantage or disadvantage. Would it have been so unimaginable for Capriati to decide that *that* was her business?[80] This is a notion aligned with an amateur ethos and not, therefore, unideological. But even, in fact especially, when technology offers more reliable or accurate evaluations and representations, it is important to question the cost of employing technical solutions to solve social and cultural issues. These technologically efficient solutions often paper over, displace, cement, or even naturalize the deeper problem—the problem of imagination that Rankine reveals for us.

I am not offering an argument against using Hawk-Eye. Colins et al. miss how—in the context of court games—the removal of ontological authority from umpires and referees can, in some instances, counter deeply seated historical and cultural biases. In Williams and Capriati's match, the network's use of Hawk-Eye's Shot Spot made the unjust refereeing so clear—so "transparent," to use Collins et al.'s term—that the umpire in question was suspended and Williams received an apology from the tournament officials. Hawk-Eye was approved for use in major tournaments the next year and introduced to tournament play in 2006. By 2020, on fifteen of seventeen match courts at the US Open, Hawk-Eye Live's electronic line-calling and video replays replaced human lines people who had worked on the court. Their labor is replaced by teams of four who work the system from a commentary booth during each match. If judged by the decrease in mistaken calls within the strict confines of the game, this is an improvement of the prior system. But still, there is some thinking to do about what gets lost with the adoption of the "line-calling technology that took the seeing away from the beholder."[81]

I agree with Colins et al. that presenting the virtual as a more accurate and unquestionable representation of reality is problematic. Hawk-Eye is currently presented as a more robust, ontological umpire. But there are other ways to represent the data produced by this system. Following arguments made by Catherine D'Ignazio and Lauren F. Klein in *Data Feminism*, I would push for Hawk-Eye's representations to include the system's and subjects' uncertainties—for the animation to more accurately represent what the technology does and does not make knowable. Taking the uncertainties of the system—and of the balls themselves—into consideration, visual representations might present these for the judges (i.e., viewers) at home to see. A fair response to calls made within the system's margin of error (which would be some small number of the total occasions of its use) might be to *replay* the point itself. A replay of a point is a different kind of parentheses than the replay of television footage or animated representations of a point. It creates a different kind of pause—a sort of hiccup in a game's progress. This would change the rhythms of games and matches in ways that might push against the pressure that television places on broadcast sports to be temporally contained and predictable. And that would in turn become part of play, the way that let serves already are. I would also like to suggest that we pause to consider what comes of players and judges relieved of the responsibility of delivering just results in the symbolic context of sport—where it should, hypothetically, be far easier to pursue relational processes of justice than it is in the messier, less clearly bounded world at large.

What kinds of *I*'s and *we*'s are produced in a world where sports themselves desire, demand, and presume that the bounce of the ball is so utterly predictable, so accurately knowable and known? Instead of pushing harder and harder for the objects, images, and officiating of the game to be true, and instead of focusing our concern on the umpires' loss of epistemological privilege and ontological authority, maybe it is possible to flip the script. To wonder how these new kinds of replays might lead the game to its next transformation is to consider what new kinds of play—between bounce and boundary—might become possible with an expanded frame. At the end of the day, a bounce cannot be true or untrue. Bounce is simply a way to describe a kind of impact. A bounce only tells us how things are. We might wonder, instead, how well these things are lined up against our individual and collective expectations—our imaginations of how they might be.

Conclusion: From the Matter of the Ball to Virtual Bounce

In part I of *Bounce*, I have shown how tennis changed from a game of ricochet, which put chance on display, to a game organized around and perhaps even obsessed with true bounce. Looking across these histories of ricochet and true bounce, it is not just the sight of players but that of spectators that has mattered.[82] They have been in the loop from the start: first peering through the netted court tennis galleries to see if they made the right bets, and later gathering around glowing boxes, following and judging the path of invisible, optically optimized, and computer-generated imagery (CGI) animated balls. This points to a larger story—a burgeoning social logic of collective participation in judgment and experience—that tracks across televisual and computational developments. The culminating logics of physical bounce, spectatorship, and authority draw heavily on virtual forms of animated and computational bounce. Tracking, estimating, and animating bounce, like that created by Hawk-Eye, is the product of long histories in animation and computing.

If balls are special weavers of intersubjectivity, as the materialities and modes of circulation of these quasi-objects (quasi-subjects) shift, then how does intersubjectivity change? How does bounce work in animation and CGI? What do these modes of representation have to do with bounce? And what does all this teach us about individual and collective forms of play? As fields, animation and computing first became intertwined with the rise of computer graphics and in recent decades, have been increasingly hard to tease apart. While bouncing balls and ballgames have been central test objects for both fields from their beginnings, the two fields had very different understandings and approaches to producing bounce, at least initially. The chapters in part I have focused on the physical ball and the physics of bounce. The chapters in part II look at the translation of physical balls into virtual bounce and what this shows us about the operation of representation and simulation in the domains of animation and computing.

Figure 2.15
A tennis ball, coffee cup, and red tennis shoe found on the corner of Potomac Ave and Jefferson Ave, Los Angeles, 2023. Photograph by the author.

Figure 5.10

Virtual Bounce

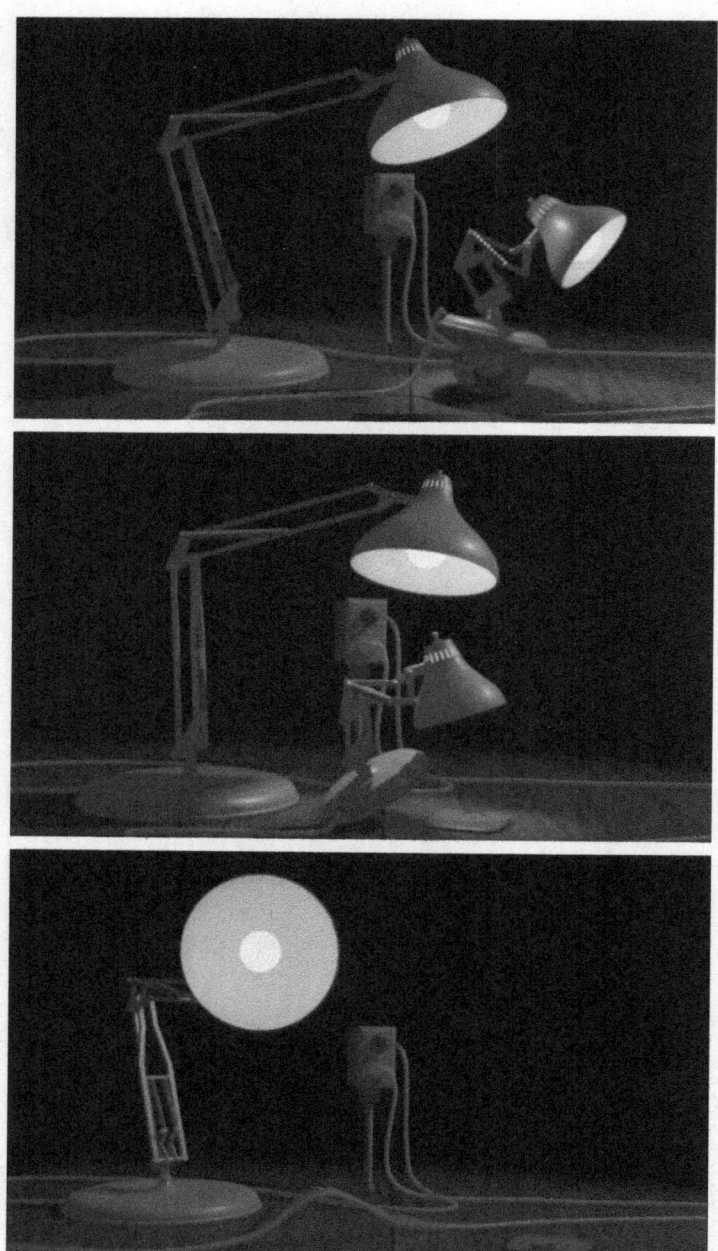

Figure 3.1
The baby lamp pops the playing ball in Pixar's *Luxo Jr.*, 1986. Screenshots by the author.

3 Squash and Stretch

"Motion Practices of Modernity"

A gray Luxo task lamp rests on a spotlit wooden surface, plugged into a protruding outlet.[1] A second long cord trails to another electric object offscreen. A yellow ball—with a blue stripe around its middle and a red star on either half—rolls into view and rebounds off the base of the lamp, with the clanging sound of a somewhat solid object striking metal. Roused by the impact, the Luxo lamp creaks into action, spotlighting the ball from a few angles before knocking it back offscreen from whence it came. The ball quickly returns, and again the lamp knocks it back. On its third appearance, the ball rolls past the lamp and out of frame. A second, smaller—and altogether springier— gray Luxo task lamp hops on screen chasing after it. The junior lamp looks up at the senior lamp, wagging with pleasure at the fort/da game, before bouncing off in pursuit. Returning with the ball, he hops on top of it, bouncing up and down faster and faster until it pops. (We have been directed to gender the little lamp by the film's title *Luxo Jr.*) The ball slowly deflates, until it lies completely squashed. As the little lamp noses the flattened ball, the larger lamp disapproves with a shaking head, leaving Junior similarly deflated. He hops heavily offscreen, only to return, reanimated, chasing a beach ball twice his size. The larger lamp directs its lightbulb at the camera—breaking the fourth wall to ask the audience to share in its bemused frustration with the little one's play—before a final round of resigned head shaking.

At the annual meeting of SIGGRAPH, the international Association for Computing Machinery's Special Interest Group on Computer Graphics and Interactive Techniques, in 1986, the debut of Pixar's two-minute digital animated short *Luxo Jr.* was met with a standing ovation. The short demonstrated

the importation of a specific set of two-dimensional (2D) animation principles into the field of three-dimensional (3D) animation.[2] The character
names "Luxo Jr." and "Luxo Sr." drag the scene between bemused indulgent
larger lamp and eager energetic smaller lamp into a normatively gendered
exchange between father and son, but the otherwise ungendered desk lamps
resist this reading. They could just as easily be father and daughter, mother
and son, mother and daughter, parent and child, adult and child, or just
larger and smaller animated beings. Light sources, the lamps cast their electric
gazes here and there—moving, shaking, and hopping with different degrees
of creak and spring after smaller and larger inflated balls. Bounces—of lamps
and light—emerge as the main characters of the short film and of animation
writ large. *Luxo Jr.* represents a semiotics of gesture based in the squash-and-
stretch animation techniques developed at Disney Studios—grounded in
notions of elasticity, spring, inflation, and deflation. With *Luxo Jr.*, the Disney
principles made the leap (a short hop, really) into digital animation, colliding
with other conceptions of bounce developed in computing.

This chapter and chapter 4 address the fundamental role of bounce in
the histories of animation and computing, respectively. Anyone who regularly looks at screens has probably followed a bouncing ball—or a ball-like
object, be it an angry bird or Sonic the Hedgehog—as it bounced, hopped,
flew, boinged, or careened across a glowing surface. Today, we are saturated
by this virtual and visual experience: *Pong* and its countless variations;
other video-game simulations of physical ballgames ranging from pinball,
marbles, and bowling to soccer, basketball, and tennis; bouncing ball, DVD,
and other logo screensavers; cartoon characters like Goofy, Wile E. Coyote,
Marvin the Martian, and SpongeBob swatting at misbehaving baseballs and
bouncing off walls or canyon floors themselves; a ball, bouncing character,
or bright color guiding the timing of karaoke lyrics; or just one more ad
populated by a hopping sprite selling something as it dances across the
small and large light-emittting diode (LED) screens that line highways, bus
stops, malls, and elevators—broadcast in public and private space alike.
These digitally animated balls tell the stories of the fields of animation and
computing. Each field uses bounce as a basic training tool for beginners
and builds from there. The resulting simulations coordinate interaction,
model physical motion, publicly demonstrate technological potential, and
produce collective performances.[3]

In the last four decades, animation and computing have become deeply entwined. Their entanglement is seen most clearly in computer graphics, and the video game industry is the obvious example. But there are others closer to hand: screensavers, wallpapers, and background or browser animations for personal computers and phones. All are replete with animated graphics. As artist and writer Lev Manovich points out, the loading animation for Netscape Navigator—stars moving across the night sky—was, for a moment, "the most widely seen moving image sequence ever."[4] Apple has a host of patented animations for its operating systems. These include the dreaded "spinning beach ball (or pinwheel) of death"—a disheartening signal that the machine is too busy computing to respond to other requests—and the bounce scroll, which responds to scrolling beyond a document's boundaries ("overscrolling") by bouncing back to the nearest display area.[5] Then there are the older forms of animation such as studio production logos, following in the endearing hops of Pixar's mascot, Luxo Jr.[6] And, of course, there is the wide world of screens beyond personal computers and phones: television monitors, tablets, printers, automatic teller machines (ATMs), liquid crystal display (LCD) billboards—the list goes on. Bounce animates them all.

In part I, the cases of court tennis and modern tennis demonstrate that when balls are bounced—or kicked, or otherwise passed around—they distribute subjectivity and enact collectives. Quintessential quasi-objects, balls contain a certain strangeness from the start. But an additional strangeness emerges when these quasi-objects/quasi-subjects move into cinematic and electronic environments. There is no expectation of being able to pick these balls up—to throw them, puncture them, or kick them into the stands. A particular filmstrip, videotape, or computer graphic may be commodified, collected, or archived, but the media object is considered the work, rather than the objects represented and simulated by that media. In all these ways, animated balls have seemingly no equivalence with the tennis balls that my father kept in his pocket to play catch with me and my brother, or the baseball signed by Jackie Robinson and the rest of the Dodgers that my grandmother's boss, the Doc, somehow procured for my father to cheer him up when he was in the hospital for six months after a car crash at age thirteen.

Simulated objects do not carry an aura in the same way that physical objects do. Neither do they accrue and carry economic value in the

Figure 3.2
The feared "spinning beach ball (or pinwheel) of death," which takes the place of the hand or arrow pointer to indicate that an application in Apple's macOS system is too busy to respond to further user requests.

same manner. We can generate emotional, psychological, or intellectual attachments to them, but we cannot take a ball off the field and out of play, or bear witness to how it inhabits space and carries marks of time and handling. The mechanism to create an object's aura, therefore, shifts. It makes the barest kind of sense to write about animated and computer-generated balls in the same way that one would write about a ball that can be served across a net, tucked in a jacket pocket, or chewed by a dog. Yet the shift into cinematic and electronic space enacts an intensifying—a doubling—of the logic of the quasi-object. Both animation and computing call heavily on the history of the physical ball, and yet the animated digital ball is discernible only as a simulation through the mechanics of bounce. Balls are quasi-objects in virtual spaces by way of quasi-actions— or interactions—by way of their bouncing movements. This changes how they pass subjectivity around and where collectivities are constructed through their passing.

How did physical balls become digital, and what happened when they did? Animation and computing have become difficult to think apart but they began in fundamentally different places. The chapters in part II disentangle these two histories, showing how they began with different relationships to representation and simulation—different understandings of what it means to produce liveliness. Importantly, the bounce and balls that initially appear in animation are different from those that first populate the world of computing.[7] A ball is the first object that an animator learns to bring to life in both classic, hand-drawn cel animation, which was made by drawing sequential images on transparent celluloid sheets, and in computer-generated animation, where software is used to generate animated sequences. This is because it is the simplest kind of character—helpfully inanimate, but lively and amenable to being squashed and stretched in all directions. The bounce in animation emerges from a properly squashed, stretched, and positioned sequence of drawings of a single ball, which is then rounded out with properly timed sound effects. By contrast, in computing, bounce programs precede and produce the ball. Bounce programs made their initial appearance when computing machines were first attached to oscilloscope screens. They quickly became standard, a kind of program often included in the user manuals for early analog and digital computers as part of the suite of demonstration programs. And they remain a common tutorial for introducing people to many programming languages today.

Taken together, we can perceive a countervailing logic bouncing between animation and computing: while early animators moved from ball to bounce (from object to action) in their quest to create lively characters, early engineers and programmers moved in the opposite direction, from bounce to ball (from action to object) and the bounce logics, programmed into frameworks of early graphical interfaces to do the job of registering, measuring, simulating, and representing different kinds of collision, were used to turn computers into real-time machines and televisions into lively, active instruments.

The rest of this chapter chases down animation's bouncing balls. In the most general sense, animation can be described as a set of practices designed to trick the senses by setting a series of successive images in motion rapidly enough to exceed the eye's threshold for maintaining the distinction between them. Once beyond this ocular threshold, images appear to move, change shape, and come to life seamlessly. Techniques for capturing motion and creating moving images dates as far back as drawings of animals with multiple legs found in Paleolithic cave paintings. During the Han dynasty, an engineer named Ding Huan made a device that Joseph Needham describes as "a nine-storied hill censer . . . on which many strange birds and mysterious animals were attached. All . . . moved quite naturally . . . presumably as soon as the lamp was lit."[8] In the last two centuries, this effect has been achieved through a wide range of techniques and technologies, from mechanical toys and flipbooks to hand-drawn cel animation and computer-generated imagery (CGI). Over this long stretch of time, bouncing balls became animation's beating heart.

The central role of the bouncing ball in animation points to deep historical and theoretical connections between animation and sport, games, and play. In *Animation, Sport and Culture*, Paul Wells argues that modern sport and animation emerged hand in hand in the early nineteenth century, each calling on the other to articulate themselves as "the new motion practices of modernity."[9] Wells identifies three properties common to animation and sport: "the shared propensity for metaphor, the technical modes of execution that result in what might be termed bio-mechanical verisimilitude, and a similar status as vehicles of hyperreality."[10] In other words, both sport and animation are practices of building symbolic worlds that offer rule-bound frameworks within which to set both bodies and stories in motion. To think through the stakes of contemporary motion practices, just follow the bouncing ball.

Philosophical Toys

In the early nineteenth century, a wealth of optical toys and magic lantern technologies were made and patented in Europe and the United States, motivated primarily by scientists' desire for a device capable of "capturing the world in its own image" and by entertainers compelled by new modes of producing spectacle and illusion.[11] The names given to these devices demonstrate a similar practice of using Greek roots to construct a genealogical claim to that revered civilization, as we saw with Thomas Hancock's naming the process of stabilizing rubber "vulcanization" and Major Walter Clopton Wingfield's naming lawn tennis *Sphairistike* (after initially calling it *The Major's Game* in the very first published edition of his new rules for the game). The thaumatrope (wonder-turner), phenakistoscope (view- or eye-deceiver), daedaleum (named for Daedelus, who built wings for his son, Icarus—with tragic results), zoopraxiscope (life-by-practice viewer), and zoetrope (wheel of life) were all popular entertainment devices that displayed sequential images of successive phases of motion. They each moved at a speed that surpassed the capacity of the human eye to distinguish between individual images, producing the illusion of continuous movement. Because movement was the overarching subject, many examples displayed on these devices were drawn from physical culture, like balls and ballgames.

The self-proclaimed inventor of the thaumatrope, John Ayrton Paris, was a British physician who attended to the intersection of entertainment and scientific pursuits.[12] Ayrton Paris situated his invention—whose name translates as "wonder-turner"—within a broader project to promote play and wonder as methods for encouraging scientific education. His 1827 book *Philosophy in Sport Made Science in Earnest* presents toys as pedagogical objects—"instruments of *Philosophical Instruction*"[13]—through which real understanding can be achieved.[14] (The word "sport" here is used in a way that is closer to the meaning of "play" than the common use of the word today.)

The book features Mr. Seymour (Ayrton Paris's author surrogate) welcoming his son, Tom, home from his first term at school with a series of toys, beginning with a simple "game at ball":

> I would give the boy some general notions with regard to the properties of matter, such as its gravitation, vis inertiae, elasticity, &c. What apparatus can be required for such a purpose, beyond some of the more simple toys? Indeed, I will undertake to demonstrate the three grand laws of motion by a game at ball; while the composition and resolution of forces may be beautifully exemplified during a game of marbles . . . We will commence with the ball; which will illustrate the nature and phenomena of *elasticity*, as it leaps from the ground;—of *rotatory motion*, while it runs along its surface;—of *reflected motion* . . . as it rebounds from the wall;—and of *projectiles*, as it is whirled through the air; at the same time the cricket-bat may serve to explain the *centre of percussion*. A game at marbles may be made subservient to the same purposes, and will farther assist us in conveying clear ideas upon the subject of the *collision of elastic and non-elastic bodies*, and of their *velocities and direction after.*[15]

To test his son's knowledge, Seymour peppers him with questions about the natural principles that each object demonstrates. Over the course of the book, different balls appear to illustrate various points: a small ball, a pith ball, a cannonball, a rebounding ball, a bandy ball, a football, a handball, Isaac Newton's tennis ball, and an India rubber ball painted with the face of the vicar.[16] These variously composed, shaped, and weighted balls are the key objects for instilling an early love of science in Tom because they offer "the rare privilege of philosophy" through play.[17] *Philosophy in Sport* is addressed to the young schoolboy and, peripherally, to his sisters, his mother, and (sometimes scoldingly) his teachers. While Mr. Seymour tests Tom, the boy's sisters and mother and the vicar serve as spectators and foils, cheering Tom on while proposing contradictory theories that are quickly proved incorrect. The vicar serves as a device for providing historical and etymological grounding that will tie the toys to the classical tradition.

Here, Ayrton Paris adds another valence of the ethos of muscular Christianity and the historical construction of masculinity. In his opening note to readers, he nods to two writers—the popular science writer Jane Marcet and the author of adult and children's literature Maria Edgeworth, both of whom were concerned with women's education—but the author whom he actually quotes is Jean-Jacques Rousseau, on how "the characters of men are often determined by the earliest impressions."[18] Ayrton Paris's book is concerned with the construction of the character of the schoolboy through impressions (note the physicality of the word).[19] But Ayrton Paris's Tom—in contrast to the Tom in *Tom Brown's School Days* by Thomas Hughes (perhaps the most famous schoolboy story of the era)—does not develop his character through vigorous participation in ballgames. Paris's Tom is, instead, redirected to play with toys—from balls to thaumatropes—for the purpose of the mind's exercise and engagement with fundamental questions of scientific discourse.

For Wells, Ayrton Paris's thaumatrope is perhaps "the first formal evidence that sports and games were used as vehicles to explore the idea of motion through devices seeking to create the illusion of a moving image." He identifies many similar devices that display motion practices from sport, games, and dance.[20] Joseph Plateau used revolving dancers in his phenakistoscope, while Thomas Talbot Bury included circus horses in his. Wells writes, "By the time Horner's zoetrope appeared commercially in the 1860s, zoetrope strips produced by Britain's most successful production house, the London Stereoscopic and Photographic, featured 'Foot ball', 'Base ball', 'The Gymnast', 'Steeplechase', 'The Skipping Girl' and 'The Sportsman.'"[21] Film historian Tom Gunning calls these philosophical toys of the Victorian era "technological images" to emphasize the mechanical nature of their manipulation of human perception. "Both manual and perceptual," he elaborates, "they not only united amusement with education, but also employed a mechanical device to manipulate human perception by coordinating the hand and the eye. Modes of representation and narration become radically revised through new interfaces with the processes of perception and the precision of technology."[22]

For Gunning, the appearance of these optical toys disrupted the entrenched history of static pictorial representation. This revision of representation and narration was due to the suddenly unfixed, composite nature of the image, which could be perceived only when the device was "properly in motion."[23] They also disrupted, or at least redirected, the use of ballgames to train moral character through vigorous physical activity. These devices unfixed the image while fixing the viewer into a far more constrained range of body positions as the condition of their viewing. The philosophical toys and magic lanterns of the nineteenth century tricked the senses and retrained the body.

The Body Elastic

The second half of the nineteenth century, the same moment that saw the rise of vulcanized play discussed in chapter 2, saw balls, ballgames, and ballplayers all serve as subjects for works of pre-cinema and proto-cinema.[24] In this sense, animation served as a site of another kind of vulcanized play. In both the production of still and moving images and the production of scientific knowledge, photography introduced a new central role for bounce. Techniques for capturing the bounce of light, sound, and other signals abound in science. At their most basic, all optical technologies depend on the registration of light that has bounced off a surface, or set of surfaces, to create an image. This indexical registration of light's bounce on paper, glass plate, celluloid film, or image sensor—the remainder of light's second bounce off the material of registration—underpins photography's various claims on truth and the many long-standing debates around these claims. In her history of the cinematographic desires of nineteenth-century scientists, Jimena Canales identifies French physiologist and chronophotographer Étienne-Jules Marey as one of many scientists who dreamed of the creation of a single machine capable of "capturing the world in its own image," independent of a fallible human operator.[25] The notion of "capturing the world in its own image" might be read as a pursuit of photorealism, but Marey was not interested in shepherding realist pictorial traditions of representation into cinematographic technologies. He was interested in devices that could record and verify empirically true results about the range, rhythms, and mechanics of bodily movements. These were for the sake of physiological norms, and he repeatedly used rubber as a material assistant in this pursuit.[26]

In histories of photography and cinema, Marey is commonly recognized alongside English photographer Eadweard Muybridge for their early use of high-speed photography to capture movement. Muybridge is best known for serial images displaying the spectrum of human and animal locomotion. He photographed university athletes swinging bats and rackets or throwing, catching, hitting (and occasionally accidentally dropping) balls used for football, baseball, cricket, and lawn tennis. Muybridge's use of ballgames and other university sports as subject matter shows how, as these forms of play became institutionalized, they came to represent and exemplify the range of human movement. Using mechanical techniques similar to other optical toys, he made serial images—and animated these images with his zoopraxiscope.

Marey's chronophotography, in contradistinction, displayed progressive intervals of a body's motion through space in a single frame. He used this method to compare the intervals between the body's captured positions, representing bodies solely in terms of duration and process, which resulted in what media historian Lisa Cartwright describes as a shift from "the observed and analogically classified body to the experimented upon and digitally ordered body."[27] Marey used this approach to create a significant body of knowledge "about the function of the human body in the military and in sport; to find a way to streamline and capitalize on the energy exerted by the human body in motion; and to make the body conform to physiological standards."[28] As a result of a specialty in cardiology and explicit interest in sport, Marey did more extensive work with athletes than Muybridge, asking them to demonstrate their movements in front of the camera. He created several early devices for data capture that used rubber tubing, shoe soles, and other elastic components to measure gait, circulation, and more—including a wearable sphygmograph, which rendered the circulatory pulse as a waveform (a predecessor to the oscilloscope, which would become the first progressive display to be attached to computing machines).Wells sums up Marey's research as servicing "emergent sports science, kinematic enquiry about velocity, the accurate graphic transcription of animation within the image, and as a continuity of movement articulating the animation from frame to frame."[29]

Marey's ultimate object of study was the movement of human and animal bodies, but like Ayrton Paris, Marey begins his discussion of movement in his 1894 book *Le Movement* with a ball. The first photographic plate of

Figure 3.3
"Photography of the Movement of a Falling Body," in Étienne-Jules Marey, *Le Movement*, 1895.

the chapter presents a white "indiarubber ball of 30 grammes weight and 11 centimetres in diameter" dropped in front of "a black-velvet curtain." "The photographic plate," as he describes, "receives a series of images of this ball, showing the positions it occupies at each successive exposure."[30] In this way, all the necessary elements are obtained for determining "the laws of motion."[31] The image, titled "Photography of the Movement of a Falling Body," presents the rubber ball as the most simple and basic body from the start.[32] After the vertical drop experiment, more balls follow—test objects illustrating how the chronophotographic apparatus works and, later in the book, serving as stand-ins for the human body.[33]

To deal with his less ball-like human subjects, Marey developed techniques for stripping bodies of their material messiness and isolating their patterns of movement. He applied reflective tape to certain parts of subjects' limbs. From the series of timed flashes, light bounced off the reflective pieces and was otherwise absorbed by the subjects' black clothes and the black backdrop. The resulting body images were rendered as sequences of dots and lines, arranged in sets of discrete intervals. As Cartwright explains: "His focus on movement required an elision of those aspects of the body not implicated in duration, motility, and temporality . . . Marey effectively reorganized the body to make it embody its own status as an object subject to laws of temporality and duration."[34] All bodies are always inevitably slipping in and out of objecthood, and a body rendered as "an object subject to laws of temporality and duration" demonstrates this clearly. Bodies that are constituted as much by the intervals between series of dots and lines as by the marks themselves are explicitly quasi-objects. Marey's method seems at first glance to untrick the eye—a scientific demonstration of Xeno's paradox of an arrow in flight. But of course, it is just a different trick, in which the repeated absence and presence of the flash produce and disarticulate continuous motion into discrete intervals.

Addressing how to measure the forces that govern human locomotion, Marey compares elastic balls and human bodies most explicitly in a later chapter of *Le Movement*. As a physiologist, one of Marey's primary goals was "to gain knowledge about the function of the human body in the military and in sport; to find ways to streamline and capitalize on the energy exerted by the human body in motion; and to make the body conform to physiological standards."[35] He turns to an elastic rubber ball's storage of

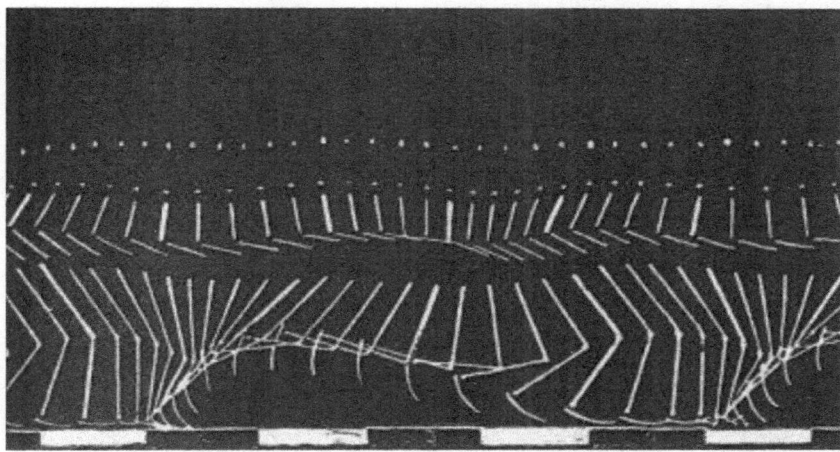

Figure 3.4
"Image of a Runner Reduced to a System of Bright Lines for Representing the Position of His Limbs," in Étienne-Jules Marey, *Le Movement,*1895.

energy across multiple bounces to compare this to "the work performed in walking":[36]

> To make our point clear, let us take the case of an elastic ball falling on a hard surface from a certain height. Let us suppose the ball weighs 100 grams, and that it falls through a distance of one metre. As the ball reaches the ground, the action of gravity will have produced work to the extent of 100 gram per metres. But let us go a little further. The ball in consequence will be flattened against the ground, and rebound to a certain height, .60 metres for instance. When the ball has reached this height of the energy which 1ms accrued from the effects of gravity, only 40 grammetres have been expended, because on letting it fall down again we shall find that there are 60 grammetres of energy left. These 60 grammetres, then, were recovered by the elastic force of the ball, which had stored them up. Is there anything analogous, when at the end of a movement the antagonistic muscles tend to stop it; and will these muscles contribute anything towards the succeeding movement in the opposite direction? The following facts lead us to believe that there is such restitution.[37]

Clarifying how the human body stores and expends energy while walking, Marey's use of a rubber ball uses the same logic that would later underpin Disney's principles of animation.[38] Here, the regularity and elasticity of a rubber ball—a spherically regular, elastic body in motion—allow the human body to be imagined as just a slightly more complex example of it. Both physiologists and animators are interested in understanding the mechanics of the gait, albeit for different reasons. Whereas for the former, the gait communicates fundamentals of biomechanics, for the latter, the gait is a key site of cultural expression. Twentieth-century animators would later capitalize on gait's semantic potential—its capacity to perform and communicate a wide range of cultural, historical, and biomechanical meanings—particularly in their quest for the comic.

Working from the position of a philosopher rather than a physiologist, Marey's colleague at the École Normale Supérieure in Paris, Henri Bergson, drew heavily on both the physical and social mechanics of elastic and inelastic movements for his theorization of the comic.[39] In his 1900 work *Laughter: An Essay on the Meaning of the Comic,* Bergson elaborates that bodies can be described on a spectrum ranging from elastic to nonelastic. He argues that "tension and elasticity are two forces, mutually complementary, which life brings into play."[40] Setting the elastic against the mechanical, Bergson positions the elastic as the key metaphor for vitality, life, and spirit, while the inelastic or mechanical is the metaphor for all that is automatic, inorganic, and machinelike. The unexpected slips between the two categories serve as the core switchboard of comedy.[41] For Bergson, the comic is "any arrangement of acts and events . . . which gives us, in a single combination, the illusion of life and the distinct impression of a mechanical arrangement."[42] This definition drives home how the mechanics of comedy—mechanics that animated cartoons will depend upon—involves a continual play with perception, both of the lines dividing the elastic and inelastic and the living and the nonliving.

Rubber balls occupy a strange place in Bergson's schema as an example of objects, for which humans can be mistaken. He writes:

WE LAUGH EVERY TIME A PERSON GIVES US THE IMPRESSION OF BEING A THING. We laugh at Sancho Panza tumbled into a bed-quilt and tossed into the air like a football. We laugh at Baron Munchausen turned into a cannon-ball and traveling through space . . . The first time, the clowns came and went, collided, fell and jumped up again in a uniformly accelerated rhythm, visibly intent upon affecting a crescendo. And it was more and more to the jumping up again, the rebound, that the attention of the public was attracted. Gradually, one lost sight of the fact that they were men of flesh and blood like ourselves; one began to think of bundles of all sorts, falling and knocking against each other. Then the vision assumed a more definite aspect. The forms grew rounder, the bodies rolled together and seemed to pick themselves up like balls. Then at last appeared the image towards which the whole of this scene had doubtless been unconsciously evolving—large rubber balls hurled against one another in every direction.[43]

In this scene, the clarifying moments come when the clown's bodies' become balls that have their own motion—bodies that are able to "pick themselves up like balls." What appear as chaotic rubber balls are humans who have passed out of humanness and into thingness. From bodies to balls, from human to thing, these slips remind the reader of their own bodied thingness—an absurd yet "unconsciously evolving" image of humans as largely out of control, colliding elastic objects. Rolling from figure to figure and from being to ball, Bergson's more fundamental argument is that all form emerges from movement: "What is real is the continual change of form: form is only a snapshot view of transition."[44] Transitions always hold the potential for deviation from whatever expectations are embodied by prior "snapshot views." For Bergson, when a trajectory is interrupted (this is the potential moment for bounce) and then an expectation of a direction of movement is disappointed (there is some failure to bounce), the liveness of the body in motion (a liveness signified by the elasticity of a rubber ball) is thrown into question. The grounds for ableism are visible here (and throughout the essay)—the discomfort with difference, with disappointed expectation. In the face of such disruptive discomforts, we resort to the social gesture of laughter.

Bergson developed his theory of human perception in dialogue with how Marey's images contained sequential phases of movement appearing in a single frame: "In just the same way the thousands of successive positions of the runner are contracted into one sole symbolic attitude, which our eye perceives, which art reproduces, and which becomes for everyone the image of the running man."[45] According to Canales, Bergson and Marey shared a deep disappointment with early cinematic technologies because these technologies split the apparatus for recording phenomena (understood to be the point of analysis) from the apparatus for projecting the recording (understood to be the point of synthesis). This supported the production of spectacle and illusion rather than the production of verifiable truth.[46] As Canales articulates it, Bergson's complaint was that "the cinematographic camera captured an illusory form of movement; it did not capture real movement . . . What the cinematographic method inevitably missed was the 'perpetual creation of possibility' that was, for Bergson, the undeniable mark of the living world. When directly compared to how it appeared on the screen, the world seemed more alive than ever; its future ever more mysterious."[47]

Cinematographic methods offered a kind of captive motion practice—all but missing the possibility of play.

Similar to the separation between the apparatus of production and the apparatus of projection (or reception), there lay a stark distance between the way that material technologies, such as rubber, were produced in European colonies and how they were worked and played with in the metropoles. The ability to understand living bodies in terms of elasticity derived, in part, from rubber becoming a common material. The turn to elastic bodies, material and metaphorical, developed in direct relationship to this relatively new access to rubber itself. In the moment when Marey and Bergson were working and writing, rubber had become a common, even if still relatively new, material in Europe. The massacres in Congo and Putamayo were both in full swing. News of the abuses at these sites of wild rubber extraction was beginning to circulate, although Roger Casement's comprehensive reports would not appear until 1904 and 1914. One feature of the Congo massacre was the severing of body parts—most commonly men's hands and genitalia—as punishment for not gathering a great enough quantity of rubber. These starkly gruesome substitutions of parts of the human body for "missing" quantities of rubber suggest another reading of Bergson's scene of cavorting clowns transmuted into "large rubber balls hurled against one another in every direction."[48] In this alternative reading, the scene evolves into an unconscious articulation of the everyday mechanisms of dehumanization that enable the conflation of human body parts with specific quantities of raw rubber. The comic, then, reappears as a mouthpiece for describing the ugliest and harshest truths of humanity, ruthlessly setting them next to the most glorious elements of life. Laughter breaks apart resistance to holding these contradictory truths unresolved.

Gestural Semiotics (Step in Time)

"What is the outline? . . . It is not something definite. It is not, believe it or not, that every object has a line around it! There is no such line."[50] This is the way that famed physicist Richard Feynman articulates a key understanding offered by the field of physics. Objects in the world are not so clearly and neatly cordoned off from each other. But classic animation relies on outlines—and on changing the shapes of the lines bounding an object—to

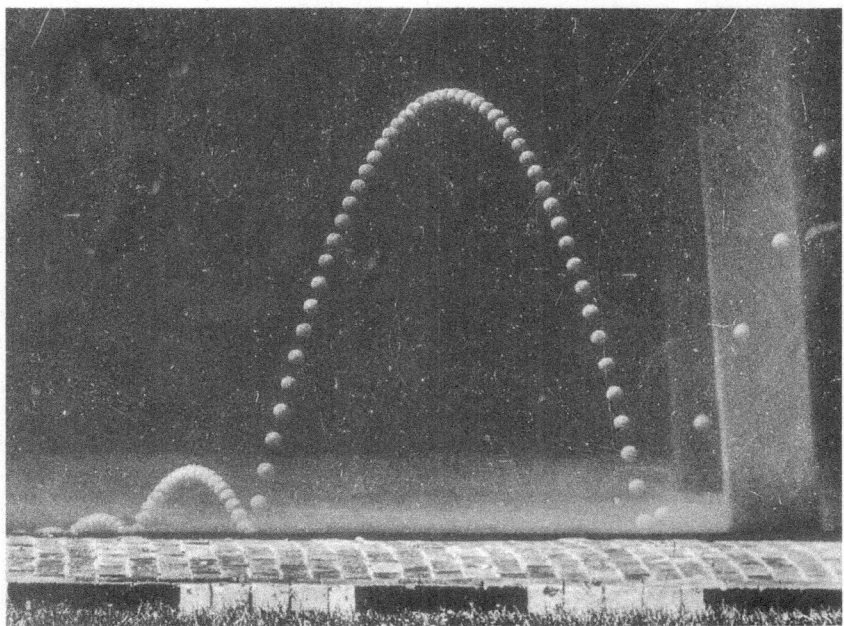

Figure 3.5
Étienne-Jules Marey, *Balle Rebondissante, Étude de la Trajectoire* (Bouncing Ball, a Study of Trajectory), 1886.[49]

represent the way that objects behave in the world. While animation as a medium is not necessarily comedic, much of the animation made in the twentieth century was comedy. This was in part because, as Donald Crafton argues, the burgeoning craft of character animation drew on the traditions of captioned illustrations and flipbooks, as well as on the widely published motion studies of Muybridge and Marey.[51] Many of the earliest animators in the United States began as illustrators and cartoonists. They brought their approach to characters, narratives, and humor to their new medium— developing character animation in the direction of the cartoonist tradition, which employs humor as one of its main methods of communicating. What animation offered cartoonists was a new opportunity to produce humor with representations of physical comedy. The clowns careening around as rubber balls that Bergson describes could now be drawn on celluloid with the slips between body and ball moving in the other direction. With the rise of 2D character animation, the bouncing ball became the foundational object for

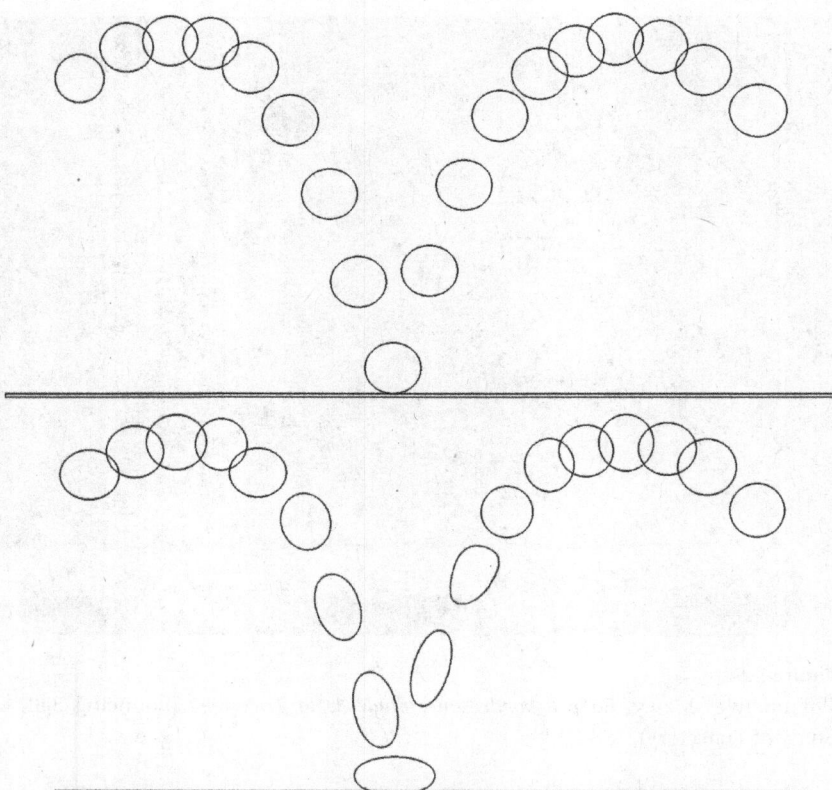

Figure 3.6
Illustration of squash and stretch based on a pair of drawings in *The Illusion of Life: Disney Animation*. In their explanation of the drawings, the authors describe the same phenomenon that Marey captured with his chronophotography. The first shows how "in moving the circle (representing the ball) down and back up, it was discovered that the ball would seem to have more weight if the drawings were closer together at the top and spaced farther apart at the bottom." And the second explains that "if the bottom drawing was flattened, it gave the appearance of bouncing. Elongating the drawings on each side made it easier to follow and gave more snap to the action. Thus the beginnings of Squash and Stretch." Illustration by the author.

training animators' gestures. The first thing that most beginning animators learn how to do—then and now—is to make a ball bounce. Building off this basic exercise, students go on to learn how to render sequences of shapes toward the creation of lively characters.

Over the course of the twentieth century, the most dominant force in the field of animation was Disney Studios, founded as Disney Brothers Cartoon Studio in 1923. Walt Disney helped establish hand-drawn 2D cartoons as the commercial form of animated cinema. Like the optical toys before them, emerging at the start of the twentieth century, 2D animated films were comprised of sequences of images, now called "frames." In the case of hand-drawn (so-called traditional) animation, each frame in the sequence was drawn on a cel, a transparent celluloid sheet used by most animators at the time. Transparency allowed them to stack pages so they could use the prior image in the sequence as a guide for the one that they were working on. Like Marey's images, the transparent stacks made the interval between instances of a body's movement visible. In 2D animation, the illusion of life and construction of character are created through the redrawn body from frame to frame.[52]

Walt Disney had encountered the conjunction of the athletic and the animated early on, when he checked out two of Muybridge's motion studies from the Kansas City Public Library.[53] What Marey took as an analytic task for the purpose of furthering scientific knowledge, Walt Disney and others took as creative tasks for producing mass entertainment. Wells argues that for Disney and other twentieth-century animation studios, "sport was a vehicle by which animation as an evolving medium could readily test itself technically, aesthetically, and socioculturally."[54] In this context, sport appeared as both a site for sorting out body mechanics and a source of narrative content. From its early days onward, Disney animators used the bouncing ball as a standard object for learning how to produce liveness, character, personality, and the illusion of life.

The classic text in the field of 2D character animation is the 1981 *Disney Animation: The Illusion of Life*. Written by Frank Thomas and Ollie Johnston—two animators who cut their teeth during the early days of Disney Studios—the book names what the field, at least as defined by Disney, considered the goal of animation.[55] The word "illusion "is as important as the word "life." From the start, the first animated films, as Donald Crafton puts it, "were concerned with making objects appear to move with a mysterious life of their own."[56] The problem of how to animate is the problem of how to "breathe life into a Golem."[57] Animation embraces illusion. Its goal is neither to make or create natural or artificial life in the world, nor is it to develop tools for acting on and reacting to the world. For Disney, animation was about "making drawings that appear to think and make decisions and act of their own volition."[58] For the purposes of this chapter, the principles laid out in *The Illusion of Life* are important because they demonstrate the central role that bouncing balls play in the task of training animators. From the early days onward, Disney animators used bouncing balls to learn the trick of imbuing drawings with liveliness, character, personality, and the illusion of life.

According to Disney, there are twelve principles of animation, the most important of which is "squash and stretch," a method for representing the elasticity of bodies. This conveys liveness by showing, and even exaggerating, how bodies change shape as they encounter the physical forces of their cartoon environments (gravity, weather, and other settings) and encounter other harder or softer bodies. It grants the drawn object a sense of weight, mass, and dimensionality.[59] Squash and stretch begins with animators learning to move from ball, the ur-object of animation, to bounce. The ball starts as a sequence of "simple circles."[60] This literal "bounding" technique, which constitutes objects through its outline, is how 2D character animation moves from ball (the representation of a static object) to bounce (the representation of an object through its relation to the world). By carefully calculating the spacing, strategically squashed and stretched circles create a projection of a single, lively ball. The diagrams of the basic bouncing ball show how the distance between the ball's positions on the page should change from one cel to the next. At the height of the bounce, the positions of the circles overlap from one cel to the next. This creates the impression that the ball is hanging in the air. As the ball begins to descend, it picks up speed, an effect achieved by placing more distance between the circles from one drawing to the next. Since every animation cel occupies a fixed interval of time that is known in advance, the timing of the scene is achieved by manipulating distance.

At Disney, animators took the behavior of one specific kind of ball as their starting point: a ball that was already one of the most iconic industrial objects of play. Thomas and Johnston describe the very first assignment given to beginning animators:

> The assignment was merely to represent the ball by a simple circle, and then, on successive drawings, have it drop, hit the ground, and bounce back into the air,

ready to repeat the whole process . . . It seemed like simplicity itself, but through the test we learned the mechanics of animating a scene while also being introduced to Timing and Squash and Stretch . . . We were encouraged to change the shape of the ball in the faster segments of the bounce, making an elongated circle that would be easier to see, then quickly to flatten it as it hit the ground, giving a solid contact as well as *the squashed shape of a rubber ball in action*. This change at the bottom also gave the feeling of thrust for the spring back into the air, but if we made an extra drawing or two at that point to get the most out of this action, the ball stayed on the ground too long, creating weird effects of hopping instead of bouncing. (Some tests looked more like a jumping bean from Mexico than any kind of ball.) . . . However, many of the circular forms just seemed to take off as if they had a life of their own.[61]

Although it is only an aside in Thomas and Johnston's text, there are instructive ironies in the jumping bean, posed as the wrong kind of motion. The motion of a jumping bean is caused by an actually alive larva moving about inside the bean. It is thus too lively and unpredictable to serve as an initial model for animators.[62] The rubber ball, on the other hand, was a kind of ur-object: an ideal for animators to hold in mind. By translating this ideal onto the page, they tackled the mechanics of animation by drawing circles spaced and shaped in a manner that would make a ball appear to bounce when the pages were set in motion. Rubber balls in action—with their elastic bounce—become the best examples by which to learn how to represent the liveness of all kinds of bodies in motion. The underlying goal is for animators to learn how to make things behave "like themselves," or rather how to construct all thingness and selfness through visible movements. So, while drawing a jumping bean may be desirable at some point in the future, mastering the more regular bounce of a ball comes first. Its reliable behavior allows animators to get their own bodies' motions under control. They use a regular object to develop their own reliable, embodied technical practice of representation. The industrially elastic rubber ball, with its true bounce, becomes the presumption from which all variations are made: the basic unit that animators use to create characters with visible personalities.

Balls, as Thomas and Johnston note, are comparatively simple objects, especially when rendered as hand-drawn circles. Working with a rubber ball, as an ideal elemental body unit, led Disney animators to understand all bodies in motion as existing on a spectrum of elasticity. After they learned to represent the simple motion dynamics of a regular rubber object, animators moved on to apply squash and stretch, timing, and other principles to more complex characters. To find examples of more complex representations of body elasticity, the animators at Disney turned to the sports section, where they found endless examples of bodies captured in full extension:

> On the sports pages of the daily newspapers we found a gold mine that had been overlooked. Here were great photos showing the elasticity of the human body in every kind of reach and stretch and violent action. Our animation principles were clearly evident in the bulges and bumps that contrasted to long, straight thrusts. Mixed in with these contortions were examples of the whole figure communicating joy, frustration, concentration, and all the other intense emotions of the sports world. These examples opened our eyes and started us observing in a new way.[63]

Sport was an arena for demonstrating the human range of motion, and sporting bodies were understood as performing at the edge of both physical capacity and emotional expression. The photographs chosen for the sports section would have been those that conveyed the most dramatic moments—the photo finish, for example—but athletes engage in a performance that is directed only secondarily at the camera. This record of involuntary physical expressiveness made the sports page a rich archive for Disney's animators: a trove of what felt, bodily physics look like. Along with the other assembled principles, they could then use squash and stretch to project the motion and emotion of the athlete forward and backward in increments—from the instant of the photographic image to the photo finish.

While Disney animators turned to the sports pages to learn what full bodily extension and emotional expression look like, they also turned to sport more generally as a vehicle that possessed "ready choreographies and embedded narratives that echoed and reflected sporting technique and execution, and a social reach that included professional sportspeople, invested amateurs and committed fans and spectators."[64] Directed by Jack Kinney, *How to Play Baseball*, a series of short animated films that featured Goofy as the ultimate sportsman presented the most extended examples. Wells describes the roll of the ball in one film: "The ball displays excessive agency, and as such, apprehends rhythm, speed, directionality in and of itself; the ball in many ways ceases to be a ball but a visual phenomenon that embodies how animation controls motion, making fast and slow, now and then, past, present, and future, the stuff of manipulation beyond the contextual space in which it occurs."[65] Here, the representation of a particular ball bouncing, flying, spiraling, and spitting in Goofy's face serves as a wink at an audience that knows how this kind of ball is supposed to behave. It lets them in on the open secret

Figure 3.7
Goofy faces a spitball in Disney's *How to Play Baseball* (1942). Screenshot by the author.

that the animator is the ultimate athlete, "a corporeal manipulator," demonstrating virtuosity by way of the ball's carefully calculated wildness.[66]

Constructing the illusion of life requires animators to use the difference between sequential images to construct the physics of cartoon worlds. The way that the ball bounces tells a story about how the techniques of animation are used to create cartoon physics—the chosen rules of any given animated world. The bouncing ball became the foundation of Disney Studios' system of representation—one based on elasticity as the ultimate sign of life. Disney's principles are directed toward the central task of making a character's character—personality—visible and audible on screen. To construct characters, animators have to deconstruct movement and reconstruct it with meaning—a semiotics of gesture. Disney animators base their semiotics on elasticity, a signifier of kinds and degrees of life, and they use this signifying system to grant (and withhold) character to any and all (animal, vegetable, or mineral) by way of visual and aural representation.

In *Donald Duck Joins Up*, Richard Shale writes that "Disney creates cartoon figures which the public can care about . . . he applies the concept of personality to cartoon characters by treating the characters like actors."[67] If in sixteenth-century court tennis, we saw a beautiful game that was thought to make a player's true character legible to those with eyes to see, and if with nineteenth-century muscular Christianity, we saw physical activity taken as a method for building desirable character in a person, then with twentieth-century cartoon animation, we see this idea stripped of its moral imperative and physical enactment and made into a purely technical assignment of representation. Where muscular Christianity proposes that motion practices can be deployed to make people act ethically in the world, animation uses motion practices to endow nonhuman things—drawings, objects, machines, and others—with the illusion of life. Cultural expressions become a library available for symbolic manipulation.

All Together Now

Nineteenth-century optical toys, Muybridge's motion studies, Marey's chronophotographies, and the earliest 2D films were visual technologies. They do not speak (at least not overtly) to the extent to which the illusory realities created by twentieth-century animators relied on sound as much as they relied on image. In particular, cartoons rely on the juxtaposition of sound and image to generate mood and humor. In *How to Play Baseball,* the ball spits, screeches to a halt, and chops through a bat with the buzz of a chainsaw. Much of the humor in early Disney cartoons comes from the matching and mismatching of sounds to bodies (e.g., a ball sounding like a spitting ballplayer, or Goofy all twisted up and then sounding like a metal spring as he releases the ball). These strategic mismatches of sound and image are acted out within the safe confines of an overarching rhythm—one that keeps the characters' body movements bouncing in time with the beat of the music. The bouncing balls that appear in the history of synchronized sound are different from the balls whose physical bodies were understood to be demonstrating natural laws of motion. Here, the bounce is beat; it is tempo and timing. By pulling disparate components into harmonious relation, these bouncing balls actively organize the interaction. Where the visual motion practices of animation turn to histories of sport—to understand how to coordinate a sequence of drawings into believable bodies—the sonic motion practices of animation look to the history of scored and scripted music to coordinate sound and bodily movement, both individual and collective.

Figure 3.8
Paramount screen song stock poster, c.1930. Courtesy of the Academy Museum.

Before synchronized sound, there were silent "bouncing ball" cartoons: cartoons that audiences and musicians sang and played along to by following the bouncing ball. Max and Dave Fleischer invented this bouncing ball in 1925 for their series of *Song Cartoons*.[68] The effect was adopted and adapted in the years and decades that followed, and it is still commonly used for karaoke today. In 1925, the effect was achieved by using high-contrast film to shoot an image track. While the image track was projected with printed lyrics, an employee would hold a stick topped with a luminescent ball, physically moving the stick up and down to make the ball appear to land on the top of each word or syllable as the moment came to sing it. The high-contrast film made the dark stick invisible, allowing the glowing ball to appear to bounce in time of its own volition. In the theater, the ball would bounce to the rhythm from left to right, landing on the right word at the right time, communicating the tempo to the musicians and helping the audience feel the time and sing along. The controlled bounce coordinated interaction to produce a collective, guided performance. As the method developed, animated balls and animated characters took the place of the physical ball on a stick.

The success of the bouncing ball cartoons served to strengthen the push toward synchronized sound. With the bouncing ball cartoons, the privileged moment of interaction and coordination was the moment when the film played in the theater and the audience and musicians sang and played along with it. The development of synchronized sound moved this coordination and interaction backstage: out of the theater and into the production studio. Figuring out how to synchronize the image track to the soundtrack meant figuring out how to plan the individual drawings and the notes of the music together instead of improvising a score to match a finished cartoon, or vice versa. Thomas and Johnston write that "Walt insisted there must be a way the two could be worked together and be controlled and built upon and changed. What kind of graph or chart or score could be devised that would bring the music and picture together?"[69] Wilfred Jackson supplied two answers: a metronome and a graph that he called a "bar sheet." Jackson "reasoned that if the film ran at a constant speed of 24 frames a second, all one had to do was determine how much music went by in a second."[70] The metronome quickly became an essential tool for the animators at Disney: "A visitor walking through the halls would hear the scattered ticks and tocks coming from several rooms at the same time, as the animators listened and acted, considered and timed."[71] The combination of the metronome and the bar sheet allowed the relationship between sound and image to be controlled with a new precision and efficiency.

The bar sheet, also called a "dope sheet," was a diagram that allowed measures of music to be visibly lined up with drawn actions. After the initial attempt to sync the sound for the first Mickey Mouse film, *Steamboat Willie*, failed, Disney devised a method for scoring motion pictures based on the bar sheet. As in the bouncing ball cartoons, this method used a bouncing ball as a visual cue for musicians, but now the bounce was addressed only to the musicians recording the score by means of a special "score film" version of the picture. *Steamboat Willie* was released in 1928, and on October 16, 1928, Disney filed a patent for the method:

> The score film . . . consists of a continuous motion picture film, successive frame lengths thereof bearing a ball or other mark at different distances from the frame lines of the film . . . Such a score film is then projected so as to render the ball or other mark visible to the musicians who are to produce the sound negative. The motion of the ball or other mark (relative to the frame lines of the picture area) visually imparts to the musicians the required tempo. If, for example, a vertically movable ball or mark was shown on the score film, the tempo that the musicians utilize in playing is governed by the rapidity or frequency of the rise and fall of the ball which they observe in the projected image.[72]

The ball replaced the earlier synchronizers, such as "the leader of an orchestra or director of any sound,"[73] who was too humanly fallible, and the metronome, which was too regular and too audible to guide the recording of the final score. The bouncing ball was a collaborative instrument allowing the director, musicians, and animators to create a cartoon world that operated somewhere between clock time and lived time. It got human error out of the way while allowing for carefully controlled (frame-by-frame) variation. Animators were now able to draw to, and for, the beat. Animated characters took on an almost magical liveliness, and audiences and musicians were no longer required, or encouraged, to make their imperfectly timed and tuned sounds in live cinematic environments.

The technique led to the style of scoring that Disney Studios became known for, "Mickey Mousing." This style consisted of making sure that every animated action was accented, amplified, or echoed by an accompanying sound. The most used tempo was "a setting of 12s (twice a second) . . . which meant that a beat came every 12 frames. This just happens to be the tempo of all marches and offers a good alternative when no metronome is handy . . . Milt Kahl once proclaimed in a lecture, 'Everyone walks on 12s—unless there's something wrong with them!'"[74] A strange flip has happened here: a marching tempo that happens to fit neatly into the standard frame rate for film of 24 frames per section, is rearticulated as the only right way to walk. But, of course, not all bodies walk to the same beat. People walk at different tempos, in different tempos, with different cadences, based on culture, context, mood, moment, and capacity. They may walk irregularly, even erratically. Some do not walk at all. No one walks to begin with.

June 6, 1933.   R. O. DISNEY   1,913,048
METHOD OF AND MEANS FOR SCORING MOTION PICTURES
Filed Oct. 16, 1928

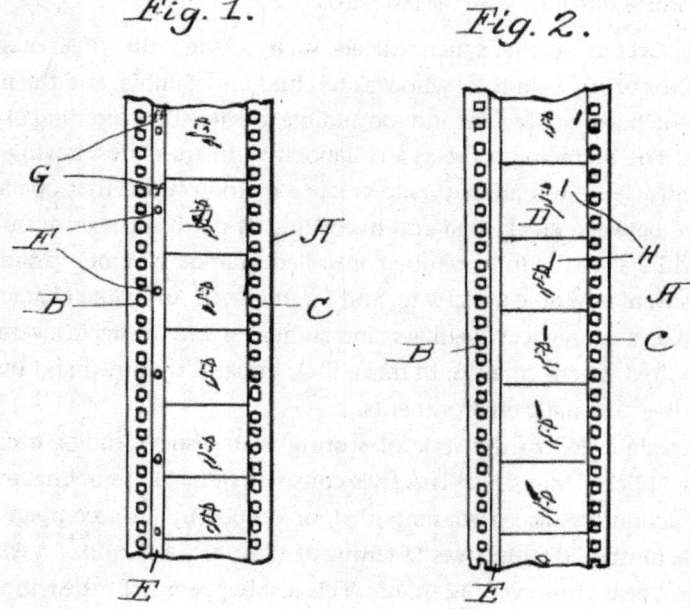

Fig. 1.   *Fig. 2.*

INVENTOR.
Roy O. Disney,
BY *Bradbury & Caswell*
ATTORNEY*S.*

Figure 3.9
Disney's patent for a "Method of and Means for Scoring Motion Pictures," 1928.

Disney Studios, however, is the culture industry writ large. Talking about the same era, cultural historian Fred Turner describes Theodor Adorno hearing the goose-stepping of German soldiers to the rhythms of American popular music.[75] Disney Studios takes this a step further, perfecting techniques for linking sound to image in a manner that produces "rhythmically obedient" bodies.[76] The bodies in question are just cartoon characters. But, as Jackson said in an interview years later, "Nobody had ever seen a drawing make a noise."[77] So the noises that cartoons make and the beats to which they walk matter because they tell the public about ideal forms of comportment—what counts as good and bad ways of carrying oneself. This is the ideological power of a representational practice—one that rests on the simulation and control of frame-by-frame and beat-by-beat coordination and interaction.

Conclusion: Winks and Nods

I must have been about eleven when I first played *Sonic the Hedgehog*. SEGA released this answer to Nintendo's *Super Mario Bros.* in 1991. Sonic first burst onto the scene as the star of the best-selling video game of 1991, giving SEGA a mascot to outstrip Nintendo's Mario. Soon after, together with millions of other players around the world, my brother and I were captivated by the speedy blue, spiky-haired, red-sneakered hedgehog who whizzed through a 2D green, hilly, roller-coaster world. One moment he sprinted along upright on his two feet, and the next, he rocketed into a dizzying blue blur of a ball. We trained our hands to follow the game's rhythms, repeatedly navigating the animated hedgehog sprite through the array of stationary and moving obstacles—our hops, leaps, and jumps timed to the spatial and temporal choreographies of the 16-bit game world. I remember the flying fox, Tails, joining the fray at some point, so I must have also played *Sonic 2*.[78] But it was that original game that I went back to years later when I had just turned thirty, during the summer when my mother was in treatment for a serious illness. There was just something about the movement—something about temporarily becoming an impatient, impeccably coordinated, ball of blue hedgehog, one traveling at "sonic" speeds, skirting all dangers, and destroying any and all enemies that crossed my path. On the second floor of the home where we were house-sitting, my partner at the time set me up to play via an emulator, and I proceeded for the first time in my life to methodically, obsessively master a video game.[79]

Almost a century after the introduction of squash and stretch—along with the dope sheet that coordinated moving sounds and images—and thirty-four years after *Luxo Jr.* teleported Disney's principles of animation into the world of interactive computer graphics, the 2020 Paramount Pictures film *Sonic the Hedgehog* briefly became the highest-grossing video game adaptation of all time. This was until it was topped by its sequel in 2022, only to be surpassed by *The Super Mario Bros. Movie* the next year. The film follows a CGI Sonic the Hedgehog as he navigates live-action loneliness and life on Earth while being pursued for his special powers. The Sonic character in the film is the product of a $5 million redesign by Marza Animation Planet. The initial design resulted in a Sonic judged to look too big, too human, and not enough like a video game character. For the film's explicit mash-up of live-action and CGI to work, the character had to be differentiated enough from the live-action world that he was traversing, recalling viewers, instead, to his video game character and our illusory sense of his video game world.

Half an hour in, viewers find the bright blue hedgehog desperately camouflaging himself among a pile of sports balls in his friend Sherriff Tom's attic to hide from Dr. Robotnik and his badniks. Wedged between an Adidas Tango 12 Adidas soccer ball, a Spalding JAM Session basketball, and other various volleyballs and American footballs, Sonic uncontrollably monologues, "OK. I'm a ball. Just a normal ball. I'm blending in like a ball. Shhhh stop talking! No, you stop talking! Be quiet Sonic! No, you be quiet Sonic! Ohhh, I hope they aren't scanning me with x-rays. I had kind of an embarrassing lunch."[80]

Figure 3.10
Sonic hides from Dr. Robotnik's badniks in Paramount's *Sonic the Hedgehog* (2020). Screenshot by the author.

Right as the badniks' lasers are about to reveal him, the hedgehog bursts from the pile, sending the other balls flying. Still tightly balled up, he rolls and bumps down the stairs. The intonation of his rolling commentary is emphasized by the acoustic Mickey Mousing of each bump—"Why Don't You Have your staircase Car-Pe-Te-D?"—concluding with a light crash into the legs of the kitchen island.[81]

Similar to *Luxo Jr.*'s reflexive ball play, *Sonic* reads as a running, winking homage to the countless bouncing, spinning, and ricocheting balls that populate the histories of sports, games, animation, and computing. The trailer compiles some of the most explicit moments: Sonic using a washing machine as a perfectly circular treadmill; Sonic playing Ping-Pong with himself (on a table constructed from a street sign with a line of soda cans as a net); Sonic playing baseball with himself in a scene reminiscent of Goofy in the Disney classic *How to Play Baseball;* Sonic popping out of the car to visit the World's Biggest RubberBand Ball, returning a second later with three rubber band balls and a classic wooden paddle ball toy. Many of Sonic's bucket list goals are sports related, including scoring a basket, which he achieves by becoming a ball to be shot and scored in a basketball arcade net. Even Dr. Robotnik's drones look suspiciously like flying soccer balls. Ricocheting like a pinball off obstacles and through video games, film, television, and other media forms, Sonic embodies the role that bounce and ball have played in the histories of both animation and video games.

What initially made *Sonic* stand out from other games of the era was the speed of the console's processor, which allowed faster gameplay than other 16-bit games. Game and media scholar Nick Bowman writes, "The 'sonic' speeds that the titular character moved at represented a core attribute of the SEGA Genesis console, which was marketed under the name 'blast-processing technology.' Blast processing allowed the Genesis console to display one image while instantaneously processing another. . . . which resulted in faster gameplay."[82] The console's capacity for speed shaped the development of both the game and the character of Sonic. As the game itself was getting faster and faster, the game designer, Yuji Naka, was searching for some feature to give the character power over his enemies, and he recalled a character that he had thought up years before, "who could roll himself into a ball and slam into enemies."[83] Naka and artist designer Naoto Oshima started with a rabbit, but hedgehogs are known for rolling into balls, so the switch was made—the path was cleared for the development of Sonic's trademark "spin attack."[84]

A team member's offhand comment that the little blue hedgehog looked supersonic supplied the name.[85] Who doesn't want to travel at the speed of sound? Hedgehogs, however, are not known for their speed. To explain Sonic's speed, he was given red sneakers, signifying speedy movement, and an impatient foot tap if he stood still for too long, signifying the intolerableness of stillness. The red, rubber-soled shoes assist players' belief in Sonic's swift, smooth, sonic movement, telling them that they should always be on the go. Vulcanized rubber reappears here to authorize a new kind of otherwise unbelievable electronic speed.

Initially, bodies and bounce appear differently in animation than they do in computing. While in this chapter, I have shown how animated bounce and lively bodies became aligned with the material of rubber, and a body's stretchiness was the mark of an elastic vitality. In chapter 4, I will show how, in computing, bounce aligns first with electric signals, which take the place of rubber and elasticity, and point to the ascendance of the nervous system and signal processing being understood as the basic sites and signs of life and liveness. Even though they have different starting points—from ball to bounce versus bounce to ball—where animation and computing overlap is in their dependence on bounce to both facilitate and demonstrate forms of virtual and sonic coordination in the service of collective, interactive process. Audiences had, for the most part, been shepherded out of singing along to movies after the invention of synced sound, leaving the labor of coordination hidden behind the screen. Yet, remarkably, the real-time interaction of computer users and video games players brought them back into the loop— once again requiring participants to harmonize actions and expressions to the medium at hand.

4   From Ping to *Pong*

Okay . . . Here's how your brain works. It has lots of little processing elements called neurons. And every so often a neuron goes ping. And what makes it go ping is that it's hearing pings from other neurons and each time it hears a ping from another neuron, it adds a little weight to some store of input that it's got. And when it gets it, when it's got enough input, it goes ping. And so if you want to know how the brain works, all you need to know is how the neurons decide to adjust those weights that they have when a ping arrives. That's all you need to know. There's got to be some procedure used for adjusting weights. And if we can figure it out, we know how the brain works.

—Geoffrey Hinton, therobotbrains.ai transcript

When you create a simulated world, nothing falls to earth unless you remember to turn the gravity on. Nothing bounces unless you tell it exactly where and when to bounce.

—Brian Hayes, "Computing Science: The Way the Ball Bounces," *Scientific American*, 1996

Sleight of Hand (Men at Play)

During his two years as the head of the Information Processing Techniques Office at the Advanced Research Projects Agency (ARPA), J. C. R. Licklider—known to friends as J. C. R. or "Lick," funneled the US Department of Defense's massive resources into the development of the graphical user interface and ARPANET, the precursor to the Internet. Prior to his work at ARPA, Licklider led the human factors team for the Semi-Automatic Ground Environment (SAGE) project at the Massachusetts Institute of Technology (MIT). There, building on the Whirlwind computer technology created by Jay Forrester

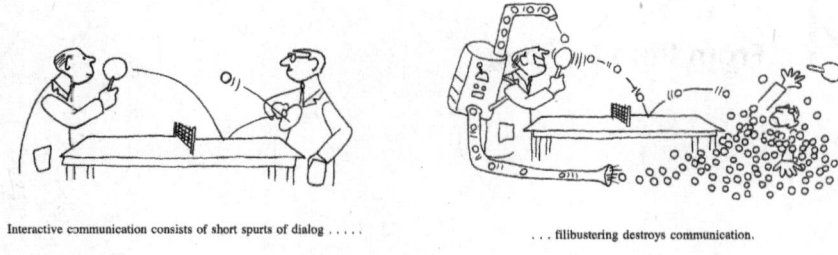

Interactive communication consists of short spurts of dialog . . . . .

. . . filibustering destroys communication.

Figure 4.1
Two drawings made by Rowland Wilson to illustrate J. C. R. Licklider's "The Computer as a Communication Device," 1968.[1]

and others, Licklider and his team coordinated information from multiple radar sites and produced a single image of a large amount of airspace. Afterward, in the thick of the Cold War in 1968, he published "The Computer as a Communication Device," a foundational essay that laid the groundwork and preliminary plan for how computers would be used to create the universe of networked communication.

Among the ten illustrations that cartoonist Rowland B. Wilson drew for Licklider's essay, a pair of drawings show two men sporting lab coats playing table tennis.[2] By no means a normal practice for science and technology journals at the time, the cartoon's illustrations provides readers with the kind of "manipulable models" for which the paper argues.[3] This turn to illustration points toward how modeling, computational or otherwise, always rests on representation.[4] If in chapter 3, animation proved a representational practice that relies on simulation and interaction, then in this chapter, computing, while relying on representation, appears as a practice that foregrounds interaction and simulation. Wilson's drawings stage two divergent futures for the nascent world of networked computing as two scenes of table tennis—or Ping-Pong, the name given to the game by the Hamley Brothers toy shop in England and trademarked by the Parker Brothers board game company in the United States.

Whereas bounce in animation developed in step with the late nineteenth-century vulcanization of play, bounce in computing develops as part of the feverish technological pursuits of the Cold War. A "cold" war meant no direct fighting between the two superpowers—but were that to occur, they shared a vision of bombs lobbed back and forth: mutually assured destruction. In this context, Wilson represented the scene of networked computing as two scientists batting balls to and fro across the table, conjuring the face-off between the United States and the Soviet Union—and their scientists locked in this race. The first of Wilson's drawings shows a friendly rally between colleagues—the ideal to and fro of dialogue. Licklider, recognizable as the player on the right by his glasses, hair, and strong nose, plays with one hand casually tucked in his lab coat pocket. Both players have small smiles on their faces as they hit a single ball back and forth. In the second image, the tone has changed: the opposing player's smile is now large and gleefully maniacal. The two-channel ball machine that he wears is relentlessly firing a barrage of balls across the table at Licklider from both above and under the table. This is neither dialogue nor dissemination, but pure bombardment—the complete destruction of communication.[5]

At the center of the utopic and dystopic scenes are two conceptions of bounce that are central to the history of computing: signals imagined as projectiles that bounce off surfaces (*ping*) and the world imagined as a rule-bound game space within which different models—or simulations—of bounce play out (*pong*). In both drawings, Ping-Pong balls stand in for messages which, given the shared materiality of celluloid balls and celluloid film reels, is a perhaps inadvertent joke about the impact that electronic communication and information storage will have on these technologies. But the cartoon's more direct humor is Licklider's physical slip into the sea of celluloid balls. This physical slip creates a visual slip between body and ball. Licklider's head, hands, and paddle are rendered as separate bits and pieces, bobbing just barely above the surface: a few mildly misshapen, outsized balls lost atop the growing pile.[6]

Licklider imagined and advocated for a world of networked communication that would look like the ideal dialogue in the first cartoon, but Wilson's second drawing cautions that, when some players have more powerful equipment than others, there is no level playing field (or table). When this happens, people slip from being players to being objects of play. The cartoon's unconscious slip, or premonition, figures networked computing machines as the introduction of dramatically unequal, asymmetric power. The only machine that appears in the cartoon is the ball machine—not something that was part of the utopic world of networked machines that Licklider was envisioning. But the ball machine's barrage in the second image is sadly more evocative of the world of networked communication in the early twenty-first century: a world of continuous data gathering in the service of surveillance capitalism. Facilitated by endless personal pings, pokes, pushes, nudges, and notifications, this environment has people fighting to keep their heads above water—to not become mere bits and bobs amid massive oceans of information.

The Ping Pong balls illustrating "The Computer as a Communication Device" are just one instance of this point. The history of computing is bursting with bouncing balls. Bounce programs made their initial appearance the first time that a computing machine was attached to an oscilloscope screen. From there, different kinds of collisions became essential to a range of computing contexts: from robotics and molecular modeling to video gaming and computer animation.[7] The bouncing balls that appear in hardware demos—and as introductory exercises for software suites—are too numerous to count. Perhaps the most famous is the Amiga Boing Ball, which first debuted in 1984 at the Consumer Electronics Show. Demonstrating Amiga's graphical capabilities, the ball would become a symbol of the Amiga brand—adapted and responded to by other brands like Atari, Commodore 64, and Sinclair QL to show that they too boasted graphics displays capable of such feats.[8] Initially created in 1989 to prevent burn-in on cathode-ray tube (CRT) monitors, the screensaver suite *After Dark* displayed bouncing balls among its early animations, along with doodles, cans of worms, a twinkling starry night, and (most beloved) flying toasters. Later animations included marbles, geobounce, globes, and gravity.[9] Other software suites like Flash, Java and JavaScript, Python, Unity, HTML5, and scalable vector graphics (SVG) all include bouncing ball demos for learning the software.

This proliferation led to many different approaches to the problem of collision detection and the programming of bounce. In his 1996 article "Computing Science: The Way the Ball Bounces," Brian Hayes lays out the four most common approaches, giving special attention to the lattice method, which makes both space and time discrete by mapping ball trajectories onto a grid. Employing a rule that only one ball can occupy any one point on the grid at any one point in time, this method offered the most efficient collision detection and required less processing power. Technically, these balls hop along discontinuous trajectories, but as with the discrete frames in animation, our eyes do not register their motion as discontinuous. And, given the bounded area of the lattice—the restricted frame is the key—a game of ball can model a universal computer. Hayes writes that once a lattice is put in place:

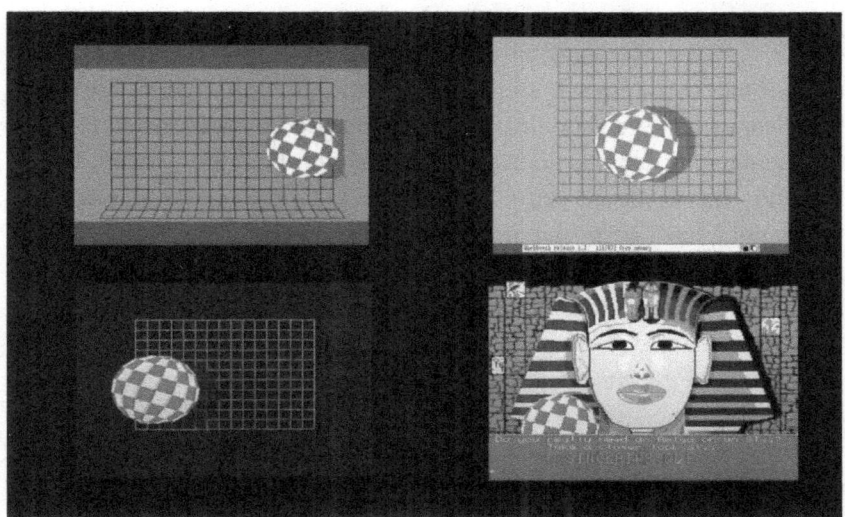

Figure 4.2
"Boing Bouncing Ball Demo Amiga/Atari/Commodore 64/Sinclair QL Comparison |
Nostalgia Nerd Extra," created and uploaded to YouTube by B:\Nostalgia Nerd. Courtesy of B:\NostalgiaNerd.

> Not only can this primitive computer simulate billiards; what's more remarkable is
> that the simulated game of billiards can in turn simulate any computer! It turns
> out that billiard balls bouncing around a tabletop can be interpreted as implement-
> ing the AND, OR, and NOT functions of Boolean logic. A place on the table where
> two trajectories cross—where a collision will take place if two balls are traveling
> those paths at the same time—represents a logic gate. The gate has a logic value
> of TRUE if the two balls collide and is FALSE otherwise. By properly arranging many
> such gates one can build a universal computer—a machine that can emulate any
> other digital computer. Thus a PC might emulate a cellular automaton emulating a
> billiard-ball computer emulating a PC.[10]

By describing this special ability of billiards to simulate any computer, Hayes
demonstrates that "much of what goes on inside a simulation has no counterpart in the real world."[11] Given that simulation, as Hayes argues, had been
accepted as "a third way of doing science," joining deduction and empiricism, his conclusion was a call for caution and acutely critical evaluation of
the knowledge produced via simulation.[12]

By running simulations, computing machines—in concert with real-time
display screens—changed the gamut of possible questions and answers.

In chapters 1 and 2, we saw how crucial it is in empirical science to recognize how the experimental method and laboratory environment affect what comes to be known and how. Likewise, part of understanding any knowledge produced via simulation is understanding the different methods of manufacturing interactions in computing and how these methods, like laboratory environments, affect what comes to be known. To consider how interactions have been manufactured in computing, this chapter offers a prehistory of the proliferation of bounce programs, following the bounce in early computing from its origins in radar's ping through the iconic and endlessly reiterated video game *Pong*.

What do we gain from thinking about the kinds of bounce and versions of ballgames that appear in computing? Approaching the history of computing via histories of play, game, and sport locates the computer as not simply the next installment in a longer history of painting, photography, and cinema, but also as part of an equally long history of rule-bound, real-time, physical-symbolic interactions: a technical and technological history of motion practices and motion capture. In this chapter, I show how bouncing balls served as training tools and object lessons for computer programmers from the field's earliest days. Whereas animators, as we saw in chapter 3, moved from ball to bounce—from object to action—in computing, it worked the other way around. In computing, bounce precedes the ball. Engineers and programmers first had to program bounce logics into the frameworks of early graphical user interfaces to make balls bounce across their screens. In doing so, they turn computers into real-time machines and televisions into active instruments. If in animation, the problem was how to coordinate animators' drawings with musicians and actors' voices to produce lively believable bodies—animated characters operating within paradigms of visual realism or the hyperrealism of animation—the problem in computing was how to build systems for real-time communication and coordination both within and between bodies—especially between humans and machines.

Ping (Knowing Through Sounding)

I'll ping you. It's on my calendar to ping you this morning but I am on a conference call so may not get to pinging you until this afternoon. Ideally, I could ping you before my afternoon meeting but I might have to wait until after the meeting to ping you. I might even just say "screw it" and ping you during the actual meeting.

I may have to postpone the ping until evening. Additional unforeseeable delays might mean I can't ping you until tomorrow. But I'm going to ping you. I hope I can ping you today but if not, rest assured that a ping is nigh.

—Gary M. Almeter, "I'll Ping You," *McSweeney's*, March 26, 2018

The modern world is marked by ever-increasing pings. Phones and smart-watches offer sonic and vibrational notifications and nudgings; microwaves, dishwashers, refrigerators, coffee-makers, toasters, and washers and dryers announce their completion (or other needs) with preset tones; fire alarms with waning batteries chirp (and occasionally shriek). Meanwhile, packets of electrical signals shoot to and fro, performing the continual background work of networked computing—checking for existence, for connection, for openness to requests, for latency. As Vivian Sobchak observes, rather than "the stream of interior consciousness that used to temporally 'co-here' as one's subjective identity . . . today what seems, for many, to hold identity together is coherence of another kind: the ongoing *affirmation* of constant cell phone calls, electronic pages, 'palm pilot' messaging—these standing less as significant communication than as the exterior, objective proof of one's existence, of one's 'being-in-the-world.'"[13] Identity now "co-heres" through the ongoing receipt of various pings and pongs.

"Ping" figures communication as sets of isolatable projectiles. Derived from the Latin verb *pung*—meaning to jab, poke, prick, puncture, or mark with points—ping first carried those meanings into Old English and Middle English before acquiring an onomatopoetic aspect in the seventeenth century of "a short, resonant, high-pitched (usually metallic) sound, as that made by firing of a bullet, the ringing of a small bell, etc.," and as a name for this kind of sound itself.[14] In the twentieth century, this form was adapted to name both the ultrasonic sound of sonar and the audible signal that represented this sound to the sonar operator. When radar screens met computing machines, "ping" became the name for a type of packet—a test signal—used to prompt automatic responses from networked computing devices and a name for the degree of latency, or delay, between two devices on a computer network.[15] In these contexts, ping became a method of producing empirical knowledge: by testing for response time, pings measure the difference between how long it takes one signal versus another signal to bounce back. This, in turn, identi-fies how far away an object is or how synced two machines are. With the rise of smart devices, the meaning expanded to describe messages—electronic or otherwise—between people as well as machines, while still carrying the implication of physical impact: "Ping me," "I'll ping you."[16]

Ping first became a method of knowledge production with the invention of radar in 1935. In radar, all objects appear by and through bounce: electromagnetic waves (radio waves) are emitted, and the reflections of these waves determine the location of distant objects. In "Genealogy of the Computer Screen," media scholar Charlie Gere describes the radar screen as "arguably the most indexical technology of representation ever devised, presenting as it does nothing more than a visual trace of the objects it tracks."[17] Early radar was continuous: it simply communicated presence or absence—*there* or *not there*. Developed later during the interwar period, pulsed radar sends out electromagnetic energy at regular intervals. The introduction of the interval collected additional kinds of information, like location and velocity. The US navy calls pulsed radar "echo-ranging."[18] In his work on ping, communication scholar John Shiga argues that pulsed radar was part of a more general shift to what he calls "active acoustics," a shift from listening to projecting. Active acoustic systems, as Shiga describes them, "project a sonic or ultrasonic pulse ('ping') into the water, listen for echoes, and use the delay between ping and echo to calculate the distance between the search vessel and the target."[19] This shift, he argues, led to the development of an "electro-acoustical discourse of ping" based on "analogies between the transmission of electrical power and the mechanical acceleration of projectiles."[20] Pulses shifted the framework of underwater acoustic transmission from one of passive listening to one of projective force.

Projective pulses of electromagnetic or sonic waves through air or water or the ionosphere is one of the kinds of ping involved in radar and sonar: ping as signal. Representations of these signals were initially displayed on oscilloscopes, progressive CRT displays showing continuous wave phenomena in real time. Oscilloscopes visualize variations in signal voltage, usually on a two-dimensional (2D) plot that represents signal variation as a function of time. In the case of radar or sonar, as they are received, oscilloscopes display the echoes of the pings that the system itself has emitted. Made familiar by countless representations of radar and sonar scopes in movies and video games, this second ping is where and how the word transitions from referring to a projected signal to referring to a received result.

Computers changed fundamentally when computing machines integrated with the pings of real-time display. Gere summarizes this as follows: "It

Figure 4.3
A Project Diana QSL card, produced by the US army for signal reception reports, illustrates the results of the "moonbounce" technique, which involved bouncing a radar signal off the moon and receiving the return signal and was first completed on January 10, 1946.

is only when radar and the computer are combined that the modern device we call the computer is able to emerge."[21] By visualizing dynamic results as they progress across the screen, computing machines could now receive and respond to input in something called, and experienced as, real time. In other words, computers did not begin as real-time machines: they became them. Converting the oscilloscope into a display technology for the output of computing machines meant a number of things. Unlike earlier display technologies for computers (paper, tape, film, and cards), the oscilloscope allowed the results of computational problems to be observed as they changed over time. The echo of the initial ping of projected electromagnetic or sonic waves was replaced by the transmission of electrical signals within the computer, which now became responsible for the ping appearing on the display. These electrical signals, which brought computers into a new relationship with real time, were in turn imagined as projectile-esque pings.[22]

The desire for real-time interaction—and, more specifically, real-time command and control capacity—was crucial for the branch of computing tied to the development of aviation, information theory, and cybernetics.

The first computer to be attached to an oscilloscope was the Whirlwind computer at MIT, a machine designed for the purpose of operating in real time. As project member Robert Everett explains: "Whirlwind was a special machine; the motivation for it was also rather special. It was designed by engineers for engineering purposes and for controlling real-time devices. Real-time control was the motivation behind Whirlwind."[23] A US navy–funded attempt to build an analog computer to act as an airplane stability control analyzer, the Whirlwind project started in MIT's Servomechanism Laboratory in 1946.[24] In this context, figuring out how to simulate bounce was a first step toward simulating more complex interactions between pilots, planes, and projectiles. The Whirlwind had visual displays of one kind or another from the start. But the oscilloscope was the first progressive display (i.e., the first that updated continuously in real time). Real-time control required putting and keeping people "in the loop" of communication—it required getting all the humans, machines, and other actors in sync.[25] Although the original intention was to use the device in flight simulators, long before its first public demonstration in April 1951, the project director, Jay Forrester, shifted away from that goal to focus on building a more general-purpose digital computer for aircraft tracking and scanning.[26] This was fundamentally different goal from those that analog and digital computers were tasked with at the time. Scanning and tracking aircraft required machines that could solve differential equations in real time and display these results as they were being produced.[27] By presenting the computed results, and, in the process, testifying that all the parts of the computer were communicating "in harmony," the oscilloscope display became a kind of result in and of itself.[28]

For mathematicians and physicists, differential equations offer the most accurate way to describe a world in motion in all its conditions of possibility. A language of nature, differential equations are a heuristic for the dynamics of matter. Connecting oscilloscopes to computing machines made it possible not only to observe wave phenomena in a repeatable context, as in the case of radar, but also to write programs that modeled wave phenomena on continually updating displays. This is the shift from *representing* signals received from the external world to *simulating* phenomena with computer programs—something that Hayes will pick up on fifty years down the road. Attaching the computer to the oscilloscope made it possible to both model and operate on nature: using a computer makes it

Figure 4.4
Ramona Ferenz at the oscilloscope display at Whirlwind test control in the Barta Building, 1950. Courtesy of MIT Museum.

possible not only to describe the world as it exists, but to model and make new ones.

Programming Bounce

The first bounce programs were written for the Whirlwind computer at MIT in 1949 as one of four kinds of programs used to test the new connection between the computing machine and the oscilloscope screen. These programs brought together two distinct notions of bounce—bounce as signal and bounce as simulation. In the context of computing, they stand as early examples of the now-omnipresent practice of simulation as a form of knowledge production. Hayes puts it this way: "When you create a simulated world, nothing falls to earth unless you remember to turn the gravity on. Nothing bounces unless you tell it exactly where and when to bounce . . . you come to expect a one-to-one mapping between processes in the physical world and algorithms in the simulated one. But much of what goes on inside a simulation has no counterpart in the real world."[29] Unlike in the larger, more unruly world, computer objects will not collide unless they are programmed to do so. Once set in motion, they will not interfere with each other unless the program includes rules of interference (rules of physics). Bounce is one of the answers to the question of what happens when two objects collide—one possible rule of interference.

At its most basic, a bounce program is a simple differential equation. Whereas in the case of radar, bounce describes the dynamic topography of the world, in the case of computing, bounce programs mathematically model conditions of interaction, producing virtual space. In mathematics, the differential equation for measuring bounce is called the *coefficient of restitution*, also known as *Newton's experimental law*. By dividing the relative speeds of the objects before their collision by their relative speeds after the collision, the equation measures restitution—how much kinetic energy remains—after the collision between two objects, which deforms them; the assumption is that the collision results in some loss of energy, as either heat or work done. This usually generates a positive real number between 0 and 1.[30] When the coefficient of restitution equals 1, the objects are said to have collided elastically, which means that each maintained its prior speed. (Sports equipment such as rackets, golf clubs, and balls are sometimes labeled with a coefficient of restitution specification, a ratio that describes a relationship between two colliding bodies. When a particular object is said to have a coefficient of restitution, an assumption is being made about what that object will be colliding with.)

In his history of computer-aided design (CAD), computer graphics and CAD pioneer turned journalist, David E. Weisberg, recalls how in 1949, Charles Adams "wrote a short program that displayed a bouncing ball on the display. This was done by solving three simultaneous differential equations. A little later, probably in late 1950, Adams and John Gilmore wrote the first computer game. It consisted of trying to get the ball to go through a hole in the floor by changing the frequency of the calculations."[31] In Adams and

Gilmore's game, the ball appeared as a moving spot on the scope. It bounced from left to right. Its movement across the screen described the path of an object based on a mass, initial velocity, and coefficient of restitution named by the bounce program's differential equations. Whenever it encountered the positions that had been designated as the ground, a sound was produced. Initially just for Whirlwind's programmers, the game was played not by manipulating a joystick or controller but by manipulating the equations in the program until they struck on a combination that resulted in the ball going through the hole in the floor.

Adams and Gilmore's program created a ball (conditions for an arcing vector), a floor (an x-axis from which the ball would rebound), and a hole in the floor (conditions under which that x-axis would change, allowing the ball to pass through). The Whirlwind's bounce programs are early examples of how rules of interference became the basic foundations of world-modeling in computing: important initial steps in the branch of computing concerned with generating physical spatial relationships for virtual worlds. There are multiple ways to write this kind of rule and to produce bounce interactions. With his bounce programs, Adams was concerned only with producing one body and a ground that it detected and collided with. His approach to programming bounce was algebraic, reflecting what he described as his ideal programming procedure, "in which the mathematical formulation together with the initial conditions can simply be set down in words and symbols and then solved directly by a computer without further programming."[32] The algebraic approach to programming bounce has the advantage of accuracy, something important for a project like Whirlwind, which simulated the movement of things like planes, rockets, and missiles. But an algebraic approach does not allow feedback; it requires all the information to be present and accounted for up front.

Bouncing Ball was first demonstrated to the public during an Open House Day in April 1951. The program that was presented had three settings, each of which was meant to simulate the gravity of a different environment. A switch chose one of three "gravity" options, initiating one of three options in coefficients of restitution: either the ball bounced extremely elastically (the moon), like a standard rubber ball (Earth), or like a heavy, dead object (Jupiter). Game historian Van Burnham describes *Bouncing Ball* as perhaps the first computer-CRT demonstration program, qualifying that it does not quite earn its place in the category of game: "It certainly didn't do much.

A dot appeared at the top of the screen, fell to the bottom, and bounced, with a loud 'thok' from the console speaker. It bounced off the sides and the floor of the displayed box, gradually losing momentum until it hit the floor and quietly rolled off the screen through a hole in the bottom line. And that was all."[33] The algebraic approach to computing bounce made for a dull game. For anyone who did not have the ability to tinker with the equations in the program, there was no to and fro of an ongoing exchange. Play started—and ended—with the first push of the button. Sitting somewhere between shot-put and ball-toss carnival games, the program still earns its place in the history of video games because it was a kind of game for its creators. Moreover, it was a key technological precursor for later games.

Beyond its position in a genealogy of electronic games in game history, the demonstration program had some more interesting aspects. *Bouncing Ball* was not a one-off program written for an open house at MIT; it was one iteration of a *kind* of test program for simulating the motion of objects in physical space, one of the central goals of the Whirlwind project. Not long after Adams and Gilmore's initial program, writing a bouncing ball program became an introductory problem—first for new members of the Whirlwind team and then for students trained in the brand-new field of computer programming. In a report from February 2, 1951, Hrand Saxenian writes: "The new men have also been writing programs of their own to run in test storage on February 2nd. O. G. Aberth has a program for displaying the path of motion of a bouncing ball using a variable coefficient of restitution."[34] (Aberth's program may have been the one used for the April 1951 Open House.) By 1953, students in a MIT computer programming course (course number 6.537) were using "WWI [Whirlwind I] to finish *the* bouncing ball exercise," and then they proceeded to work on term problems.[35] One of these students, Harold H. Seward, went on to make a program that converted "programs punched in standard Flexowriter form, store the program in electrostatic storage, and start at the address designated."[36] Another report states that a "'bouncing ball plot' (2589 pl) was converted successfully."[37] Bouncing ball exercises became a way to demonstrate the successful completion of more complex programs later. Finally, they became one of four standard demonstration tapes and were used for things like testing new scope decoders.[38] Both before and long after they were deployed to communicate the machine's potential to visitors and potential funders at the Open House Day, bouncing ball programs were written by programmers who were learning both how to get a handle on the interaction between computer and screen and how to model interactions in physical space.

By 1958, bouncing ball programs had become standard enough to be included in the instruction manuals for analog computers, alongside instructions for programming trajectories of other kinds of projectiles. In the manual of a Donner Analog computer, one of these sets of programming instructions gave William Higinbotham, then director of the Instrumentation Division at Brookhaven National Laboratory, the idea of building "a tennis game for two to play, that displayed the court, the net and the moving ball on a cathode ray tube"[39] for the lab's annual visitors' day:

> The instruction booklet that came with this analogue computer described how to generate various curves on the cathode-ray tube of an oscilloscope, using resistors, capacitors and relays. Among the examples were the trajectory of a bullet subject to gravity and wind resistance, [intercontinental ballistic missile (ICBM)] trajectories and a bouncing ball. The latter suggested the tennis game. Four of the operational amplifiers were used to generate the ball motions and the others to sense when the ball hits the ground, or the net, and to switch the controls to the person in whose court the ball was located.[40]

None of these bounce programs, including the game of computer tennis by Higinbotham, prioritized *looking like* what they were understood to represent (tennis balls, planes, missiles, etc.). They did not pursue visual realism. Even if they had wanted to do so, computing power at the time did not offer that option. Instead, these programs focused on *simulating* the ways that balls, planes, or missiles moved through the air and encountered hard surfaces. Earlier in his career, Higinbotham had worked at MIT's Radiation Laboratory on airborne, ship-borne, and land-based radar, leaving a few years before the Whirlwind project started. He describes his creation of a computer tennis game as a "natural progression" from his prior work, which had "involved designing means to present the echoes returned from distant targets on cathode-ray tubes, in angle and distance, not far from the problems involved in the tennis game display."[41] Modeling a bounce in the game was imbricated with modeling radar echoes on displays, and both models presented similar computing problems.

Higinbotham made *Tennis for Two* a demonstration vehicle for the high school students, college students, and general public attending Brookhaven's annual visitors' days, which, in 1958, were held on three successive weekends.[42] Unlike the *Bouncing Ball* demonstration at MIT, Higinbotham's game allowed back-and-forth play. To turn the bounce program into a game of computer tennis, Higinbotham generated a long horizontal line at the bottom of the oscilloscope screen to represent the ground and a short vertical line jutting up to represent the net. He also built a pair of handheld controllers. Tennis, as Higinbotham knew it, was already a fairly abstract form of gameplay. The modern tennis court only obliquely refers back to the courts of the Early Modern era, but while the irregular walls, chase lines, and asymmetrical architectural interruptions had long since fallen away, it maintains the rectangular play area. What remains, too, is a mathematically simpler and symmetrical court with a net across the middle. This mathematical simplicity and regularity made tennis a good candidate for translation into a new game space.

The controllers that Higinbotham built let the players alternatingly control the motion of the ball. Each controller had a knob that allowed the player to choose one of three angles (up, down, or level) and a button to initiate the hit. If the ball made it over the net, then the other player could select an angle and hit the ball either before or after it bounced. They had to return the ball before it reached the end of the court.[43] Higinbotham created wind resistance with a 10-megaohm resistor, and he set up the bounce program so some energy was lost with each bounce. The racket was not represented, so there was no visual representation of impact other than the ball changing directions—a streak of light moving back and forth—but whenever the ball was hit (or bounced off the floor or net) the machine's switches sounded. It was not possible to change the velocity of the hit or to miss the ball—if the ball reached the end of the court without being hit, it would simply "freeze in the serve position"—but it was possible for the ball to hit the net and bounce back, ending the point.[44]

Interviewed for the documentary *When Games Went Click*, Robert Dvorak Jr., the son of one of Higinbotham's colleagues, recalls being brought in to test the game of computer tennis: "It's amazing looking at today's technology that you could be fascinated by this little dot moving across the face of the oscilloscope, but sure enough, you could actually see a tennis ball, and there was a little net in the middle and you would bat the ball back and forth across the screen."[45] What does Dvorak mean when he says that he could "actually see a tennis ball"? What made a moving spot of light, leaving a trajectory trail akin to a comet, identifiable as a ball? And what made it

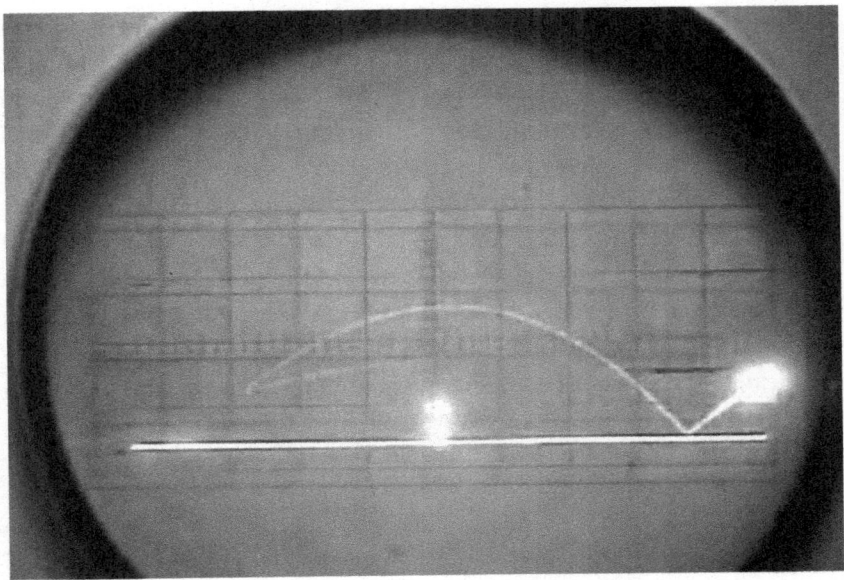

Figure 4.5
A re-creation of the original *Tennis for Two* game constructed for the fiftieth anniversary of its original debut at Brookhaven National Laboratory in 1958. Footage of the re-created game is included in a short documentary posted to the Brookhaven National Laboratory YouTube channel in 2008, as well as in a longer documentary *When Games Went Click: The Story of Tennis for Two*, completed in 2013. Screenshot by the author.

identifiable as "a tennis ball" in particular? First, the computer manual had called it one; and second, Higinbotham had created controllers that allowed players to hit the dot to and fro. It was a ball because it was called a ball, but more emphatically because it behaved like a ball. This computed behavior passed the *I*—the ability to take an action—from player to player. Finally, it was a *tennis* ball because Higinbotham presented the game as computer tennis, representing and programming an elastic bounce that resembled that of a tennis ball—clearly distinct from the bounce of, say, a hacky sack, baseball, or football.

Tennis for Two was presented to the public just one year after *Sputnik*'s 1957 launch. The context of the space race shows up in a similar way as it did with *Bouncing Ball*. The second version of the game, presented in 1959, had a larger screen and three gravitational settings (similar to three settings on the Whirlwind's *Bouncing Ball* program), which allowed players to play "variations of tennis on the moon, with low gravity, or on Jupiter, with high gravity."[46] More and less gravity were represented by changes in the bounce of the ball. Despite being the display of a continually updating result, the ball was imagined as a stable physical object, whose behavior was changed by an external environment. In both 1958 and 1959, people waited in lines for hours to play. After 1959, the game was put away in favor of new projects and demonstrations. But over the next decades, as computing reshaped communicative and cultural life—expanding from the military-industrial complex to the military-entertainment complex, and from there to culture at large—bouncing balls populated computer programming course assignments, tech demos, design software, and the growing industry of video games.[47]

"Electronic Coordination"

The stories of the creation of the earliest and most iconic video games and platforms have been told by a wide range of game scholars.[48] *Pong* was the first massively commercially successful video game, and Magnavox Odyssey was the first commercial home video game console. As was the case at MIT in 1951 and at Brookhaven in 1958, both *Pong* and the Magnavox Odyssey were made by men with backgrounds in engineering and radar. *Bouncing Ball* and *Tennis for Two* were made in the context of laboratories developing technologies for missile defense and nuclear energy—explicitly intended for potential funders and self-selecting publics—but Atari's *Pong* and the Magnavox Odyssey console were packaged and promoted as commercial products for mass distribution. These early bouncing ballgames were sold as ways to develop the general public's "electronic coordination."[49]

The creator of the Magnavox Odyssey, Ralph Baer set out to make games for the television-viewing public. His primary objective, as described in a 1971 patent, was to make "an apparatus and methods for displaying video signals upon the screen of a television receiver where some or all of the video signals [are] both generated and controlled by the viewer."[50] In contrast to *Bouncing Ball*, where the primary site of play was programming equations, the physical control of symbolic objects on the screen now occurred through control units connected to an everyday domestic object: a television receiver. Baer strongly believed that televisions should be "active rather than passive

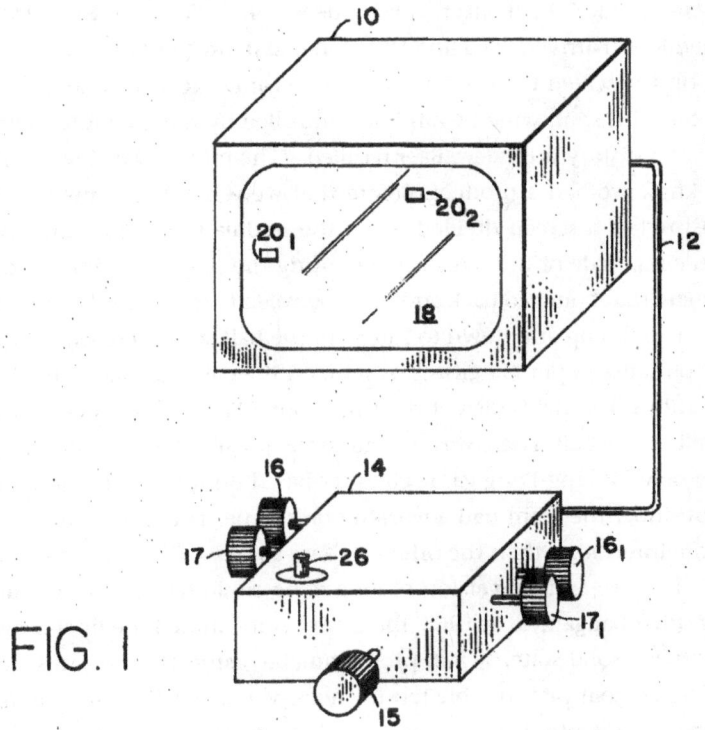

Figure 4.6
Drawings of a television display and hand controller from Ralph Baer's 1973 patent for a "Television, Gaming, and Training Apparatus."

instruments."[51] He imagined the television-viewing audience becoming players, participants, and partners in waveform generation and manipulation.[52] Baer considered watching a passive and, thus, less desirable activity. This assumption about the passivity of sight still exists in much of the discourse around interactivity and virtuality. Visible (measurable) body gestures that control what occurs on screen are seen as "active," and thus more desirable.[53] Interaction, in this context, meant people using televisions as instruments for playing with both the machine and with each other.

Baer's training was in the field of television engineering. While working at Sanders Defense Associates, he worked on how to make games for television. He brought fellow Sanders employees—Bill Harrison, a former air force radar specialist, and Bill Rusch, an electrical engineer from MIT—onto the team in 1967. When Rusch arrived, Baer and Harrison had figured out how to control two spots on the screen to create a set of chase games. Rusch came up with the idea of a third spot, which would be controlled by the machine itself: a ball. In an oral history interview, Baer recalled, "The minute we played Ping-Pong, we knew we had a product. Before that we weren't too sure."[54] The game consisted of a screen divided down the middle by a white line, two squares on either side of the screen representing the players' paddles, and a smaller square that bounced back and forth between them. The players could move their paddles up and down to intercept the ball, and there was a knob that they could use to put "English," or spin, on the ball. Burnham describes the implications of Rusch's idea of the third spot: "Ping-Pong. Hockey. Tennis. Football. Volleyball. There was no end to the possibilities. Harrison engineered the new TV Ping-Pong game circuitry based on Rusch's initial design and by November the team had a working prototype. The new architecture was far more interesting than the original chase games—and infinitely more fun."[55] The basic to-and-fro version of Ping-Pong could be iterated into any number of other ballgames. In 1968, the team demonstrated the final Brown Box prototype: a solid-state, switch-programmable game of forty transistors and forty diodes that played table tennis, hockey, volleyball, skiing, tennis, football, soccer, roulette, baseball, horse racing, chase games, maze games, and target-shooting games with a light rifle.

Baer and Sanders Defense Associates' 1971 patent for a "Television Gaming and Training Apparatus" was sweeping: The invention was presented as an apparatus "for the generation, display, manipulation, and use of symbols or geometric figures upon the screen of the television receivers for the purpose of training simulation, for playing games, and for engaging in other activities by one or more participants."[56] Patenting games was not new. In the middle of the nineteenth century, the British patent office began publishing an index of patents called "Class 132: Toys, Games, and Exercises."[57] As we saw in chapter 2, Major Walter Clopton Wingfield's patent for a "New and Improved Portable Court for Playing the Ancient Game of Tennis" in 1874 introduced a simpler and more portable version of tennis that ended up being adopted around the world. The Hamley Brothers in England and the Parker Brothers in the United States patented and rigorously prosecuted their 1900 trademarks on the name *ping-pong*, which is why the game that Magnavox Odyssey eventually released was called "Table Tennis," not

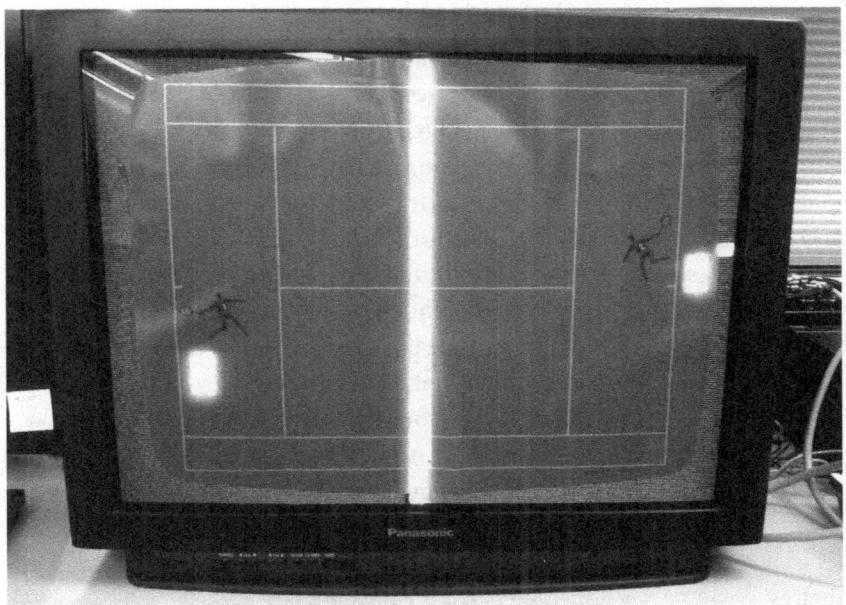

Figure 4.7
Playing tennis with the plastic overlay on the Magnavox Odyssey during a 2013 site visit to the Strong Museum of Play. Photograph by the author.

"Ping-Pong." Patenting is its own kind of bounding practice—one that aligns with the original mean of bound: to mark off territory. Baer and Sanders Defense Associates were not claiming the invention of any one game, but rather the invention of *any and all* games that involved players using controls to move symbolic objects on television screens. They would later argue that this patent gave them exclusive rights to make all ball-and-paddle video games. They had used the relocation of ball play to electronic environments as an opportunity to claim ownership of one of the largest genres of play in human history.

The television companies to whom Baer pitched the product were not convinced that a mass audience wanted their televisions to be "active instruments."[58] Finally, Magnavox agreed to take on the project. Baer had hoped to convince television stations to broadcast overhead footage of sports fields and courts to serve as backdrops for the games, but he did not succeed. When Magnavox released the first Odyssey system, it came with twelve translucent mylar overlays that players placed on their television screens to change the game. Placing and replacing the overlays required the player to touch the screen and feel the static that was making them adhere to it. Just as the signals had become manipulable, the screen had become a touchable surface.[59] The Odyssey's overlays, more generally, made the television into a machine with which people imagined interacting or playing. Underlying the patented ambition, the promise or fantasy is that games, simulated training programs, and other activities would help people develop the capacity for good interaction with these new machines. The patent does not go into detail about the type of training that a person would undergo, but when the Magnavox Odyssey launched as the first home console in 1972, the tagline that accompanied the table tennis game was: "Table Tennis: The basic Odyssey game that develops your electronic coordination."[60] Coordination involves the bodily capacity for good timing. Table tennis here is imagined as a vehicle for importing the hand-eye coordination of ballgames into the context of electronics. As the tagline suggests, the television screen is the new sports field, and the ability to play on this field—figured as the ability to move "symbols or geometric figures"—will be developed by playing ballgames.[61]

Pong: The Ball Is Square

The same year that the Magnavox Odyssey consoles made their way into living rooms, Nolan Bushnell's company Atari launched the arcade version of *Pong*, which would become perhaps the most iconic video game of all time.[62] The word "pong" is the counterpart to "ping"—a general name for short, resonant, low-pitched sounds. And while the bounce games made at MIT and Brookhaven had sonic elements, this was the first commercial game that contained sound. The name *Pong* called up associations to Ping-Pong while effectively skirting the Parker Brothers' trademark on the name of that game. Multiple books on the history of video games carry a subtitle along the lines of: "from *Pong* to . . ."[63] In the decades since its launch, thousands of adaptations—commercial or otherwise—have been created for arcades, living rooms, auditoriums, personal computers, mobile phones, tech demos, and laboratories. Versions have been made with analog and digital computers, dedicated chips, software programs, atoms, electrons, and cortical neurons. Despite Baer's patent, when scientists design electronic ballgames for pigs or chimps to learn, these games do not call back to the Odyssey's tennis game. They are, instead, inevitably compared to, if not simply called, *Pong*.[64] Like

Figure 4.8
A mother and son playing *Pong* in Germany c.1976, and a set of Atari Paddle controllers. The controllers were called "paddles" in part because the first game that they
were used for was *Pong*. Photograph on the left courtesy of Alamy; photograph on the
right, courtesy of the Strong National Museum of Play.

tennis during the Scientific Revolution, in the second half of the twentieth
century, *Pong* became a popular object to think with—ready made for translating scientific ideas to the public.

The original game is famously simple: "Avoid Missing Ball for High Score."
Bushnell's first game, *Computer Space*, had failed, and its failure had been
blamed on its complexity. In the aftermath of that failure, and after attending a demonstration of Baer's Ping-Pong game, Bushnell asked his new engineer, Al Alcorn, to make an arcade game based on tennis: just a ball, two
paddles, and the score.[65] Alcorn was new to designing games, but Bushnell's
plan was to give him a simple training exercise that would ease him into
electronic games.[66] In an interview with computer historian Henry Lowood,
Alcorn recalls understanding the task as "Let's just do the most simple game
to save time."[67] Lowood says that Alcorn was initially asked to make a game
based on Ping-Pong, but in his interview with Lowood, Alcorn maintains
that the simplest game he thought of was tennis. Regardless of which was the
referent for Alcorn's designs, tennis and Ping-Pong—both already simplifica-
tions and abstractions of court tennis—shared a set of formal properties that
met the desire for a simple kind of play. The forms and mathematical logics
of one-on-one, back-and-forth ball play in a symmetrical and divided space
were simple enough to translate across scales and materials.

As balls go, the *Pong* ball is an odd one. For starters, it is rectangular,[68] but
it does not bounce the way that one would expect a rectangular object in
the world to bounce—with some kind of wild, off-kilter ricochet. Instead, its

motion is regular, more like that expected of a sphere. And yet its trajectories are missing certain kinds of curves and arcs. The ball's rectangular shape was a product of the original game's raster graphics. Visually, the ball was a graphic index of the smallest possible interval of time that it could take to adjust the intensity of the CRT's electron beam.[69] The ball was a pixel. And, unlike the first Magnavox Odyssey games, it not only bounced off the players' paddles, but also off the "walls" at the top and bottom of the screen. In his history of computer animation, Tom Sito describes *Pong* as "one of the greatest break-throughs in motion graphics." He added, "You didn't need a science degree or pages of detailed instructions . . . just plug it in and anybody could play *PONG* in minutes."[70] Computer graphics, like traditional animation, began with bounce. But it was a very different kind of bounce—one, at least at the start, that was highly elastic, yet devoid of all squashing and stretching.

Sonically, the ball was an amplified index of voltage shifts. Van Burnham explains, "The image seen on-screen was a reflection of the shifting ON/OFF patterns in the circuitry, drawn with a pulsing raster scan. As the patterns changed, they created a shift in voltage that, in turn, generated wave forms, that—when amplified—would make the most fantastic sound. *PONG*. It was perfect."[71] In fact, there are three different "pong" sounds in the game: the sound of the ball contacting the wall (duration 16 ms, frequency of 226 Hz), the sound of the ball contacting a paddle (duration of 96 ms and frequency of 459 Hz), and the sound of a ball being "missed" and a point being added to a player's score (duration 257 ms, frequency 490 Hz), which fall in the bass (60–250 Hz) and low-mid (250–500 Hz) frequency ranges respectively.[72] Similar to Baer's initial desire to integrate live television broadcasting into his electronic games, Bushnell had originally wanted recorded sound clips of ball bounces and crowd cheering from professional tennis matches. Yet when he played the game that Alcorn had built, he "forgot all of his other project ideas."[73] A former carnival barker turned electrical engineer, Bush-nell recognized a game that would keep people playing. At least in its earliest enactments, the ball's sound, indexing every moment of collision and point conclusion, was the sound of the system.

On top of its square shape and alternating harmonic and disharmonic sounds, the *Pong* ball is strange in some other ways. The simulation of envi-ronmental conditions, present in prior projectile simulations (from *Bouncing Ball* and *Tennis for Two* to Bushnell's own *Computer Space*), is gone. The *Pong* ball is not subject to three-dimensional (3D) inconveniences like gravity,

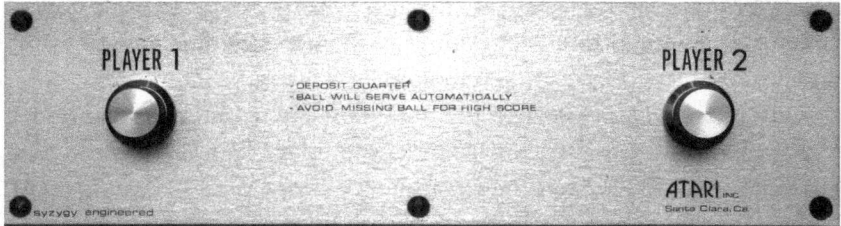

Figure 4.9
Pong instructions as they appeared on the first arcade cabinets. Courtesy of the Strong National Museum of Play.

friction, wind, or spin; it suffers no loss of energy. It appears to be perfectly elastic (meaning that it has a coefficient of restitution of 1). Unlike the early Magnavox Odyssey ballgames, the player can control its motion trajectory and velocity only indirectly. Players do not hit the ball so much as block it. *Pong*'s single instruction—"Avoid missing ball for high score"—drives this home.[74] Hitting a ball usually implies direction, aim, and at least some attempt at placement. *Pong* players are relieved of these complexities.[75] All you have to do to keep the ball in play is to avoid missing it.

The simplicity was partly due to encoding the game logic in hardware. In his telling of the complex institutional, technical, and social history of *Pong*, Lowood emphasizes, "The original game ran not one line of program code. It did not use a microprocessor or a custom integrated circuit; rather, it was a digital logic design made from components familiar to a television engineer who thoroughly understood the various ways pulse waveforms could be generated and manipulated."[76] Lowood goes on to quote Alcorn: "[Bushnell] designed out the need for the computer, because the computers were so slow at that time . . . So there was this brilliant leap that Nolan made about how he could get rid of just a little bit of logic [and still] do the same thing the computer's going to do, just much, much faster, so he didn't need the computer."[77] *Pong* was an example of computing without a computer. By disposing of the gravity settings and other complications,[78] they realized two key things: games did not need to simulate planetary physics—and, as a corollary, balls did not need to appear round—and computing processes did not belong exclusively to any one type of machine or material. As video game historian Raiford Guins puts it: the *Pong* prototype operates as "a liminal artifact . . . seemingly outside of television history, despite being a television and game of television engineering, while claimed as an 'iconic object' in the history of computing to recognize predigital innovations in electronic game design and anticipatory practices for a (then) soon-to-be video game industry."[79]

In its moment, *Pong* was a smash hit. Almost overnight, Atari went from a $750 investment to a $2 billion "wonder company."[80] Arcade sales skyrocketed. Bushnell decided to produce a home console version. At this point, Magnavox sued, claiming that Bushnell had stolen the Ping-Pong idea from Baer after playing a demo version of the Odyssey in May 1972. *Pong* had allegedly violated the 1971 patent. Although Bushnell denied copying Baer's game, his presence at a demo of Baer's game in California bolstered Magnavox's case. They won the suit, and Atari was forced to release the home console version of *Pong* exclusively through Magnavox. Baer's company, Sanders Defense Associates, and Magnavox went on to make millions of dollars from the patent. Later, those looking to challenge Baer and Magnavox's sweeping patent on all "video games using bouncing balls" brought Higinbotham's computer tennis game into the history of electronic games as an example of an electronic ballgame that predated the patent and rendered it void.[81] Writing about why it did not occur to him to patent his own game, Higinbotham says, "At the time it did not appear to me to be more novel than the bouncing ball circuit in the analogue computer book."[82] Bushnell was also not so caught up in the game of firsts: "I feel in some way that I didn't invent the video game—I commercialized it. The real digital game was invented by a few guys who programmed PDP-1s at MIT."[83]

Pong is important in the story of electronic bounce not only because of its stunning initial success, but also because its logic of bounce became the foundation of a slew of games in the second half of the 1970s. Atari failed to register a trademark in time to protect the name *Pong*, and as a result, like-named knockoffs sprang up. Although this hurt the company economically, it contributed to the development of the game's iconic status: both name and form were repeatedly repurposed. In 1975, General Instruments produced Atari's design for the AY-3-8500 microchip—a.k.a. "Pong-on-a-chip." In *Racing the Beam*, Nick Montfort and Ian Bogost describe how Pong-on-a-chip "allowed even companies without much electronics experience to bring *PONG*-like games to market, and many did just that. Martin Campbell Kelly writes that there were seventy-five *PONG*-like products available by the end of 1976, 'being produced in the millions for a few dollars apiece.'"[84]

If the games themselves gave arcade and living room players a new relationship to their televisions, Pong-on-a-chip gave companies a new relationship to production. Viewers were able to generate and manipulate video signals, and companies with electronics expertise built systems for this kind of manipulation through the standardization of the foundational logics of bounce.

Pixelated Participation

"Now's your chance to be a pixel in a crazy, first time anywhere experiment, consisting of you, reflectors, lights, video cameras, frame grabbers, computers, and lots o' software."[85] On a summer night in Las Vegas in 1991, during the annual SIGGRAPH conference (whose acronym stands for "Special Interest Group on GRAPHics and Interactive Techniques"), Loren and Rachel Carpenter debuted a new interactive audience participation system by way of a projected game of *Pong*. The gathered crowd of computer graphics professionals collectively played the game by using red-and-green-tipped wands they had found waiting for them on their seats.[86] People eagerly held up their individual wands to be read by the Cinematrix camera's sensors in order to help move their side of the room's *Pong* paddle to the right position on the screen to intercept the ball's trajectory.[87] To fulfill *Pong*'s iconic single instruction—"Avoid missing ball for high score"—each group had to quickly and collectively display the mix of red and green wands that would move the paddle in the right direction at the right speed to intercept the glowing, white, square ball. Each person was "a participant and a pixel in

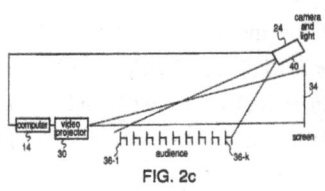
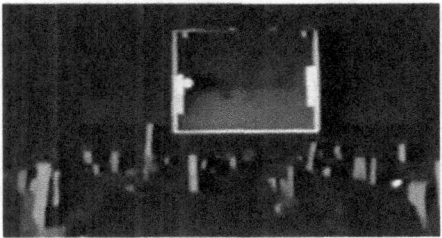

Figure 4.10
A diagram of the Cinematrix system from Loren Carpenter's patent for a "Video Imaging Method and Apparatus for Audience Participation," 1994, and a still of documentary footage from the Cinematrix demo at SIGGRAPH.

the world's largest videogame."[88] After a few minutes, when both sides of the room got the hang of how the distribution of red and green correlated with the direction and speed of the paddle, the speed of the ball increased. "The crowd faltered but then compensated, playing better than before."[89] Caught up in the collective challenge of keeping the ball bouncing back and forth, they became effervescent.[90]

At the time, the Carpenters were already known quantities at the conference, which was attended annually by thousands of industry experts in the field of computer graphics. During the prior decade, Rachel had curated art exhibitions and animation screenings for the conference (serving as an art show administrator in 1985 and art show committee member in 1989), and prior to that, in 1980, Loren had been the first to use fractals to generate the graphics for a computer-animated film. He was also the co-inventor of the Reyes rendering algorithm: the core rendering engine in Pixar's RenderMan software, used to render every Pixar film, as well as a host of non-Pixar animated films and visual effects for many live-action films. Ed Catmull, Loren Carpenter, and Rob Cook won an Academy Award of Merit for RenderMan in 2001—the first software package to get an Oscar. In 1991, the Carpenters incorporated a company to promote the Cinematrix for use at corporate theater events.

Director Adam Curtis uses footage from the Carpenters' initial presentation at SIGGRAPH to open and close "Love and Power," the first segment of his three-part series *All Watched over by Machines of Loving Grace* (2011). Curtis's documentary is critical of how computing machines have reconfigured fundamental ways of seeing and being in the world. The SIGGRAPH scene first appears intercut with an interview with Loren and Rachel. In the clip, Rachel nods silently in agreement as Loren describes the ecstatic response of the crowd. She had developed an analysis of the Cinematrix as a master's thesis in social and cultural anthropology titled "Techno Tribe," which addressed "the human experience of new communication technologies."[91] The Carpenters understood the Cinematrix as revealing a mysterious, amoeba-like capacity of the crowd to come together to play the game successfully. They, in turn, interpreted this as an instance of how a nonhierarchical society of free individuals connected by machines might be able to produce its own order and stability.[92] This apparent desire of free individuals to be part of the effort to keep a *Pong* ball bouncing back and forth, they argue, models how we might use systems of machine-assisted collaborative control to produce collective attunement at the scale of society. The desire to participate and meaningful participation are two different things. Behind the collective effervescence, estimations done by the algorithm were smoothing out the inevitable errors of the crowd. Michael Scroggins, a performance-animation and virtual-reality artist who attended the demonstration, remembers accidentally showing green when he meant to show red and vice versa, and realizing that it did not actually matter all that much. He recalls Loren Carpenter saying at the time, "If you think you are on the left you are, and if you think you are on the right, you are."[93]

The footage from the SIGGRAPH demonstration returns for the final shot of "Love and Power," with Curtis's voice intoning: "The original promise . . . was that the computer would liberate us from all of the old forms of control, and we would become Randian heroes in control of our own destiny. Instead, today we feel the opposite, that we are helpless components in a global system, a system that is controlled by a rigid logic that we are powerless to challenge or to change."[94] For Curtis, the *Pong* demonstration drives home his critique of a misdirected social model. As we watch the projected paddles struggle to get in place just in time to deflect the *Pong* ball, the narrowness of this vision of freedom is palpable. It is a vision in which freedom is reduced to the freedom to choose whether to hold up the red or green side of a wand—whether to be a red or green pixel.

In 2013, a new iteration of the Cinematrix system was presented as a demonstration of an experiment on crowd behavior for the BBC show *Bang Goes the Theory*. The setup of the system—and its facilitated interactions—were essentially the same as they had been twenty-two years earlier, all the way down to the red- and green-tipped paddles. The most dramatic difference was visual: the classic black-and-white *Pong* game was now a full-color, animated, watery blue scene of a shipwreck underwater. The collectively controlled animated paddles were bicolor, green on the top half and red on the bottom, to indicate what direction a participant will be contributing to when they hold up the green or red side of their own paddles. This animated undersea layer—overlaying the classic *Pong* display—embodies what Bruno Latour calls the "drawing together" of histories. Here, computing and animation are drawn together such that "the weakest, by manipulating inscriptions of all sorts obsessively and exclusively, become the strongest."[95] The object bouncing to and fro across the projection screen is neither a sphere nor a square. The ball is now a friendly cartoon shark.

Figure 4.11
Interactive *Pong* game demonstrated for an episode of *Bang Goes the Theory*, BBC,
2013. Screenshot by the author.

Conclusion: Cooperative Models

Because Curtis uses Loren as a strawman to tell the story of the rise and fall of
California ideology—a combination of radical individualism and cybernetic
utopianism—he does not dwell on the more basic assumptions underwriting the Cinematrix demonstration. How did a collectively played game of
Pong come to stand in as the model, par excellence, of interaction with and
through machines? And, more strangely, where did the idea that a simple
game of bounce makes a good model for a complex society come from?
Across these two chapters on animated and computer bounce, my broad
argument is that bouncing balls were used as the building blocks of representation, interaction, and simulation in both fields from their earliest days.
From Étienne-Jules Marey's *Le Movement* to the *Bouncing Ball* and *Tennis for
Two* games presented at MIT and Brookhaven's open house to the Carpenters'
interactive game of *Pong*, demonstrations of virtual bounce have continuously underwritten innovations and advances in animation and computing.
And as such, via virtual bounce, it is possible to track a burgeoning social
logic of collective participation in judgment and experience across computational and cinematic developments. These in turn create the possibility

for both collective and individual gaming experiences—and for the global spectatorship of worldwide play.

I want to conclude by returning to the two scenes of table tennis as models of good and bad communication. Licklider's central argument in "The Computer as a Communication Device" is that "modeling . . . is basic and central to communication." Communication can be defined as "'cooperative modeling'—cooperation in the construction, maintenance, and use of a model."[96] His vision was that "a particular form of digital computer organization . . . constitutes the dynamic, moldable medium that can revolutionize the art of modeling."[97] For Licklider, the structure of digital computing involved cooperation between people and machines in every part of the process—the construction, maintenance, and use of models. Formed through processes of extending bodies and minds into computing machines, this would lead to fundamentally different *I's* and *we's*. He put it this way during a 1962 panel discussion with J. R. Pierce, Claude Shannon, and Vannevar Bush on "What Computers Should Be Doing":

> Computers give us for the first time the tools with which to come to grips in a serious way with intellectual processes. We have never had them before . . . I should like to reinforce what Rosenblueth said about developing a real feeling of rapport with the computing machine . . . anybody who does not feel the extension of the human body into a machine does not really have a basis for feeling what the intellectual extension might be. Some people who have spent hours at tennis can hardly believe that the racket is not alive. People who have spent hours with computers have comparable difficulty.[98]

Here, the computer is not a Ping-Pong table, but rather a tennis racket: an instrument that, if one learns to grip, handle, and hit with it properly, one "can hardly believe . . . [it] is not alive."[99] In computing, like animation, there is an illusion of life, but it is of a different sort. While in animation, the illusion is produced through coordinated audiovisual effects for a viewing audience, in computing, the illusion is produced for a person using—playing with—the machine: first through an experience of bodily rapport with, and extension into, the machine, à la the relationship to a tennis racket, followed by an experience of intellectual extension and expansion. Licklider understands computers as machines used for the simulation of complex systems. While simulation, as a rubric, usually assumes a separation between the real world and a less-than-real digital simulation, Licklider is not imagining separation here. Instead, he imagines a coextension through a structure of networked computing—one that will fundamentally change the way that people exist with each other. It is a vision of an *I* that has been expanded from individual, sonic, and virtual coordination by way of bounce, until suddenly everyone is collectively participating in real time via digital means. Indeed, the collectivity that Licklider posits has been fully naturalized in our era of networked communication, having moved from a possibility to a capacity to a requirement for engagement in a multitude of forms of social, political, and economic life.

But as the Carpenters' interactive game of *Pong* shows, it is necessary to question the naturalization of the extensions of body and mind into computing machines. The problem with the Cinematrix's collective game of *Pong* is one shared by many iterations of programmed bounce: it dramatically idealizes modes of collision. All too often, programmed bounce presumes that both objects and environments are simple, are mathematically regular, and have a durability akin to that of billiard balls, when in fact, very few things in the real world—be they human bodies or electrical signals—actually interact in this way. The world is made through many kinds of more qualified interactions. For collective actions to produce an average of successful play, not only do those actions need to be reduced to binary logic (yes or no, 0 or 1, red or green), but a tremendous amount of effort is also needed to make environments and objects highly predictable. While this may seem innocuous in the context of a game of *Pong*, it becomes less so when considering pursuits such as the transformation of built environments to accommodate a world full of autonomously operating vehicles. Here, it is crucial to remember Brian Hayes's warning that "much of what goes on inside a simulation has no counterpart in the real world."[100] Otherwise, the desire to bring collective bodies in sync risks underwriting more overarching moves to make our worlds as regular as our most abstract and simplified electronic ball courts.

If the story of the move from the industrial to the informatic is often told in terms of the rise of open networks, the kinds of bouncing balls and bounded games that appear in computing call attention to new kinds of enclosures—and new kinds of togetherness—that have emerged through these informatic transformations. Our notion of interaction has thinned to accommodate them. There is a presumed synchronicity and uniformity—electric and electronic coordination enabled and animated by computational and animation logics—and the virtuality of this interconnectivity and interactivity is just that: virtual. And, more than this, this virtual conception of collectivity and uniformity entails certain social-political judgments and frames.

The chapters in part III, the final section of this book, take up the question of bounded spectacle to show how this happens. Against the drive for collective synchronicity and virtual coordination, it's worth remembering the awkwardness of individual entrainment, the gendered and racialized differences that disrupt and glitch uniformity, and the exceptions of forgotten play that serve to counter culturally hegemonic forms of sport.

Bounded Spectacle

5   Bounce Feel

Familiarity is what is, as it were, given, and which in being given "gives" the body the capacity to be orientated in this way or in that. The question of orientation becomes, then, a question not only about how we "find our way" but how we come to "feel at home."

—Sara Ahmed, *Queer Phenomenology*

In the information age, fans are the new experts and athletes are objectified as data, becoming sets of statistical profiles and avatars.

—Astria Suparak and Brett Kashmere, "Introduction: A Non-Zero-Sum Game."

Sitting Down to Play

The first time I sat down to play *FIFA* in the fall of 2016, I felt distinctly ungainly.[1,2] My ungainliness was partly beginner's awkwardness. As soon as I sat on the couch and picked up the Xbox 360 controller, I was acutely aware that my fingers had the wrong muscle memory. The last time that I had played video games with any real regularity was back in the 1990s, during the reign of Super Nintendo Entertainment System controllers, when Electronic Arts (EA) published the very first version of the *FIFA* series. EA's *FIFA*—first launched in 1993 and renamed in *EA Sports FC* in 2023, after the collapse of EA's long-standing licensing deal with the eponymous (and infamous) football association—is a global entertainment software franchise. It is the world's most popular and profitable sports video game.[3] At any given moment of any given day, the game is being played on the gaming platform Steam by anywhere from 25,000 to 110,000 players.[4] That a soccer-themed game is the most popular sports video game in the world is perhaps not

Figure 5.1
Portrait of my character during shooting practice. Photograph created from video documentation of my first time playing *FIFA '15*.

surprising since soccer—or "football," as it is known in most places—is called "the world's game":[5]

> If there is anything close to a global game, it is association football. Also known as soccer, football, fussball, fútbol, fotball, fitba, and futebol, the game is played in well-manicured and wealthy parks of gleaming metropolises and on patches of asphalt, grass, or dirt commandeered by players in cities and towns round the world. At any given moment, someone is kicking a soccer ball down a field somewhere, fueled by and fueling a multibillion industry that promises entertainment and drama pass after pass, goal after goal.[6]

These are the words that Henry Lowood and Raiford Guins and I came up with to describe the mythic scale of soccer in the introduction of the volume we co-edited on *FIFA*. Thousands of soccer-themed games populate the history of video games.[7] Some of these simulate management and strategy, some center on individual star players, some offer adaptations of the game—featuring extraordinary aggression or expressive cars or dolphins zooming around as the soccer players—and some, like *FIFA*, emphasize realism, claiming to allow players to *feel the game*.[8]

The final pair of chapters in *Bounce* take on two mythic games—*FIFA* and the Mesoamerican ballgame *ulama*, respectively—as examples of bounded spectacles. At first glance, *FIFA* and *ulama* appear incongruous. Their forms, scales, and logics of play are drastically different, but both depend on self-sustaining myths about participation, initiation, a logic of demonstration, and simulated play. The myth of *FIFA* borrows directly from the game and professional organization that it simulates to build a spatial myth of global spectacle and global capitalism covering and claiming "The World" as its field of play. In contrast, the myth of *ulama* is temporal, countering the myths of capitalism and colonialism with claims of epic historical duration and cosmology. *FIFA* is an open and universalizing bid for total virtual control and representation of "The World's Game."[9] The Associación De Juego de Pelota Mesoamericao (AJUPEME) and others working to revive, sustain, and grow *ulama* are participating in an anticolonial effort to create an opening for perseverance and preservation in a closed world.

My past life as a professional women's squash player made the methods of training for *ulama* familiar. Even though I was new to the game, I could jump right in. But my years of squash did not prepare me for *FIFA*. Or rather, they prepared me to feel awkward, uncoordinated—ungainly. The pleasure that I have taken in sports has historically been grounded in familiar feelings: muscle burning, sweat dripping, satisfying contact, wind hitting my face. The lack of this dense sensory experience is precisely what makes simulated sports strange to me.[10] The shift of a tangible, full-body activity into an animated audiovisual field always pulls me up short. In sports video games, the feel of playing a sport is translated into interactive audiovisual media. Animated graphics and well-timed sound effects stand in for experiences of proprioception and physical contact.

Often, it is not sports in general but professional televised sports in particular that are used as models for video games. As Abe Stein writes in "Playing the Game on Television," *FIFA* and other sports video games "borrow and build on, and reference, many of the established conventions of mediated sports, not least of which is the televisual." The relationship between sports and sports video games is dialogic and intertextual.[11] A regular viewer of televised sports games has as many, or more, skills to bring to playing a sports video game as a highly skilled athlete. Regardless of whether the represented activity is organized around eye-hand coordination (as in the case of tennis), eye-hip coordination (as in the case of *ulama*), or eye-foot coordination (as in the case of soccer), hand controllers are the primary sites of contact for most video games, requiring most of a player's body to be fairly stationary so their eyes, ears, and hands can get attuned to the action.[12]

Legs crossed on the couch, I began virtual shooting practice. My avatar, *FIFA 15*'s default Lionel Messi, stepped up and shot the ball from its set position. It missed, rebounding off the top bar of the goal. Some Avril Lavigne-esque pop music blared from the speakers as play reset. I did it again, and I discovered that I could repeat this exact miss with ease. Again and again, and again and again, I sent the ball flying off the top bar. And just like that, I was drawn in, charmed by the unearned consistency, the impossible accuracy.[13] This was a different kind of bounce than the one that I knew so well from squash—different as well from the kinds of bounce found in *ulama*, court tennis, and modern tennis. While athletes and sports manufacturers focus on making their bodies and equipment perform as reliably and consistently as possible, video game designers work in the other direction, putting an immense amount of effort into manufacturing interactions that convincingly simulate the complexity—and the resulting planned chance, or just contained unpredictability—of the real world. This is part of what makes finding these kinds of edges pleasurable. I moved on to a practice match, scrolling through an array of options for game duration, weather, pitch pattern and wear, and ball choice (options included balls from Adidas, Mitre, Nike, Puma, and Umbro, as well as one carrying EA's brand). I again went with the defaults. Two men's teams took the field, wearing blue and red uniforms. (There was no option to play as a women's team or a coed team in *FIFA '15*.)

At kickoff, another problem presented itself: the ball was nowhere to be seen. Play commenced. As I watched, a red arrow jumped from avatar to avatar following the path of an invisible object. I determined that this indicated which player I controlled at any given moment. As I fumbled with the controller, the opposite team scored a goal. And the game went on in this way: the arrow and the artificial intelligence (AI)–driven players' clustering movements pointing me approximately—and belatedly—toward the ball's location. I made small, ineffectual attempts to direct any of my players to make contact. At some point, I lost the match. This play had not been particularly fun. But, like hitting the crossbar, the invisible ball glitch and the bouncing arrow captured my attention. I had been searching for something compelling about this distinctly unathletic activity—something that would counter my resistance to sitting down to play. Like the charm of impossible

Figure 5.2
The character that I am attempting to control dribbles an invisible ball as an AI opponent comes in for a slide tackle. Photograph created from video documentation of
my first time playing *FIFA '15*.

accuracy, this accidentally invisible ball and the arrow that chased it did
just that: they compelled some pursuit. In part, my resistance to *FIFA* had
been grounded in an expectation—a desire, really—for the game to give me
a familiar feeling, for it to make me feel like I was playing a sport. The way
through this resistance was to stop wanting the game to be something that
it was not. I had to become curious about what, in fact, it was.

Despite the centrality of the ball and its feel to the game, it was sometimes an afterthought for the game's designers. In 2012, while touting the
excellence of the series, then–general manager for EA Football Matt Bilby
said: "Part of the benefit of going through the process of releasing the best
games in the world is that we've got very good and . . . those 'oh my god
we forgot to put the ball in the game' [moments] don't happen. And we did
actually have a year when we forgot to put the ball in the game. But that
was a long time ago."[14] Bilby gave this assurance that the days of forgetting
to put the ball on the pitch were in the past, three years before my encounter with the invisible ball glitch. Online threads testify to versions of this
issue showing up in *FIFA '15*, *FIFA '16*, and *FIFA '17*.[15] Raiford Guins reports
encountering a related glitch in *FIFA '17*: a trophy inherited ball mechanics

and could be dribbled, passed, and shot as a ball! *FIFA '20* had a floating ball glitch that consisted of one or more of the ball options not casting shadows. These glitches are not errors of expectation, but rather technical mistakes and oversights. As Henry Lowood reminds us in his work on *FIFA*'s game engine, "Software does not always work the way it is supposed to work."[16]

What interests me about Bilby's overconfident proclamation is the acknowledgment that—even though the ball is the very thing that players train their eyes on and direct their movement toward, unlike the game's other interactive objects (the avatars of men's professional team players)—the ball can be overlooked and left out of the game completely. The object that gives players the felt sense of the game's spatial model—its underlying common sense—is so fundamental that it can be, well, forgotten. The ball in *FIFA* offers a new kind of common sense for what balls can and cannot be. They cannot acquire wear and tear. They can be invisible, or missing their shadow, or shaped like a square, a fish, or a trophy. They are a key site for the construction of game feel, but they are not the object of the game. When my eyes, with all their prior training, fixed themselves on the bounce of this invisible ball, a world of densely layered interactions between balls—and other bouncing and bounced-off bodies—was revealed.

This chapter follows the bounce in *FIFA* from ball to body, exploring the relationship between bounce, feel, and the game's bounded spectacle. This exploration touches on the construction of game feel alongside questions of whose game this is—and who gets the feel for this game. I begin with the ball in *FIFA*, building on chapter 4's argument about bounce programs, to argue that they operate as a common sense of computer graphics. The movement of this animated interactive programmed ball, in other words, creates a kind of common sense and palpable feel for the game. From there, I turn our attention to the hovering arrow hopping from avatar to avatar. The player indicator arrow is the visual representation of the game's individualizing mechanism—one that collapses a team sport into one-on-one play. This mechanism serves as a transition from bouncing balls to bounce and the body. The final move in the chapter completes the jump from ball to body by shifting to discuss another kind of bounce that is foundational to *FIFA*: the continuous bounce of light off motion-capture suits used to produce the game's animated avatars. This foundational aspect of the game opens onto the discourse around the 2016 addition of professional women footballers to the game. Throughout, glitches appear that shake up established sense and sensation.

The Common Sense of Computer Graphics

The ball in *FIFA* is strange, as balls go. As I describe in chapter 3, this strangeness is shared by many animated and digital simulations of recognizable, tangible objects. There is no expectation of being able to pick this ball up, puncture it, or kick it into the stands. Thanks to the rise of non-fungible tokens (NFTs), a ball in *FIFA* technically could be set aside, sold to collectors, or acquired by the Smithsonian after a particularly auspicious goal, but it will not be.[17] Three-dimensional (3D) graphical objects have an additional strangeness. As Jacob Gaboury explains, "Graphical objects exist in their totality—as a collection of coordinate points, image files, and object databases—prior to their manifestation as a visible image."[18] For Gaboury, graphical objects are in this sense "nonphenomenological" because "in order to simulate our perception, graphics must not only calculate that which is to be seen but also anticipate and hide that which is known yet should not be shown, that which must be made hidden and invisible."[19] The invisible ball glitch extends this point by making evident that even when the graphic fails to appear, the object can continue on its course. In the world of *FIFA*, the position, speed, and trajectory of the ball continually cue the game's AI to determine the range of possible players' positions and actions—and to then select from that set. Whether or not its graphic representation appears, the ball serves as a constantly moving center around which relations are programmed and produced.[20]

Because it is what players' avatars' interact with most directly, the ball is a crucial vehicle for constructing the embodied experience of the game, what game designer Steve Swink calls "game feel."[21] Swink argues that it is "the tactile, kinesthetic sense of manipulating a virtual object"—created through a combination of real-time control, simulated space, and polish—that makes any given game pleasurable to play.[22] Think of the experience of slotting a *Tetris* piece perfectly into place, flinging an *Angry Bird* at a pile of green pigs, or moving a cursor to work on a different paragraph in word processing software. Game feel is what allows a player to absorb their use of a controller to direct an avatar to kick a ball down a pixelated field as a bodily sensation of play. By following the ball with their eyes and using their hands to initiate the kicking, dribbling, passing, and shooting of the ball, players are trained into the common sense of the game's spatial model. In the circulation of the *I* between their own and their (human or AI) opponents' avatars, they are pulled into participation. Media scholar Noah Wardrip-Fruin calls the collision, movement physics, and navigation logics that produce a game's continuous spatial model "graphical logics." He argues that these serve as the foundation of "playable models"—the procedural representations that make the rules of simulations like *FIFA* legible and palpable to the game's players.[23] The ball in *FIFA* does what balls do: it constructs intersubjectivities by shaping players' sensoriums around the game's graphical logics.

As chapter 4 demonstrates, the history of computing is awash with bouncing balls—so much so that programmed bounce operates as a kind of common sense of computer graphics.[24] Here, I am extending that argument by exploring the relationships between common sense, sense perception, and training of the senses. "Common sense" is a term used in overlapping ways by Antonio Gramsci, Stuart Hall, and Clifford Geertz to describe the kinds of everyday thinking that are received simply as knowledge or truth, bypassing any process of critical reflection. On these accounts, common sense makes conceptions of the world, along with any attendant inequalities and oppressions, appear natural. Geertz describes it as an "of course-ness . . . a relatively organized body of considered thought" that, crucially, "rests its [authority] on the assertion that it is not a case at all . . . The world is its authority."[25] For Geertz, common sense is comprised of those wisdoms, assessments, and judgments of our perceptions that are mistaken as natural. Applied to the technical realm of programming bounce, the "of courseness" of game worlds is produced from a history, and "bounce" is a cultural and ideological expression of that history. This analysis draws our attention to embodied "of courseness": to the ways that sense is always built from, and through, the training of sensation and the way that the trajectory—the course—of the ball has been part of this training. We might say that this training is a kind of programming.

The programmers who initially worked on the prototype that would become *FIFA* were familiar with the well-established traditions of programming bounce. One of the two British designers who made the first prototype, Jules Burt, describes in interviews how, as a kid growing up in Saudi Arabia, he campaigned for his parents to get him a Commodore Amiga computer after he saw the now-iconic red-and-white checkered soccer ball bounce across the Amiga's screen. First presented at the Consumer Electronics Show, organized by the Consumer Technology Association) in 1994, the ball in the Amiga Boing Ball demo is one of the most famous soccer balls in the history of computer graphics. Following in the tradition initiated by the Whirlwind computer of using bouncing balls to demonstrate new capacities of graphical

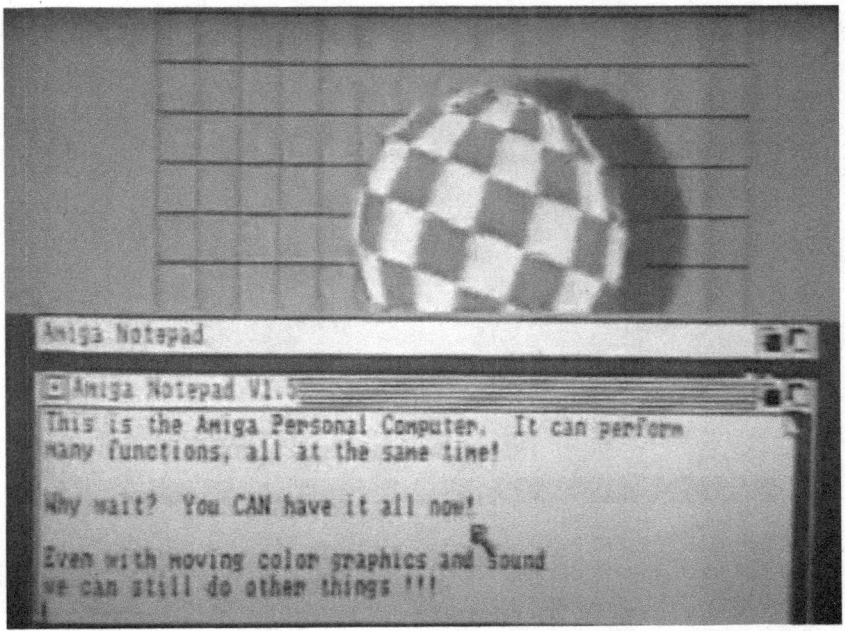

Figure 5.3
Demo of Commodore Amiga Boing Ball on the "Amiga and Atari" episode of *Computer Chronicles,* PBS, 1985. Screenshot by the author.

user interfaces, the Boing demo highlighted the Amiga's new multitasking capacities. In his history of the Commodore Amiga, Jimmy Maher writes that "demonstrators delighted in showing the ball bouncing and booming merrily away in one virtual screen, while one or more other programs ran in another."[26] But the arrival of the Commodore Amiga proved to be a disappointment to the future designer of *FIFA*'s prototype. Burt discovered that the ball's independent movement was an illusion: there was no capacity to actually interact with the bouncing, booming soccer ball.[27] As Maher explains, the viewport—or the "camera" capturing the ball—was in motion, not the graphical object itself.[28] Burt's childhood disappointment became a motivation. He went on to work for Commodore, making games for the Amiga, until he and John Law started their own studio. In 1991, they were hired by EA's Mark Lewis to make a prototype of a soccer game.[29]

There was one aspect of Burt and Law's prototype that stood out most to Jan Tian, the programmer whom EA assigned to take over development of the game series. Instead of staying glued to the avatar's feet, the ball was "knocked

Figure 5.4
Still of Jan Tian's eponymous character, Brazil's number 11 player Janco Tianno, in
@WeirdFIFA's "FIFA HISTORY 94–16" video, as an example of the isometric view.
Screenshot by the author.

forward," and then the avatars chased it.[30] Tian could recognize and evaluate
the interactions between the ball and the avatars. This was markedly different
from prior soccer video games, including *Amiga Soccer* in 1988 and *Sensible
Soccer* in 1992. What made Burt and Law's prototype "feel like real soccer," for
Tian, was how the ball could be kicked and chased—together with the "isometric viewpoint," which gave the game a sense of depth and suited "a more
real TV view" by appearing to look down on the game from one corner of the
stadium.[31] The effect of the isometric view indicates that the long-standing
co-constitution of professional ball sports and broadcast television had been
naturalized. Television spectatorship was already intrinsic to the game's feel—
so much so that it became the thing to simulate. From there, "the most difficult task to program correctly" was how to position avatars, as Tian puts it,
"in a way that resembled professional football, rather than a playground kick-
around where children swarm after the ball."[32] The interactive dynamics of
team sports are exponentially more complex than the one-on-one to-and-fro
of tennis and *Pong*. While the ball's position, speed, and trajectory cue shift in
the positions of the avatars on the field, for the game to play like soccer, the
avatars' positions also have to take place in relation to each other.

When *FIFA International Soccer* launched in 1993, the black-and-white checkered ball was propelled down the field rather than appearing glued to the avatars' feet. And when it was kicked with a reinforcing thud, a shadow accompanied the ball to create a sense of it sailing through the air. Despite this strong start, the experience of the ball's motion and responsiveness was the subject of ongoing player complaints for decades. Longtime *FIFA* fans will be familiar with the way that the game's feel was routinely compared to its main competitor, *Pro Evolution Soccer (PES)*, and found wanting.[33] A key aspect of *FIFA*'s comparatively bad game feel was the floating suspension of the ball: it hung in the air for so long that it did not believably match the behavior of a real soccer ball. In 2014, an article in *Scientific American* announced that *FIFA* had finally got "the physics right."[34] After two decades of players' complaints, engineers and animators at the company took a close look at the projectile physics, and they discovered a lurking error in the drag coefficient, the equation used to calculate the effect of the resistance from an environment. They had to calculate the air's resistance on an object passing through it—in this case, a soccer ball—so the ball "at long last, could sail smartly through the air."[35]

This constant back-and-forth about the ball's rightness and wrongness marks its importance to the game. Since 2014, almost every year's release has been accompanied by new ball physics—or refinement of the ball control system—to improve players' experience of touching the ball. *FIFA '18* touted "Real Touch" and "Strategic Dribbling," *FIFA '19* boasted an "Active Touch" feature that added animations for trapping and flicking the ball in new ways, and *FIFA '20* introduced strafe dribbling and an amorphous "new ball physics."[36] Despite having "gotten the physics right" in 2014, EA Sports has continued to revisit the behavior of the ball with each new version of the game. *FIFA '15* promised players that "each touch of the ball will affect its trajectory, and the game will even take into account how and where the ball was touched," producing a ball that "acts more like you'd expect it to." And *FIFA '16* introduced "No Touch Dribbling": "Sometimes the best touch is the one you don't take." Over the franchise's history, refining and elaborating ways for players to feel the game through their experience of ball control—through ball physics, touch, and no touch—have made each successive year's version of the game an essential purchase for the committed player.

The series also offers a sense of ball control via over fifty customizable balls for players to choose among and purchase. The balls not only look different but are also programmed to play slightly differently from one another. These variations determine, in turn, how the ball's movement changes based on which professional player's avatar has possession, what weather and difficulty settings are in place, and the expertise of the player holding the game controller. The endless fine-tuning of feel aims to meet a set of expectations about the behavior of soccer balls and soccer players that the video game's players bring to the game. All of this is part of the series' rhetoric of realism. Matt Prior, the creative director of the series, said it this way: "Until *FIFA* is indistinguishable from football in real life and plays exactly like football, we'll always have more to do."[37] This emphasis on realism, in turn, underwrites EA's argument that players need to purchase the latest version of the game each year. This pursuit of "real life" is common to many sports simulations and more broadly to many video games.[38] If taken literally, it sets up errant expectations, like the one that I experienced during shooting practice. I thought that the ball that rebounded again and again off the crossbar was not behaving like a ball; it felt like an error. But it was bouncing exactly the way that it had been programmed to bounce. The shooting practice program was establishing my relationship with this particular kind of animated, interactive bounce.

Common sense is pervasive, but it is not static. It both creates expectations and adapts to keep up with how they shift. Despite all the effort expended to make these animated interactive objects appear to fly and bounce just like real soccer balls, they are fundamentally different from their tangible referents. When the programmers remember to put them in the game, they offer a bounce with a degree of predictability that physical balls can only pursue asymptotically. To whatever degree they fail to match a players' expectations built through their time spent playing soccer—and viewing the game on television—they are, at least initially, experienced as unreliable. Nevertheless, as players train themselves via this kind of simulated bounce, they create new sets of expectations. This feedback structures our expectations and desires for objects in the physical world in turn—so we want them to also behave with more predictability.

Body-Hopping (or Always on the Ball)

The invisible ball glitch that I encountered in *FIFA* made another kind of bounce visible. As I watched my players flail after the invisible—to my eyes—ball, a red arrow hopped from avatar to avatar, following the graphically absent and yet otherwise present ball. The red arrow indicated which player was my avatar at any given moment, and which player would respond to the controls in my hands. It was always the player closest to the ball. These hopping red-and-white triangles are called "player indicator arrows."[39] While *FIFA*'s bouncing ball nests nicely in the longer history of programming bounce, the player indicator arrows are a kind of bounce that belongs more fully to video games.

The default setting in *FIFA* is for the game's AI to do the switching from avatar to avatar automatically, although it is possible for players to change this setting to control these jumps. Many more advanced players do just that.[40] In either case, bouncing from avatar to avatar is the mechanic that collapses a team sport into a one-on-one game.[41] When *FIFA* transforms the already professionalized and globally spectacularized version of the beautiful game into a primarily one-on-one video game, some of the measuring and placing of the self in relationship to set boundaries that happens in court games is relocated to the act of hopping from character to character on a player's given team. Instead of a collection of *I*'s working together—à la so much of the rhetoric of team sports, or as Michel Serres would have it, "passing the I" back and forth—to become a collective through the act of passing, each player is distributed across a set of avatars, in effect becoming a "we" unto themselves.[42] During offensive play, they pass the ball from themselves to themselves. When on defense, they hop from body to body anticipating or chasing the ball down. These jumps from avatar to avatar are more akin to the movement of a television viewer's eyes than they are to an athlete's field of vision or mode of action. This is especially so since most *FIFA* games maintain the isometric camera shot for actual gameplay, only going to cut scenes during game stoppages. In *FIFA*, players are always on the ball.[43]

This is fundamentally different from playing soccer. And it is, to redirect a phrase of Gregory Bateson's, "a difference which makes a difference."[44] A soccer player spends the majority of their time during a ninety-minute match standing or slowly jogging. One sports scientist estimates that players spend an average of seven minutes and forty seconds sprinting with between sixty to ninety seconds of ball time during the entire ninety-minute game.[45] Even the most famous soccer players—Megan Rapinoe, Lionel Messi, and others—spend only a small percentage of any game "on the ball." Indeed, one piece of soccer's common sense is that games are won "off the ball."

Figure 5.5
A close-up of my character, identifiable via the player indicator arrow above their head, dribbling an invisible ball as an AI opponent comes in for a slide tackle. Photograph created from video documentation of my first time playing *FIFA '15*.

Being aware of, but not in direct control of, the ball—pausing, waiting and watching, slipping and drifting into position to possibly receive a future pass or cut off a striker's run—are all fundamental to the game of soccer. In *FIFA*, the waiting, watching, drifting, and pausing are discarded and otherwise distributed to the AI. The bounce of the player indicator arrow that tracks the players' continual bouncing between animated bodies makes clear a set of underlying values. In *FIFA*, players occupy all the bodies, all the roles. They are always the central actor, the dominant "I." Because, as any good soccer fan has learned from hours and hours of watching broadcast matches, being on the ball is the most compelling and important place to be. This particular kind of common sense of televised sports is not necessary to soccer. One can imagine a soccer video game that simulates the temporal and spatial experience of an individual player on a team, but that would be a very different game. In *FIFA*, which simulates the experience of the television viewer rather than the experience of the player, the player becomes a kind of ball—a quasi-object, quasi-subject passed from avatar to avatar, from animated body to animated body. The player switches positions, orientations, and roles every few seconds to stay always on the ball.

Believable Bodies

The animated avatars, which players continually hop between, are produced using a third kind of bounce—one that is also foundational to the feel of the game. While I was pursuing the common sense of the ball, I happened on a photograph of four players. They were wearing gray-and-black motion-capture suits and helmets, running in a rough line across a thin carpet of green AstroTurf, and they were midstride. Bright red lights placed on the floor and on a lighting rig shone at them and at the lens of the camera taking the picture. These lights continuously bounced off the Ping-Pong ball–esque markers—attached to the players' suits at their joints and other "parts that move"—back to the over one hundred cameras in EA's motion-capture studio.[46] Motion capture, like many optical recording technologies, used bounce as a technique for producing believable bodies. The players in the photograph were all members of the US Women's National soccer team. It was a

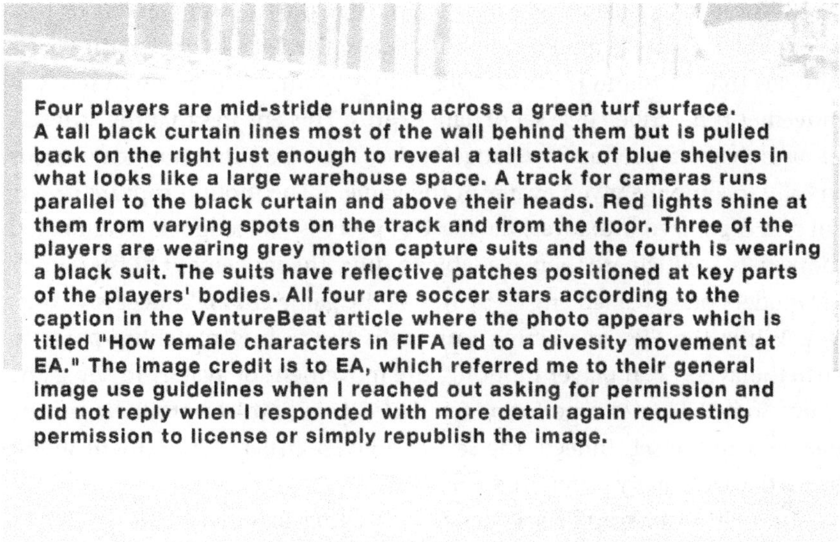

Four players are mid-stride running across a green turf surface. A tall black curtain lines most of the wall behind them but is pulled back on the right just enough to reveal a tall stack of blue shelves in what looks like a large warehouse space. A track for cameras runs parallel to the black curtain and above their heads. Red lights shine at them from varying spots on the track and from the floor. Three of the players are wearing grey motion capture suits and the fourth is wearing a black suit. The suits have reflective patches positioned at key parts of the players' bodies. All four are soccer stars according to the caption in the VentureBeat article where the photo appears which is titled "How female characters in FIFA led to a divesity movement at EA." The image credit is to EA, which referred me to their general image use guidelines when I reached out asking for permission and did not reply when I responded with more detail again requesting permission to license or simply republish the image.

Figure 5.6
A written description of the image that I saw in the *VentureBeat* article.

promotional image circulated by EA to herald the introduction of women to the game in *FIFA '16.*

Twenty-two years into the series's history, when I sat down to play my first practice match in *FIFA '15*,[47] the simulation of soccer presented in *FIFA '15* did not include an option to play on women's or coed teams.[48] Given the substantial effort expended on creating a vast variety of balls for the game, it is significant that there had not been an equivalent effort to create a variety of bodies up to that point. In that same year, the United States–Japan FIFA Women's World Cup final shattered all prior television ratings, becoming the most viewed soccer match ever in the United States. So when the shift to expand the range of bodies represented in the game finally did occur one year later, in 2016, capturing the movements of specific US Women's National team players was seen as a key component. The addition of women characters to the game was positioned as necessary for representing the game's global reach.

EA started using motion capture in *FIFA* in 1997. Their offices in Vancouver house one of the largest motion-capture studios in the world.[49] Currently, the process works something like this: An actor/athlete puts on a suit adorned with reflective balls positioned at key positions on their body. These balls operate as either active (signal-emitting) markers or passive (light-reflecting) markers. In either case, the balls bounce light to the cameras that surround the area. Each suited-up athlete performs a series of movements described by a set of dance cards. This ensures that the athlete's movements will produce the range of data that the animation team needs to fully locomote a given avatar in the game.[50] The motion capture data is cut and organized into a vast database of poses (small snippets of animated movement), which are then available to drive the movement of the game's 3D model bodies (called "rigs") by way of the game engine's successive queries. While the FIFA marketing team calls the task of matching poses to virtual players "real player motion," the movement of avatars in the game is not indexical to the performance that a given athlete produces on the motion-capture set. Indeed, the sets of a given athlete's movement might be used across many different avatars.

Through an ongoing selection and stringing together of tiny snippets of animated movements, believable bodies are composed in real time. To be believable, they conceal the fundamental modularity and multiplicity of the process. Games scholar Amanda Phillips aptly calls motion-capture

Figure 5.7
Example of a motion-capture "dance card" from Kristjan Zadziuk's great 2016 Game
Developers Conference talk "Motion Matching—'Dance Card' Breakdown." Cour-
tesy of Kristjan Zadziuk.

animation "masking technologies" that "record the data of certain bodies
and map them to others, simultaneously deflecting and reaffirming the
importance of racial and gender identity in digital performance."[51] The
process of masking the modularity and multiplicity—and deflecting and
reaffirming racial and gender identity in the process—operates differently
in *FIFA* than in games with fictional characters because the avatars are
representations of specific identifiable athletes in the world. Well before
their incorporation into video games like *FIFA*, athletes had already become
avatars of a sort.[52] As Brian Frye demonstrates, since the invention of baseball cards, professional athletes have had two bodies—one corporeal (the
body that they inhabit) and one public (figured as circulating commodified
images and statistics, and thus alienable).[53]

Games like *FIFA* that emphasize an asymptotic pursuit of realism depend
on motion-capture animation more than most other games. Like the interactive bounce of the ball, motion-capture is central to simulating the feel of
the professional game. In *FIFA*'s case, until 2023, this pursuit depended on
EA's exclusive licensing agreement with the football association FIFA, which
among other things included the right to use its name and to capture and use
professional soccer players' likenesses (their public and alienable bodies).[54]

A combination of visual likeness, movement likeness, and behavior likeness makes animated avatars' bodies believable—particularly as specific professional soccer stars. The avatars in the game are modeled on professional players both visually and, to different degrees, behaviorally. While I am not going to discuss the behavioral aspects in detail in this chapter, what is important to know is that EA employs thousands of people to watch every league, national, and international football match live and keep running tallies and ratings of players on pass accuracy, shooting, passing, dribbling, defense, and physical capacity. This data gets channeled into the programming of each player's avatar and refreshed with each annual release or update. So the motion-capture libraries provide the range of possible visual animations of movement, while the behavioral data helps to determine what from this range is available for a given avatar at a given moment. While one motion-capture crew ushers a select few stars and stand-ins through the locomotion dance cards in the large Vancouver studio, another crew travels from team to team, scanning professional players' likenesses—their heads and facial expressions—and thousands of scouts report running player statistics. Accordingly, it is the publicly perceived racial and gender identities of the professional athletes, reaffirmed by their captured likenesses, that serve to mask the variety of bodies whose movements generated the animations used by any given in-game avatar.

The video game series reflects and extends the history of the organization that it is named after and whose name it licenses. One of the consequences of *FIFA* simulating FIFA (as opposed to, say, simulating soccer) is how the game series has simulated and extended the association's historical exclusion, what Michael Pennington calls the "ritualised exclusion" of women.[55] Founded in 1904 in Paris, FIFA has driven the development of men's professional soccer for over a century. Today, it is perhaps best known as the corruption-plagued, multibillion-dollar organization behind one of the most successful global television spectacles, the FIFA World Cup. The development of the men's game has gone hand in hand with the enforced absenting of women from the field. Throughout the twentieth century, despite enthusiasm from players and spectators, women were repeatedly banned from professional play.

The English Football Association banned women from playing matches from 1921 through 1971. Brazil's ban lasted through 1981. For its part, FIFA worked to undermine the first women's world championships organized by the Federation of Independent European Female Football in Italy in 1970 and in Mexico in 1971, sending a directive to the Mexican federation prohibiting it from organizing the 1971 event.[56] Even after FIFA agreed to host a Women's World Cup competition in China two decades later in 1991—two years before the first *FIFA* video game launched—the organization continued to give lackluster support to the women's game.[57] The approach taken by EA and the *FIFA* production team for the first twenty-three years of the game's history of not including women's teams and women players in the game, follows FIFA's lead. Which, in turn, follows another kind of common sense: sport is the domain of boys and men, and men matter more and constitute a bigger, more profitable market. This kind of common sense is one that requires women to exhaustingly repeat the fight for inclusion over and over.

By way of full-body motion capture of a few especially famous players—and head and face scanning of all the team members of twelve national teams—women were introduced to the game in *FIFA '16*. The company explained the two-decade absence by pointing to the technological challenges posed by capturing women's movements. *FIFA* series producer Gilliard Lopes said, "When we tried to implement this functionality in the previous generation, we came to the conclusion that our tools were not yet flexible enough to authentically represent the physical characteristics of the athletes in the game."[58] Similarly, *FIFA* senior producer Nick Shannon said, "The key for us was when we brought it into the game, we had to bring it in properly, and we needed some supporting technology to be able to do that . . . We've been looking each year as to 'can we do it' and comparing to priorities at the time as well. Once the technologies were in place, we could do it properly."[59] While the game had been able to authentically represent men for two decades, something about women required new, more flexible technological tools to represent them properly. This refrain is not unique to *FIFA*. When asked questions about the absence of women avatars in their games, video game developers often respond by saying that it is technologically impossible, or otherwise too difficult and too expensive, to create them. As Sara Ahmed writes, "When the arrival of some bodies is noticed, when an arrival is noticeable, it generates disorientation in how things are arranged."[60] The disruption of the common sense that sport is the domain of boys and men is disorienting, and the necessary rearrangements are expensive.

The two main technological challenges initially put forward by *FIFA* producers and designers were the difference between men's and women's body movements and the question of hair length and movement.[61] The initial prototype of women players consisted of a "female head" on "a man's body." This was used to identify everything that needed to be done to introduce women into the game.[62] Katie Scott, a game designer who joined the FIFA team in 2015 and who has led a push for diversity in EA games, says, "Just in terms of the technical side, the way that women move their hips is the really big thing, and their shoulders. That's a really big difference, typically."[63] In the context of FIFA and FIFA, the range of body types, comportments, and movements that players enact is shaped first by the situation of professional soccer. With its long-defended masculinist frame, this category has already filtered out much of the range of existing and possible bodies and movement cultures when the simulation of professional soccer does a second pass of filtering.

The gendered distinctions that Scott describes appear, then, not just within the constrained frames of the athletic body, specifically the soccer body, but even more specifically within the frame of the game's existing 3D model of the soccer body. The latter was made as a one-size-fits-all form based on an average men's soccer player's skeleton. FIFA vice president and general manager David Rutter explains, "We rebuilt the animation rig to support the different dimensions and proportions of a woman's body . . . and then applied that motion capture skeleton to those very believable bodies to make sure that the standing, walking, jogging, sprinting, passing, shooting, is actually women animation rather than male animation."[64] Men do not all run alike or have identical body types. But the lack of distinguishability between avatars' movements becomes a problem that necessitates a solution only when there is a risk that avatars, previously identifiable as men by default, might suddenly be mistaken for the avatars of women players. Where the debate over the rightness and wrongness of the ball once marked its centrality to the game, here the back-and-forth over what constitutes right versus wrong representation of movement marks the centrality of gender.

When a game series like FIFA finally decides to address its legacy of representation and shift its practices, it is expensive. Along with the task of building new body rigs, spending two decades creating only "male" avatars created a massive representation deficit in the game's animation libraries. Scott describes spending her entire cinematic budget on animation to begin to even out an animation library that started with a ratio of 18,000 clips of "male animation" compared to 1,000 clips of women.[65] What warrants emphasis here is that women's bodies did not create these challenges, absenting them from all consideration for two decades prior did.

For the design team that took up the task of introducing women avatars into the game, hair presented another key site for distinction. The initial prototype "looked like a female because it had long hair, but it wasn't enough."[66] Given that many players on women's teams crop their hair short and many male players wear their hair long, this ironic identification, of course, stands in direct contrast to a wide variation in hairstyles of contemporary professional soccer players. A compilation of the top ten players in every *FIFA* from 1994 to 2000 put together by the Romanian YouTuber Shade, includes a striking number of long-haired men, including Francesco Totti, Gabriel Batistuta, David Ginola, Edgar Davids, Ronaldinho, Edinson Cavani, David Luiz, and Falcao—to name just a few.[67] The design team's conflation of long hair with "female" aligns with the cultural association of long hair with heterosexual femininity. In the context of sport, long hair has historically been used as a tool to shore up femininity, on the one hand, and overtly or covertly challenge athletes with short hair about their sexuality, on the other.[68]

The moment that women are introduced into *FIFA*, the need to uphold and forcefully enact gender segregation—the starting point for most modern sport for over a century—appears as well. It is precisely because the borders between masculinity and femininity are thin in sporting contexts that, as Jennifer Doyle has phrased it, the best female athletes are always in danger of "running out of gender."[69] Historically, gender segregation has been policed by regulatory bodies—enacted in practice almost exclusively through challenging the gender of women athletes, excluding any athletes who cannot be made to fit into cisgender categories. Already measured against notions of femininity and beauty grounded in white supremacist ideals—and forms of women's sport that developed in step with white feminism—women athletes of color are routinely policed in this way.[70] Via a similarly exclusionary logic in the United States, anti-trans sports policies have been pursued at the state and federal level in the United States and beyond, and tend to frame the argument as one about fair play and level playing fields.

As Doyle goes on to describe, "Mainstream sports culture theatricalizes the exile and abjection of the feminine, the effeminate, the queer . . . It stages gender segregation as not only natural but necessary to a sense of fairness. It does so in syncopation with a racialist logic that presents the black body especially as vitality, as raw force, as athleticism itself."[71] These methods of exile, Doyle describes, shore up the idealization of the bodies of cisgender men. Sport simulation games are modeled both on professional sports' "ideal" bodies and on motion capture's model bodies. In this context—first by only capturing the motion of men's bodies for over two decades, and then by generating a discourse around the inclusion of women, identified as a technological problem (rather than an ideological position), to be solved with the proper technical enactment of gender—the gender segregation and discrimination that are already so deeply embedded in sport is virtually recapitulated.[72]

In the discourse around hair, the intersection of gender segregation and racialization is not only a matter of length but also movement. Nick Shannon frames hair movement as another technological challenge that must be solved to bring women into the game, saying, "We had to do a lot of optimization and work to make sure that the hair movement could be seen, because you can see it while you're playing."[73] *FIFA* general manager and vice president David Rutter emphasizes, "We've even gone so far as to rejig the physics on our hair to make sure that the ponytails are more believable."[74] To bring women into *FIFA*, avatars have to have long hair, pulled back into ponytails that move in a believable manner.

What kind of movement is believable? Their image of soccer women seems largely based on the internationally successful sports star Mia Hamm.[75] Presumably Ruther and the rest of the *FIFA* team would have been familiar with *Mia Hamm Soccer 64*, the first women's soccer video game published in 2000, which featured avatars that all sported Hamm's iconic ponytail, which bounced up and down as they traversed the field. The ways that hair looks, feels, and moves—the ways that it bounces—have long been a subject of politics in the context of style and beauty standards shaped by white supremacy and patriarchy.[76] In *FIFA*, these politics are evident in forums titled "WHY 85% TO UNLOCK AFROS? LOWER THAT PLEASE" and "WHY PRO CLUBS RACIST," wherein players express frustration about the limited availability and accessibility of Black hairstyles.[77] As one player writes: "There are no black men haircuts and styles. It's all white people hair."[78] These politics are likewise evident in the kind of hair featured in EA's 2019 Full Hair Tech demonstration video—engineered by their Frostbite engine—which features a faceless, "female" motion-capture mannequin showing off long, thin, straight, glossy, bouncing hair that was capable of a "full swish."[79] Here, bounce becomes one of a set of qualities marking the technical achievement of supposedly

Figure 5.8
Stills from EA's Frostbite hair montage video. Screenshot by the author.

good hair. "Proper" animated hair movement ensures that clear distinctions will continue to definitively produce differently raced and gendered bodies. To help people visualize what they are supposed to be embodying, the game continually depends on ideologies of race and gender.

FIFA's modes of representation raise issues around gender and race that are pervasive, but as more bodies and more practices are included in this game, these modes will never be static. Circulated by EA, the image of red lights bouncing off the suited-up bodies of four US soccer stars gives viewers the sense of being, "behind the scenes": any steps taken to end the twenty years of gender discrimination in the series, the image assures us, will be done with careful attention to—and fortification of—the ongoing gender segregation in sport. Women players, the photograph tells us, will be used to make women avatars. Moreover, women's play deserves representation when (and because) they are doing the work of representing their nations. But there is, in fact, no necessity for motion-capture data itself to be gender segregated, nor are the poses intrinsically distinguished by race or nationality. Each pose is just a small snippet of movement. An avatar's movements are assembled in real time from ongoing queries of the pose database. While some movement sets are player/avatar specific, other movements turn out to generalize well. They are, accordingly, applied across a wide range of avatars, regardless of assigned or perceived gender. The transition to motion matching systems streamline the process of incorporating motion-capture data into the game, and this has made it dramatically easier to add new kinds of movement. Yet this has not been accompanied by a full incorporation of women into *FIFA*.

Women avatars were finally introduced into *FIFA Ultimate Team (FUT)*, the most popular and profitable portion of the game series, thirty-one years after the series launch (and a year after its dramatic name change) in *EA Sports FC*

'24.[80] As when I encountered the programmed bounce of the ball off the top bar, sexism and racism continually produce a sense of error—of mismatched expectations—for those on the receiving end, but they are not errors in the system. They circulate as common sense—products of the same "of course" decisions made again and again and again and again . . .

As part of its struggle to provide a uniform virtual spectacle, the glitches in *FIFA* ironically demonstrate EA's own hegemonic way of chasing history. *FIFA* pursues their hegemony by expanding representation through typical capitalistic means—adding more hairstyles, for instance, or innovating to capture more markets and new populations and consumers. Yet, as they bump inevitably against the uncontainable and irreducible heterogeneity of representation, faultiness and misrepresentation haunt their dominance.

Conclusion: Alternative Of(f) Courses

It is possible to imagine an alternative universe—different trajectories, other courses, and the making of different myths. EA could name their past actions (inaction is a kind of action) and make direct amends in the forms of real changes in their approach. They could simulate the temporalities, spatial dynamics, and corporalities of soccer in different ways, from different perspectives and positions. Or, less fantastically, they could call things by their names, speaking bluntly about the impact of market share and stock price on how they design their games, or their intentions to create a game for mostly men to play. Over the course of writing this chapter, people, mostly men, have told me their *FIFA* stories: a taxi driver who pursues his own philosophy by playing *FIFA* alone; fathers and sons and college roommates who bond through play; a friend's fiancé who bought himself a PlayStation 4 just to play *FIFA* as a way to survive the loss of live sports during the COVID-19 pandemic; the editor who did beautiful structural and line edits for this very book, who shared after editing this chapter that he had nursed himself through an extreme depression the prior winter by playing the game; my coeditors of *EA Sports FIFA: Feeling the Game*, whose deep love of the game occasioned that book and this chapter.

In all these instances, the game serves as a kind of third object, or medium, for facilitating masculine intimacy, both with the self and with others.[81] Community, closeness, and deep feeling scaffold themselves on the game. This is also part of the game's feel. This pile of stories lives next to my own very different encounters of the game as an aversive beginner, inhabiting the role of the feminist killjoy.

As a former professional squash player, I have a lived stake in how sports and movement are represented to audiences. *FIFA*'s approach to representing movement makes me mad, the same way that not being allowed to play baseball made me mad when I was a fourteen-year-old girl. In *Glitch Feminism*, Legacy Russell argues that "glitched bodies—those that do not align with the canon of white cisgender heternormativity—pose a threat to social order. Range-full and cast, they cannot be programmed."[82] Russell's words make me imagine gloriously glitched versions of this game—versions where the bounces of balls and bodies are rangeful (able to range fully together), where gravity is adjustable or intermittent, where avatars from the stands jump onto the field and join in play, or where the avatars are all just balls. Hair and hips become nonevents. In a more equitable society, this would have been an obvious solution. A memory comes of the feel of afternoon sun, stretched on a sloping hill, skin salty and legs gone to jelly after a long day of play. Game feel is ball physics *and* social physics—the feel of grass between your fingers and the feeling of being able to rest where you are.

Figure 6.1
"CA.AJUPEME.USA 🖐🏽🟫⚫🏛🟫WWW.AJUPEME.-USA.COM," @ca.ajupeme.usa,
July 27, 2022. Courtesy of Raul Herrera.

6 Pok ta Pok

California AJUPEME

I slow my scroll through Instagram and watch the latest video post from @ca.ajupeme.usa.[1] A video of players practicing *ulama* is set on a diagonal between two bands of black lines with flashing GIFs. At the top right, an animated Zapatista waves a torch up and down. Her nose and mouth are masked by a red bandana, and a baby is strapped to her body. The word "RESISTE" appears, disappears, and reappears—again and again—in red across the bottom of her blue dress. Next to her, there's an image—familiar from its place on the Mexican flag, referring to the founding myth of Tenochtitlan: an eagle eating a rattlesnake, which flips back and forth at a quick clip of one-two-three-and-four. Next to the eagle sits a black-and-white photograph of a native chief, the words "WE EXIST," "WE RESIST," and "WE RISE" winking in and out across his chest. In the bottom left, an animated skeletal figure leaps in and out of the figure, drawing and shooting a bow and arrow. At the center, a pyramid gets built from bottom to top on a loop. On the right, a black-and-white image of a fist clenched in solidarity sends out brightly colored lines, and under it, rocking back and forth, is a reclining figurine—a Maya chacmool from Chichén Itzá. I hit unmute: from Control Machete's 1996 anthem "Andamos Armados," the beat wraps the movement of the GIFs—and the movement of the players—in its tempo.

The Mexican hip-hop music and the flicker of the GIFs, together with the contrast filter and sharp cuts in the video, make it difficult for me to make out the players, but they are bouncing the *ule* back and forth across the basketball court at El Cariso Community Regional Park. A few shots in, I recognize Bede

hitting the heavy rubber ball with his hip.[2] (The names of the *ulama* players in this chapter are pseudonyms used to protect the identities of the participants, except for those who hold public roles in AJUPEME.) Then Daniel
jumps to meet it with his. Later, Blanca, using one hand for balance, drops to
the ground to redirect the rolling ball, and Miguel does the same to block it
back in her direction. Raul has cut clips together of players working in small
groups, a scrimmage, and all the players, arranged in a big circle, taking turns
aiming for the polyvinyl chloride (PVC) hoop in the middle of the court. One
clip shows Daniel doing one-on-one work with new players who came from
San Diego, Santa Barbara, and Pomona. Soon, I realize that this is all footage
from the practice that I just attended the previous Saturday. I stood on the
sidelines as Raul beat Aztec rhythms on the drum and Iris shot the video
that Raul cut and edited to make this post. The distance between the feel of
being at practice and the feel of viewing the post is simultaneously surprising
and familiar. The distance between these two experiences of the same event
makes the transformations wrought by mediation sharply apparent.

I attended my first practice at El Cariso Park in June 2021. Named after
the El Cariso Hotshots, a crew of firefighters who lost twelve of their members to the Loop Fire in 1966, the park spreads out across eighty acres of Los
Angeles County in the foothills of the Angeles National Forest on the traditional lands of the Fernandeño Tataviam Band of Mission Indians. The San
Fernando City Mesoamerican Ballgame Team meets on the park's basketball
courts most Saturdays for practice. I was there to see the hip version of *ulama*
in person for the first time, and to ask for the team's permission to write
about the game.[3] Half an hour later, I found myself holding plank position
along with the rest of the players. The command of a coach ordering everyone to do sets of jumping jacks, burpees, push-ups, and lunges was a siren
song calling me back to an earlier athlete self. I did not know how not to join
in. Usually, it is a forty-minute workout, but Daniel went easy on us because
of the new players. I held the plank with the other women at the end, while
Roberto and the other men did push-ups until one of us gave up. It was so
familiar—this mode of developing grit by pushing to the edge of and then
past capacity. After the warm-up, I was grouped with Bede, also a beginner,
and Daniel's youngest son, Mateo, who was charged with patiently bouncing
the small kid's ball to us newbies so we could learn to hit with our hips.

Ranging in age from teenagers to players in their mid-forties, the team is
made of a mix of men and women. The weekend practices are open to the

Figure 6.2
(Left) A player from the San Fernando Ulama Team teaches recruits for a new San Diego team how to get down to hit a low ball. (Right) Miguel Duran, the National Ulama Trainer for the United States and a member of the California AJUPEME delegation, goes down to the ground to hit a ball. Photographs by Raul Herrera.

public. Founded in 2019, the San Fernando City team belongs to the California delegation of the Associación De Juego de Pelota Mesoamerica (AJUPEME). The association is reviving and extending the 3,500-year-old tradition of this ballgame. Armando Uscanga, a cofounder of AJUPEME, explains, "When the Holy Inquisition came, they banned this sport, if it wasn't for them, today, this sport would have been more popular than soccer. The ultimate goal of this sport is to build a professional league, like they have in soccer."[4] Like Electronic Arts' (EA's) *FIFA*, *ulama* proponents are looking to soccer and modeling their vision in relation to what is currently *the* global game. But while EA's aim has been to simulate professional men's soccer both cinematically and structurally, replicating the dynamics of both play and profit, AJUPEME aims to build and spread alternative structures for *ulama* rooted in the game's extended historical and cosmological traditions.

Elastic Cosmologies

Sometime between 1200 and 600 BCE, in Anáhuac (Nahuatl for "land by the waters," commonly identified as "Mesoamerica" today), the Olmec (Nahuatl for "rubber people") developed processes for transforming sap from rubber trees into an elastic material used to manufacture a wide range of objects,

Figure 6.3
The court at Coba, Mexico (left), and the court at Chichén Itzá, Mexico (right). The court at Chichén Itzá is much larger than the majority of known *ulama* courts. Photographs by the author.

the first and foremost of which were balls used for ritual offerings and ballgames.[5] They built an entire cosmology around a ballgame and the astounding bounce of rubber balls.[6] Later civilizations carried forward and refined the Olmec's technology and cosmology of bounce. Collectively, these cultures covered the region with ballcourts. Across an area stretching from present-day Arizona to El Salvador, more than three thousand courts have been excavated to date.[7] Many construction methods for fabricating the balls—and many variations on the game's attire and rules—were developed, resulting in a family of games with some common architectural and regulatory elements.[8]

The most spectacular version of the game is played by two teams of players. They use only their hips to hit a solid rubber ball back and forth on long outdoor courts. The courts span an area in the neighborhood of thirteen feet wide and up to two hundred feet long. Built in the shape of a capital "I," elaborated courts are lined with slanting walls which spectators can sit atop, and which occasionally boast stone rings jutting out at the center point of the court. The scale of the courts speaks to the strength of the players and the liveliness of the ball. To commence play, the ball is tossed across the court to the opposing team. Players hit the ball with their hips either on the fly, after one or two bounces, or when it is rolling on the ground. Once a ball is rolling, players wait to figure out where and how to meet it. They then go down to redirect it, one hand on the ground and the other extended up, in a position akin to a breakdancer.[9]

Like the ballgames of so many cultures, *ulama* has been played primarily by men, but there is also a history of women's play. Maria Isabel Ramos has assembled evidence of ancient ceramic female ballplayer figurines,

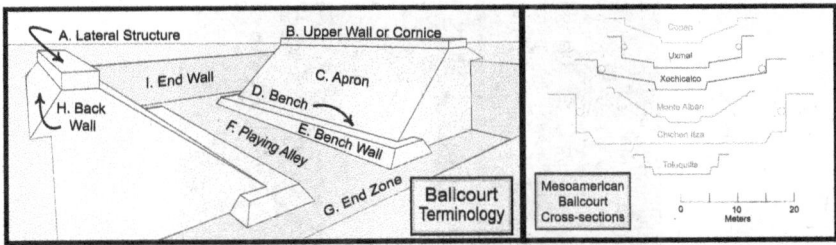

Figure 6.4
Diagram of a ballcourt and cross sections of some of the more typical ballcourts. Jacinto Quirarte has classified Copan, Uxmal, and Xochicalco at Type I, Monte Albán as Type II, Chichén Itzá as Type III, and Toluquilla as Type IV. By Madman2001—Own work, CC BY 3.0, https://commons.wikimedia.org/w/index.php?curid=3123805 and CC BY-SA 4.0, https://commons.wikimedia.org/w/index.php?curid=3156637.

sixteenth-century colonial accounts, and contemporary interviews that all point to instances of "women playing a man's game."[10] While it is true that these games have been predominantly played by men, Ramos argues that the primacy of men's play has been overstated. Many instances and traditions of women's play have been rendered invisible by their exclusion from early colonizer and Indigenous accounts, as well as from twentieth-century Mesoamerican scholarship.[11]

In Maya, the game is called *pok-ta-pok*, a name that echoes the sound of the solid rubber ball bouncing off a player's hip—*pok*—then the ground or wall—*ta*—and then another hip—*pok*. Of course, any given ball bouncing off any given wall, ground surface, or hip emits various sounds based on the specificity—all the material and environmental conditions—of that bounce. The name serves as a container. The rhythm of play is given to the reader's eye and ear, creating a space to hold the different possible tones, pitches, and intensities of impact that can inhabit that rhythm. *Pok-ta-pok* figures the ball's ongoing movement from body surface to Earth's surface and back again as the heartbeat of the game.

An essential feature of this cosmos—and its myths—was its animate character. The ball is the sun, symbolically, and the players are forces that keep the sun in motion. Art historian Manuel Aguilar-Moreno, who has done extensive work on the history and survival of the ballgame, explains: "It is a story of creation and the dynamics of the cosmos. The accoutrement is different in different places . . . but the underlying cosmology stays the same. It was a

Figure 6.5
Female ballplayer figurine, Huastec (artist), *AD 900–1400 (Late Classic—Postclassic)*, earthenware, Veracruz, Mexico. Stendahl Galleries, Los Angeles; purchased by John G. Bourne, Santa Fe, between 1970 and 1979; given to Walters Art Museum, 2013.

pan-Mesoamerican cosmology."[12] Ramos adds, "Mountains, lakes, trees, and all celestial entities were conceived as living beings . . . the human body was also a symbol for the microcosm of the universe . . . a container of life filled with supernatural powers and entities."[13] The role of the players is to keep the ball moving without stopping, continually meeting the ball and redirecting the bounce as a way of collectively keeping the universe in balance. Rendered into alphabetic writing by Francisco Ximenez in the eighteenth century as the *Popol Vuh*, the Maya creation story centers on two ballplayers, the Hero Twins. Taking their father's old equipment down from where it hangs from the ceiling, the twins go on to raise such a great racket that they get called down to the underworld to account for the commotion. They become heroes by defeating the gods at the ballgame, avoiding their father's fate (decapitation), and ascending to the heavens, where they are transformed into the sun and the moon. Their victory initiates a new age on Earth.

The English word "rubber" refers etymologically to the physical use of the elastic material—to the back-and-forth friction of rubbing. The Mayan words—such as *kik* or *quiq*, connoting blood, or the word *caoutchouc*, which is common to other Indigenous languages and translates to "tears of" or "juice of the tree"—all refer to sap in its liquid state. These names figure the material as a bodily fluid—one that requires some opening in the body of the tree in order to pour forth.[14] In Nahuatl, the primary language family of the Toltec and, later, Mixtec cultures, the game is known as *ulama*, or *ullamaliztli*.[15] The Nahuatl root "ule" or "olli," present in "Olmec," is also visible in the names *ulama* (the ballgame); *ollanqui* (the players who play with a rubber ball using their buttocks); *ules* (rubber balls); *hule* (the trees that provide the sap used to make the balls); *olli* (the sap from rubber trees), which is combined with *ololiuqui* (juice from morning glory vines) to make *ules* and other rubber objects; and *olin* (movement, wind, tremor, or earthquake).[16] The Olmec both developed rubber technologies and constructed a cosmology revolving around the concept of movement as necessary to maintaining balance and life.

Reflecting this cosmology, the game's scoring system is also organized around to-and-fro movement. Aguilar-Moreno explains:

> Rather than a linear cumulative scoring system, as is used in most Western games, the score oscillates from one team to another, with the *rayas* going up and down. This aspect of the game is consistent with Mesoamerican ideology, for it originated as a ritual practice in which there was a representation of the dynamics of the

Figure 6.6
Weekend practice at El Cariso Park. Photographs by Raul Herrera.

cosmos. The Mesoamericans believed that life is held by the balancing action of contrary and complementary forces, which are in perpetual movement . . . The scoring oscillation in *ulama* embodies that duality.[17]

Elastic movement, then, is contained as potential in the material, embodied by the ball and the players, embedded in the scoring system, and enacted by playing the game.

The largest balls are ten pounds of solid rubber. A good player can send this ball flying off their hip at up to sixty miles per hour. An anthropologist who extensively researched the ancient ballcourts, John Fox describes his first time absorbing the ball's impact as "a moment of utter revelation." After years of studying artifacts, he visited a village where a modern version of the game is still played and "felt the blow of a ball against [his] body,"[18] and finally understood why players used their hips to hit the ball.[19] Hands and feet (the instruments of choice in many ball sports today) are too fragile, too puny, to provide the necessary force or control for this kind of ball. It can seriously injure, or even kill, a player unlucky enough to be hit in the head or chest. Breaking is one of the alternatives to bouncing, and it usually brings about the end of play.

Over time, different methods and recipes were developed to produce balls and other rubber products with variable elasticity. As defined by Michael J. Tarkanian and Dorothy Hosler in "America's First Polymer Scientists: Rubber Processing, Use and Transport in Mesoamerica," elasticity is "a quantification of bounce" or "the ability of a material to store and release energy of defor-mation efficiently."[20] Tarkanian and Hosler trace the long history of the development and refinement of multiple methods for producing materials with different degrees of elasticity, durability, and strength, especially when needed to make balls, sandal soles, or adhesive and hafting bands.[21] In their survey of the archaeological evidence, historical accounts, and ethnographic studies, they identify four methods for fabricating rubber balls for ballgames: shaping a ball from a solid mass of rubber; building it with thin layers of rubber; rolling it, in a spiral fashion, from one or more thick rubber sheets; and constructing it from wound rubber strands.[22] While the cosmology was shared across regions, methods for making the balls, rules and reasons for playing the ballgames, and cultural practices and meanings associated with the ballgame were situated in specific regions and cultures in pre-Columbian times.

There was no rubber technology in Europe in the Early Modern era. When the Spanish conquistadors first witnessed the game, they found the behavior of the lively balls amazing and mystifying. Pedro Mártir de Anghiera, royal historian to Emperor Charles V and first European historian of the Americas, declared: "I don't understand how when the balls hit the ground they are sent into the air with such incredible bounce."[23] Beyond the amazing dynamic capacities of the object, the Franciscan friar Bernardino de Sahagún remarked that its "aural qualities were astonishing" as well.[24] And the Dominican friar Diego Durán wrote: "Jumping and bouncing are its qualities, upward and downward, to and fro. It can exhaust the pursuer running after it before he can catch up with it."[25] These writers would have been completely unfamil-iar with rubber bounce, and they could only have compared it to the kinds of bounce that they knew from their home lives. These would have ranged from stuffed cloth and leather balls, used to play hand and racket games, and slightly more animated, encased, and inflated bladders, used to play pallone and early forms of football and rugby. Against this backdrop of less lively objects, the rubber bounce and the body movements that it elicited were astounding.

Passing

The first recorded demonstration of rubber bounce on the European con-tinent took place in 1528. Hernán Cortés arrived at the court of Charles V of Spain with a troupe of performers. The contingent included a dozen ballplayers from Tlaxcala, a region that had allied itself with the Spanish against the Aztecs. The visit, as a whole, was not well documented, but there are some sparse records in which the names of some of the travelers were recorded, although these do not seem to include the names of the players or other entertainers.[26] The ballplayers had brought their equipment with them across the ocean: leather padding that wrapped around their waists and knees, as well as solid rubber balls to play the game that they called "batey," a Caribbean Taino name for the ballgame.[27] A series of drawings by Christoph Weiditz is the best-known documentation of the visit. A medal-maker and portraitist from the free imperial city of Ausburg, he happened to be at the court seeking royal permission to continue his profession of striking portraits of nobles out of metal. Included in Weiditz's series is a drawing of two of the Tlaxcalan ballplayers.

Weiditz positions the players on slightly sloping, grassy ground. Their backs are to each other and to us. To track the path of the ball that flies between them, they are looking over their shoulders. The player on the right has just

Figure 6.7
Drawing by Christoph Weiditz of two Tlaxcalan ballplayers, from the *Trachtenbuch des Christoph Weiditz* (1528).

hit the ball with his hip, still half-crouched, and his hands are grazing the ground. The player on the left stands ready to receive, knees bent, hip cocked. They are young men with shoulder-length, dark hair and brown, muscled chests and legs. Their heads turn toward each other, but their eyes do not track the trajectory of the ball. Instead, at the page's crease, their gazes meet at the very top and center of the image. Both the composition and the caption emphasize what would have been one of the most unusual features of the game to the European artist: "In this way the Indians play with the inflated ball, with their buttocks without raising their hands from the ground; they also have a hard leather [protector] over their buttocks to receive the impact of the ball; they are also wearing similar leather gloves."[28]

Weiditz's portrait is striking, but it does not work as a representation of the play of the game. Players do put a hand on the ground to meet balls that are bouncing or rolling close to the ground, but at the start of play, they are standing, or even jumping up, to meet the ball. He does not represent the rules, the architecture, or the team-based structure of the game, either visually or verbally. Since there is no other description of the performance at the court of Charles V, it is not fully possible to describe how that scene played out. *Ulama* is not usually a one-on-one game, and we do know that two full teams of ballplayers arrived with Cortés. Weiditz's description of the ball as "inflated" mistakes the material technology at hand. The balls were solid. He likely based this assumption on the ball's liveliness, and this presumption would have made him unable to fully understand why the players were using their hips, elbows, and thighs to direct it.

In addition, while Weiditz's sloping, uneven field of grass is an elegant visual solution for the image, it is a misleading representation. Since there was no proper court to play on—nothing akin to the massive stone courts found in every major city and town in their homeland, the players would have had to find or construct a level field of play to ensure a true bounce, without which the risk of serious injury would have been too high. Perhaps they played on one of the *trinquetes*, closed courts for pelota, a game common in Spain at the time, or on a tennis court. Charles V loved court tennis and had multiple enclosed courts at his palaces. Although Europe did not have anything like the long tradition of institutionalized ball play found in the Americas, it was amid its own ballgame craze. Early iterations of pallone, jai alai, rugby, football, and tennis were played across the continent. Like the Mesoamerican game, these games were often played on purposefully

Figure 6.8

Herri met de Bles, *Landscape with David and Bathsheba,* c.1535–1540. There are a number of Dutch paintings from this period on the theme of David and Bathsheba that
are famous in the history of sport because they include some of the earliest representations of an enclosed *jeu de paume* court, located on the grounds of a palace belong-
ing to Charles V. Roger Morgan, a historian of real tennis, suggests that the paintings
used the biblical allegory to evade the censors and offer a sly protest against Charles V,
who was hated by many for his suppression of the Protestant Reformation. Courtesy
Bridgeman Image.

cordoned off courts and fields and offered a site for high-stakes gambling by
spectators.

At the time, Europe's elite class was using balls and ballgames as a metaphor
to understand and communicate a new social and natural order of things.
Considering this, it seems strange that there is not more of a record of the
reaction to the Tlaxcalans' demonstration of their ballgame, and it raises
questions: Why didn't rubber technology make the leap across the ocean
at this moment? And what happened to the technologies and cosmologies
of bounce in the Americas after the conquest? To reframe "questions about
'how we know' to include questions about what we do *not* know, and why
not,"[29] historian of science Londa Schiebinger tells us to look at instances of
the "*nontransfer* of important bodies of knowledge from the New World into
Europe."[30] Often, ignorance is not an absence, she points out, but rather an
outcome of cultural and political struggles. Schiebinger offers the example
of a flower from the Caribbean whose seeds can be used to make a tea to
induce miscarriage. While the flower gained a Latin name, *flos pavonis* ("pea-
cock flower") and a place in the Linnean system, knowledge of its ability
to be used as an abortifacient by any woman who might not want to bring
a new life into the world did not make the journey across the ocean. The
case of rubber entails two nontransfers, played out in the realms of technical
and cultural knowledge, one across space and another across time. Neither

nontransfer was total or complete. Both dramatically and greatly shaped the trajectories of rubber technologies and the ballgame.

First, there was the nontransfer of rubber technology to Europe. The colonizers did record the Aztec process for making rubber balls, including the use of juice, from the *Ipomoea alba* species of morning glory, in the processing of latex. They failed, however, to recognize the key role that this sulfur-rich liquid played in enabling rubber objects to hold their form across time and temperature.[31] Liquid latex does not travel well, and it has a strong odor, especially when exposed to heat. So, to process it, it would be necessary to understand certain techniques. Moreover, in the Aztec Empire, rubber was a valuable tribute material, demanded as payment from conquered peoples. But the conquistadors arrived with a clear idea of what materials they wanted to amass: silver, gold, and other precious metals.[32] To them, rubber appeared unstable—again, liquid latex did not travel well and had a strong odor—and therefore undesirable.[33] In their scheme of things, rubber was neither sacred nor especially economically valuable. Likewise, ballgames were not the kind of activity that warranted serious or careful documentation and representation.

Second, in the regions where it was played, the game was suppressed. The ballgame—with its ritual significance and its astounding bounce—quickly came to be viewed as a threat to the colonizers' project. Suppressing existing technologies and cosmologies was part of the process of remaking Anáhuac into a place called the Americas.[34] Fearing the game's role in maintaining the people's connections with their gods, Spanish bishops enacted bans on play, successfully eradicating it in most regions.[35] Alongside the active suppression of the game by the bishops, the first fifty-five years of Spanish rule decimated the population of Mexico, which went from approximately 25 million in 1519 to just over 1 million people by 1575. Some were killed directly. The bulk died in epidemics of smallpox, typhus, measles, influenza, and mumps—diseases introduced to the native population by the Spanish. Death on this scale affects generations. When so many people die in such a short period of time, cultures struggle to reproduce themselves. When they also face active suppression, reproduction becomes an even greater challenge.

Direct Spanish rule lasted for about three hundred years, from 1519 to 1821. In that time, 200,000 African slaves were brought to labor alongside the Indigenous population on haciendas, the form of plantation used throughout the Spanish Empire. The independence movement of 1821 was led by conservative landowners, and life did not change significantly for much of the country's population. Between 1821 and the Mexican Revolution of

1910, the country lost a significant amount of its territory to the United States, including present-day Texas, New Mexico, Arizona, and California. (A reminder that just a little more than a century ago, the borders which there are currently such fierce calls to wall off, did not exist.) After sitting at the center of the region's cultural and political practices for thousands of years, in recent centuries, the game has danced at the edge of extinction. Nevertheless, there is evidence of men's and women's play throughout the twentieth century.

Since the Mexican Revolution, *ulama* has held an important place in the cultural imaginary. After the revolution, dramatic social and economic change was enacted, including the ending of the hacienda system and the redistribution of land, mostly under President Lázaro Cárdenas. The *mestizje* ideology understood those of combined Indigenous and Spanish descent, a product of colonialism, as central to the nation's social and economic well-being.[36] The rise of *mestizje* was popularized by writer and politician José Vasconcelos—together with his vision for *la raza* (the cosmic race). Although Vasconcelos's idea of "cosmic race" views the melting-pot theory as a necessary solution to racial discrimination, women's studies scholar Alicia Arrizón argues that "ironically, his notion of mestizaje—a positive process of miscegenation—also promoted the idea that blackness would vanish from the social fabric of Mexico and elsewhere in Latin America." We should therefore, she elaborates, "understand mestizaje as the product of a history formed by cultural encounters, colonial difference, and the 'whitening' of the Indigenous/black subordinated colonial subject."[37] Taken together, Arrizón argues that the term effectively incorporated Indigeneity into the project of the nation-state rather than centering it.[38]

The incorporation of Indigeneity into the project of the national identity engendered more support and interest, albeit intermittent, in the history and contemporary play of the ballgame. Along with ongoing iterations of play, since the Mexican revolution, the game has been the subject of documentaries; museum exhibitions and catalogs; scholarship in archaeology, anthropology, and sociology; animations; performances; children's books; and video games. The 1968 Olympics in Mexico City—perhaps best remembered in sport history as the time that an Olympic torch was first lit by a woman, Enriqueta Basillio, and for US sprinters Tommie Smith's and John Carlos's iconic Black power salutes from the medal stands—opened with an *ulama* demonstration.

Figure 6.9
Frame from Roberto Rochin's video game *Pok ta Pok* (2012), which was briefly available from Apple's App Store. Screenshot by the author.

Scholars today debate whether the game has been played continuously, or if there was a break in play caused by the conquest, after which it was revived. If *ulama* play has been continuous anywhere, it has been in the northern province of Sinaloa, where people have continued to practice, play, watch, and bet on *ulama* in four small towns with populations of just over a hundred each.[39] The two variations played therein both take place on long, level fields with no walls. Players hit a solid rubber ball with their hips in one version, and with their elbows and knees in the other. Art historian Manuel Aguilar-Moreno spearheaded The Ulama Project from 2003–2013 with California State University, Los Angeles (Cal State LA), shepherding graduate students in their study of the game at the sites where it was still played. Aguilar-Moreno speculates that game may have shed its previously spectacular architecture in Sinaloa as a strategy for staying out of sight of the church authorities.[40]

The ballgame's tradition is inextricable from working with trees and vines to gather and process the latex sap for constructing the bouncing balls that the game revolves around. As part of his 1986 film *Ulama*, which documents the history, survival, and revival of this essential cultural form, filmmaker Roberto Rochín commissioned the production of a number of *ules*. Because the vulcanized rubber that they tried to use initially had the wrong density and hardness, this commission was to preserve the cultural practice of making the balls.[41] The palpable difference between the bounce of traditional and nontraditional *ules* is a result, in part, of the divergence in the history of rubber technologies created by colonization.[42] Despite the preponderance of natural rubber coming from *Hevea brasiliensis* trees, the history of the material is neither monocultural nor monolithic. The words "natural" and "rubber" in the phrase "natural rubber" obscures the thousands of different plants whose milky sap can be made into elastic materials. Significantly, the latex used to make *ules* comes from the *olicuáhuitl* or *hule* tree rather than *H. brasiliensis*. The former produces the kind of bounce that *ulama* players know and depend on. Beginning in the middle of the twentieth century, as the game recirculated in the cultural imaginary after the Mexican Revolution, *hule* trees became difficult to find in Sinaloa because of rampant deforestation, making it difficult to make the balls for the game.[43] Different kinds of bounce require different ecologies and structures of relation. After the *ules* and other equipment made for Rochín's film were ruined or sold, he concluded that what is needed to support the

Figure 6.10
Close-up of a *Castilla elastica* tree from Roberto Rochín's film *Ulama* (1986).

survival of the game is not only the training of players—and the retrieval and re-creation of *ule* production techniques—but also a reforestation program to replace *hule* trees, materially retransmitting the practice of the game.[44] This vision lives alongside a growing number of initiatives to return to Indigenous ways of caretaking land.

"Your Hip Will Be Like Your Hand"

AJUPEME, initiated by Armando Osorio Uscanga and Reyna Puc, boasts a growing number of delegations in Central America, South America, Europe, Canada, and the United States. Uscanga and Puc have helped push to have the game presented as a cultural performance at places like Xcaret, a "water, theme, and eco-archaeological park" in Cancún, Mexico.[45] Archaeologist and *ulama* player Arturo Sanchez explains how they and other groups have been "studying ancient texts, reading what the conquest reported, and decrypting codices in order to imitate the ancient game."[46] When I was participating, the San Fernando *ulama* team was organized by three people: Raul Herrera, who served as the California delegate and vice president of AJUPEME USA–Turtle Island; Daniel, who served as the team captain; and Aztlan Tenochtitlan who served as the AJUPEME ambassador and Los Angeles Director.

After seeking permission from the Tataviam Fernandeño Band of Mission Indians to play the ballgame on their territory, the San Fernando team began recruiting people who participated in *danza Azteca*. They were just getting off the ground when the COVID-19 pandemic arrived, and quickly lost 95 percent of their participants.[47] Using the time during the pandemic to focus on developing regulations that preserve the integrity of the game, the association wishes to construct the game as a decolonial project, driven by a vision of restoring a traditional form of play, and in doing so, refiguring colonization from a totalizing event to a historical blip—as seen from a future where playing the traditional ballgame will be literally re-creative.

The team is intergenerational, welcoming all ages and genders to their practices,[48] and it is a team of beginners. Most of the best *ulama* players live in Mexico, and the San Fernando players were unable to travel during the pandemic. So the team learned by watching players in Mexico on YouTube. They shared their own practices and demonstrations via Instagram. They have a website that includes event and practice schedules, short documentary videos, a photo gallery, a sign-up, and a nascent merchandise section offering shirts, stickers, statues, and sculptures.[49] The goal is to develop and grow decolonized sport, and to introduce people, particularly young people, to the game—one played on this continent for thousands of years.

In Mexico, Indigeneity has been incorporated by the state into the national identity—often in ways that have occasioned protest and resistance. In the United States, centering national identity around white male citizens has involved, among many other exclusions, the exclusion of Mexican American history from a majority of public education curricula. By contrast, Raul envisions growing the game through schools and after-school programs, with those in the area explicitly interested in supporting a decolonizing curriculum that includes decolonized sports. Most of the team's public presentations are demonstrations at schools, holiday celebrations, and cultural events. As the pandemic eased—and people in the area slowly returned to gathering together—the San Fernando team traveled around to present demonstrations of the game at schools like the University of California, Los Angeles (UCLA), to Mt. San Antonio College, and as part of local and citywide celebrations of Mundo Maya and Día de los Muertos (Day of the Dead).[50] Once, they played in front of three thousand youth at UCLA. Raul describes being surrounded by the old *hule* trees that lined the quad and wondering whether the institution realized what it had in these trees. Exemplified by what he shares at the team's practices and demonstrations, his years of experience teaching Chicano studies classes at Los Angeles colleges has enabled him to educate participants and audiences on cultural tradition, the history of colonization, and practices of decolonization.

Players don the traditional uniform at both public demonstrations and practices. The *gamuza* is a piece of animal hide shaped like a uterus, so players can channel both male and female energies. Pulled tight around the hips, it hangs down loosely over where it is cinched around the groin. The *chimali/e* or *protectora* (protector) is a leather strap looped right under the buttocks—also cinched as tight as possible, like a corset for the butt. The third part, over top of the *gamuza*, the *faja* is a cloth belt that wraps around the belly to prevent injury and to help the ball bounce better off the body.[51] The final piece is the *bota*, a small piece of leather placed under the *fajado* to absorb impact. With their bodies serially wrapped, the players become smooth, hard, taut surfaces for the ball to rebound off. The tight wrapping restricts movement, making it challenging to walk and bend, but this degree of tautness helps the ball bounce off the body without leaving

Figure 6.11
Raul Herrera presenting at Mt. San Antonio College (Mt. SAC), April 27, 2022. Photograph by the author.

bruises. For public demonstrations, players add face paint, feather head-
dresses, and wrist and ankle pieces.

Once the players have put on their uniforms, Raul and Daniel call them
into a circle to acknowledge the six directions. First is east, which is the direction of men; next is west, the direction of women; then north, the direction
of the youth, followed by south, the direction of the elders; then up to the
heavens, the direction of the cosmos; and then down to the Earth, acknowl-
edging the land, and the Fernandeño Tataviam Band of Mission Indians, who
gave their permission for the game to be played on their traditional lands.
For each direction, Raul blows on a conch shell. Different species of conch
have different tones used for different purposes. This acknowledgment takes
place at the start of each of the team's public demonstrations and practices.
Once the game commences, Raul plays the *huehue*—the drums, which represent the old ones or ancestors, the trees. Playing the *huehue* brings the ancestors to the game. The drumming is used to give energy to the players. During
practice, Raul plays mostly Aztec rhythms, sometimes hooking a speaker up
to his phone and playing along to YouTube videos. Ads pop up here and

there—mostly for the Medieval Times dinner theater, where he was planning on taking his daughters. Eventually, he wants to find musicians to produce tracks for the practices, as well as Aztec dancers and musicians, who would hold their practices alongside the ballplayers.

To mark a delegation's acceptance in AJUPEME, they are given one or more *ules* made in Quintana Roo using traditional Mayan ball construction techniques.[52] Marco Antonio Chavez, the president of AJUPEME USA, explained the process:

> It takes six months to make a ball. Three months to make it through a long process of adding layers one by one. And then three months to dry. Before it dries it has the smell of a dead animal. You can get stopped at customs for the smell. In tournament balls, there is a piece of quartz in the middle that is the heart of the ball. Everything you make with your hands, you give your energy to. The balls carry the energy of the person who makes them.[53]

The balls made by AJUPEME are unmarked and unbranded. They cannot be bought or sold, and for the team to remain part of the organization, they must commit to not commercializing the balls or the game itself. Teams are not allowed to make money from the game by selling tickets, swag, or the balls. If they do, the association asks for the balls back. The rule aims to keep the ball from circulating as a commodity. This keeps the game apart from the consumer capitalism that suffuses, and fundamentally structures, so much of contemporary sport.[54]

The first *ules* that I encountered in person rested against the blacktop of a basketball court at El Cariso. I picked up the solid, matte black spheres, one after the other, impressed by the heft that I had read so much about. As I absorbed their weight, I remembered reading Fox's description years earlier:

> For me, absorbing the ball's impact for the first time was a moment of utter revelation. I'd written a 300-page doctoral dissertation and several academic articles on the ancient game and had lectured on the topic at conferences. I'd dissected the game's ritual meaning and political symbolism and diligently pieced together and cataloged thousands of pottery fragments excavated from the ruins of courts. But I'd never felt the blow of a ball against my body.[55]

At the time, I had gotten a rubber mold–making kit from an art supply store and made a couple of solid rubber balls, slightly larger than softballs, to try to get some sense of these blows. But those smaller, slightly misshapen purple spheres did not offer a good equivalence to the *ules* that I held now.

Their elastic density absorbs and then releases any force that they meet. The way that they looked in the light made me think about volcanic rock, asteroids, and black holes. Raul uses the black new moon emoji to represent the ball in Instagram posts. The men's ball weighs eight or nine pounds and is a bit smaller than a basketball. Newer and pocked on the surface, it's just short of round, with a lighter, almost rusty hue. The women's ball weighs around six pounds and is older than the men's ball and rounder and smoother from use. There is also a kids' ball, which is a little larger than a tennis ball, weighing in the range of two to three pounds. And there are also ten-pound balls used in practice so the eight-pound ball will seem light in competition. Because the physical cost of misjudging and mishitting a ball is so high in *ulama* (much more than in tennis), it is even more important that the bounce be true—that it be predictable and reliable and safely rebounded. Balls need to be a well-balanced sphere, and the surfaces of play need to be level, but reliability and predictability are arrived at via different routes and roots. In place of strict specifications and testing that allow balls to be easily exchanged one for the next, the longevity of any given ball allows players to get to know any particularity of its behavior.

Until 2023, the US Open used different tennis balls for the men's and women's draws in order to slow down the men's game and speed up the women's game. In the case of *ulama*, the difference in the *ules* reduces the potential force of impact and the speed of the women's game. Unlike tennis

Figure 6.12
The eight- and six-pound *ulama* balls at El Cariso Park. Photograph by the author.

balls—rapidly mass-produced pneumatic objects that are tested to ensure that they are as indistinguishable as possible and begin losing pressure when a can is opened—*ules* are solid rubber spheres that become smoother with time and use. They maintain their bounce for years. Like any other object, balls are always also processes: They are brought into form, and over time, that form shifts from impact, use, exposure, or nonuse,[56] and different processes have different temporalities.

All ballgames are played between—and, thus, play with the constitution of—objects and surfaces. The bounce of a solid rubber projectile carries force. Like any contact sport, *ulama* involves the risk of real bodily harm. In the hip version of the game, players hit the ball with their hips, upper thighs, buttocks, and bodies just above the hip. Any impact higher or lower than these points risks serious injury. Training to play is training to be an opposing force to that of the ball. It is a training out of, into, and through fear of impact. Miguel Duran, a filmmaker, first came to the team practices because he wanted to make a film about *ulama*. Within a year, he had become a member of the US National men's team, and he currently serves as the national team coach. He describes how, when they first started playing, everyone's legs and hips and sides were covered in bruises in "all kinds of colors, like a supernova."[57] Like handball, learning *ulama* requires players to endure pain and bruising. While they develop proper technique, players must be willing to bear temporary tattoos of bruises, to achieve protectively calloused skin, and a toughened psyche and soma. Learning to receive and return the bounce in *ulama* shapes everyone who steps up to meet it.

The force of contact means that sometimes when things don't bounce, they break. I reflected on this—and on Miguel's description of supernovaesque bruises—as I tended to two vivid contusions on my right hip and ribs. I tested exactly how much of an inhale my almost (or perhaps actually) fractured rib could handle. The cost of being an ungainly beginner was far higher in *ulama* than it had been in *FIFA*. I confronted the question of whether I really wanted to play this game. There is no difference in the treatment for a badly bruised versus a fractured rib. Rest. Ice. Soak in Epsom salts. Take Advil. Try not to cough, or sneeze, or laugh. Every inhalation tests the tolerance of the injured rib and strained intercostal muscles. I discovered a strange humor in the injuries, taking a perverse pleasure in the pace of my recovery. "When things don't bounce, they break" has a nice ring to it, but it is not quite right. It is not always an either/or. Something

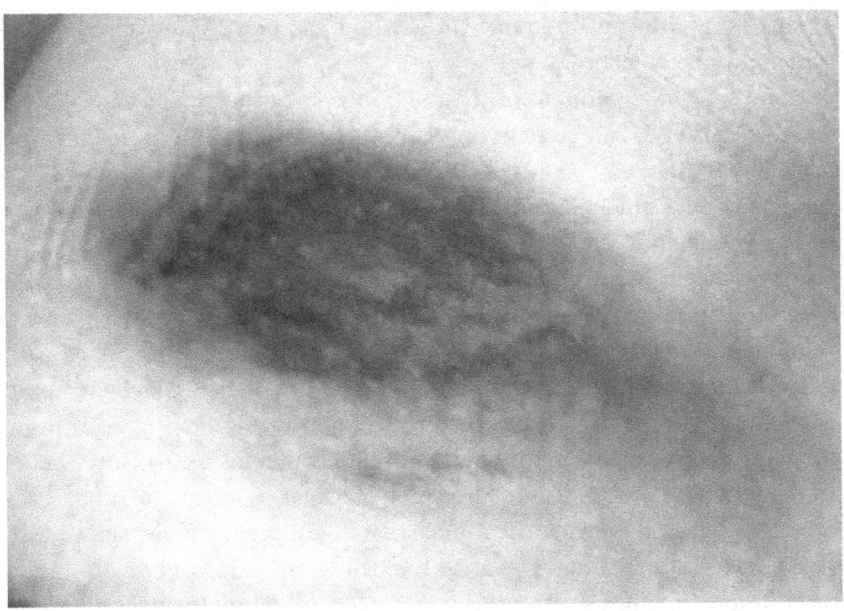

Figure 6.13
Hip contusion on day 2, September 13, 2021. Photograph by the author.

breaking will change the bounce, and sometimes things take a bad bounce and break.

The hip contusion was par for the course for a beginner: the result of Daniel tossing the ball for me to hit with my right hip over and over, as he instructed me on how to crouch and curl a bit before stepping—almost jumping—to meet the ball: "The trick is to bend your knees and then open your body up like you are greeting the sun."[58] (Here was a fitting image since, in the game, the ball represents the sun.) "Make your whole body tight and stretched like a wall."[59] During my first court tennis lesson as well, I was told to turn myself and my racket into a wall. Like that moment, I was drawn again by the instruction for the action to become something solid and still. But, of course—much more so than in court tennis—in *ulama*, players' bodies are at stake. Drawing the skin and hidden muscles, tendons, ligaments, and bone tissue taut creates a solid surface that can deflect, and perhaps increase, the ball's kinetic energy. The bodily wall refuses the ball's entry to, and impact on, the body's softer materials, while redirecting its trajectory. The *protectora* and *faja* assist in the transformation of side body into impenetrable surface, but the player must develop the underlying body technique.

The rib's bruise came from a bad bounce, a misjudgment of how and where to meet the ball brought on by pushing past my limit for fielding the ball's impact that day. As I moved and breathed as shallowly as possible, I thought about Daniel's promise that with practice, "eventually, your hip will be like your hand."[60] The image caught my ear—I pictured remapping the fine motor skills of my hand, with its opposable thumb (something so often held up as exclusively human), to another region of my body. I pictured cradling and redirecting the ball with my hip with a sense of contact akin to how it felt to catch and throw a tennis ball or baseball. Maybe it would be akin to how the squash ball used to feel on my racket strings—in those moments when time slowed and I knew that I could put the ball anywhere I wanted. Playing with time is the true trick in *ulama*, as in any to-and-fro ball-game: knowing all the kinds of bounce that might occur between body, ball, and surface and learning to wait for the ball to arrive instead of rushing to meet it, watching its travel and its spin and coordinating the right gesture—the right response—in the moment.

The ability to respond rather than react in any given moment in any context comes from explicit or implicit training. Response is an embodied practice. The question of whether I really wanted to practice—to entrain my body—into this form of play hovered in my mind. I was just past forty. I had spent much of the past decade in varying degrees of physical pain. I was in and out of physical therapy healing herniated discs and unwinding asymmetric muscle firing patterns, ones etched deep by decades of playing a racket sport. I had been unlearning the training to push past my limits without first listening and discerning whether that push was truly wanted or warranted. *Ulama* players must always use the same hip. If my hip became my hand, this would create—like handedness for most people—a new set of asymmetric patterns. We are all variously asymmetric, but some kinds of asymmetries require more support and resource to sustain.

I tried to imagine following Atzlan's suggestion that, to get my hip used to the impact, I could hang a basketball from the ceiling of my home and use it to practice. I liked the idea of living with a ball hanging at hip height in the corner of my bedroom, and I could tell that it would do the necessary entrainment. But as I considered what it would take to remake my embodiment to meet the particular kinds of risks and impacts of this game, the promise of future bruises and newly entrained asymmetries gave me pause. How would this training live next to another significant kind of embodied transformation, remaking, and risk that I had stepped into in the same moment: trying to become pregnant and have a child? Warming up and learning the game had set off some old instincts of pursuit. But with time, I discovered that I was no longer driven by this particular embodied pursuit. As I stepped toward the embodiment of pregnancy, I was ready to let this kind of training be in the past tense.

The question of how, when, and whether to approach pregnancy in particular—and parenting in general—has long been fraught for athletes with uteruses who want to bear children. The already real pragmatic challenges that many potential parents face are, then, made more complex by the risk of play to the possibility of birthing a child. This has been used as a mainstay argument for banning women from playing all kinds of ballgames professionally and, in some cases, recreationally. This argument nests inside the more general argument that women and girls who play sports risk an unacceptable degree of physical injury, and it is situated amid similarly patriarchal positions. Women who play sports will negatively affect their menstrual cycles. Or, because of the power of their menstrual cycles, those cycles will negatively affect the game and other players. Or athletic women will behave in unfeminine ways, becoming too muscular and unattractive. Or they will harm their marriage prospects. Or they will become lesbians. Or they will move in sexually explicit ways that will distract male players. The list goes on. These arguments are variously deployed in support of bans on women's play and to enforce the gender segregation enacted in most modern sports, which in turn serves as a key site for maintaining the cultural construction of gender as a strict binary.[61] Writing of the abrupt withdrawal of support for women's football in Mexico—after the country hosted the second women's world championships and cheered their national team on through the finals—cultural historians Brenda Elsey and Joshua Nadal report that Mexican *futboleras* recall being constantly criticized for playing football: "If you played women's football, it was almost as if you were thrown out of the female sex."[62]

In Ramos's interviews with girls and women who participate in *ulama* in Sinaloa, she encountered fears that the *ule* could hit the abdomen, preventing future pregnancies, or that it might hit the vagina and take the virginity of a young player. The contemporary culture of play in Sinaloa, her research shows, is deeply shaped by sharply binary gender roles and a strong culture of machismo. Girls and women pay a high social cost for daring to play the game.[63] Most girls and women stop playing or never start, in part because they do not want to confront the negative comments from their peers and family. While women pay a cost for playing, men pay a cost for not playing, or not playing well. The contemporary game in Sinaloa serves as a site for gaining, maintaining, and losing status. Ramos documents the older *ulama* players mocking the younger male players with comments like, *"Le pegas al ule como una mujer"* (You hit the ball like a girl).[64] This insult echoes the one that prompted philosopher Iris Marion Young to write her foundational work of feminist phenomenology, "Throwing Like a Girl: A Phenomenology of Feminine Body Comportment Motility and Spatiality," yet it points to something that Young does not address in her text. The presumed addressee of the insult is not a girl or a woman, but a boy or young man. He is being taught to either turn his body into a sturdy wall that can strongly deflect an eight-pound ball by enduring months of supernova-level bruising as per Miguel's description, or risk being shamefully cast out of the category of man and turned into a girl.

None of the ethnographic work from the twentieth century mentions women's play, but Ramos's interviewees report memories of women playing the game in the 1940s and 1970s. One possible explanation for this emerges from the dramatically different responses that she and a male-identified colleague received when they each separately asked the same question—whether women play the game—to men who played the game. When her colleague asked the question, the players did not admit the possibility of women playing. But when she asked the question, they acknowledged the existence of girls and women who play. Here, Ramos identifies a feedback loop akin to the ones identified by other feminist scholars of sports and video games: the presumption that certain games are "men's games." This gendered and self-enclosed loop gets reinforced by the ongoing absenting of women's play from the historical and ethnographic records due, in part, to the operation of gender on the research. This, in turn, reinforces the understanding of these games as men's games. This loop holds in place the social cost for girls and women who might want to play, ensuring that they do, indeed, stay that way.

The San Fernando City *ulama* team, following the lead of the founders of AJUPEME, welcomes everyone to play. At the team's practices and demonstrations that I attended, there were always at least a few women—mothers with their daughters, alongside other teenage girls and young women. Luna and Blanca are there most days. Iris doesn't play, but she does participate by documenting practices and demonstrations, posting on social media, and providing other kinds of support for the team. While formal competition follows the familiar gender segregation of so many sports, everyone practices and participates in the demonstrations together. There are fewer women than men who show up regularly for practices and demonstrations, and the team does not yet have enough women to field a full national team. But by welcoming players across all ages and genders—and by playing together—they offer a picture of an alternative tradition of play. If games are ways that cultures tell stories about themselves *to* themselves, this cultural revival tells a different story of a game played with the ball made from the milk of the tree.

On November 19, 2022, the team joined a group gathered at the northeast corner of MacArthur Park, in the Westlake neighborhood of Los Angeles, for the ground-breaking of the Maya Corridor Project, a street-scaping project aimed at transforming a mile-and-a-half stretch of 6th Avenue, running west from the park. To create a place of recognition for Mayan residents in Los Angeles, the street was dotted with Mayan-inspired monuments, decorative sidewalks, streetlights, and other visual culture. Raul and Miguel contributed a proposal for their vision of building an *ulama* court in MacArthur Park to the city's plans. AJUPEME is not the only group practicing *ulama* in the city; the San Fernando *ulama* team participated in the ceremony alongside another group of players from Los Angeles who are not in the association. Mayor Eric Garcetti and a number of the city's Mayan cultural leaders participated in celebrating the completion of the first step of the project.

As the city stands on the cusp of hosting the World Cup in 2026 and the Olympics in 2028, the team dreams of building *ulama* courts in public parks across the city: remaking history by working toward a future where this sport is again part of the culture's common sense. If they succeed, *ulama* courts will live next to the tennis, handball, and (most recently) pickleball courts that already bedeck the city's parks and have hosted endless hours of casual and fiercely competitive play.

Conclusion: *Rebote*

I wrote much of this chapter sitting next to the rubber tree that Amanda Perez asked me to look after when the Maravilla Handball Court, the oldest surviving handball court in Los Angeles County, was sold in 2016.[65]

Figure 6.14
The Maravilla Handball Court, Maravilla, Los Angeles, January 2011. Photograph by the author.

Pushing up through and around a tomato cage that it both relies on and has long outgrown, the plant is a tall, scraggly tangle of branches in a large terra cotta pot. Around when Amanda began organizing the Maravilla Historical Society, it was dropped off at the court by one of the Maravilla Handball Club's longtime players.[66] She pruned the small young plant, and it grew and bloomed. It grew more as she recruited family, friends, players, coaches, filmmakers, writers, and artists to help her pursue dreams of saving the court, preserving its history, raising funds to purchase the property, and turning it into a community center.[67] I came into her orbit after reading a piece by Hector Becerra in the *Los Angeles Times* about her efforts to save the court. I had been thinking about handball and other similar ball-wall games ever since artist Helen Mirra asked, "Why squash?" in a question-and-answer session after a talk that I gave on my work, and subsequently found myself researching the history of handball in Los Angeles as preparation for presenting a ball-making workshop at Machine Project. I invited Amanda to the workshop, and she in turn invited me to organize another workshop at the court, along with an art auction to raise funds for her efforts. Later, she asked me to join the board of the Maravilla Historical Society, pulling me further into the orbit of the court and its history.

Before coming to live with me, the tree lived in the shadow of a neighboring orange tree on the back porch of the small living quarters above the El Centro Grocery. The store is attached to the court. According to lore, the court's two walls—shaped like an "L," with dimensions of 20 feet by 40 feet—were built in 1928 with bricks that workers had taken from the Davidson Brick Company factory down the street. The echoes of balls bouncing off hands and walls became a regular part of the rhythm of the small *varrio*, long before it had even basic infrastructure like water, sewers, or sidewalks. In his vivid chronicle of the court's history, Sam Sweet writes that, while many of the public courts in the city of Los Angeles boast designs that merge handball's disparate lineages—Irish, Spanish, Mexican, and others—the mostly Mexican American boys and men playing at the Maravilla Handball Court "remained faithful to the uniquely Mexican game learned on the streets of El Paso: *rebote a mano con pelota dura*. Hardball."[68] The neighborhood witnessed and weathered the waves of wartime industrialization and the rise of mass incarceration over the course of the twentieth century. Throughout, the court—known simply as *el rebote*—served as a collective heartbeat for the neighborhood's Chicano community.

Rebote, the Spanish word for "rebound" or "bounce," is used as a name for both the game of handball and the wall itself. The Maravilla Handball Court's nickname, *el rebote,* shows how the wall was understood as the site and generator of bounce—like the early versions of court tennis—as much as or more than the ball. Sweet writes, "Youngsters said there was a special bounce to the wall at Maravilla—a density deep within the Davidson bricks that couldn't be replicated on any other court."[69] No matter how much effort is put into attempts to standardize sporting contexts, different walls create different bounces. Bounce is always situated, conditionally created, and situational.[70] That is, each *rebote* has its own character. It is possible to get to know a wall's particularities—the bounce that it produces.

Players also get to know particular balls. All the balls used to play handball are made of rubber, but they are not all the same.[71] The balls used to play at Maravilla were smaller and harder than the ones used for Irish derivations of handball, or the "big blue" balls (also made of rubber) used for modern racketball. Sweet reports, "They were made to order from an old man on Valley

Figure 6.15
(Left) El Centro Grocery and the Maravilla Handball Court, photographed from the street. (Right) The court photographed from above with a game in full swing. Photographs courtesy of Amanda Perez.

and Muscatel, in South San Gabriel. He eviscerated golf balls and wrapped the salvaged rubber in yard. A strip of tanned goatskin was stitched around the exterior, then soaked in blue-black ink and left to dry in the sun."[72] With its hard rubber core and leather exterior, this ball is as hard as a baseball. It stings to hit and, over time, it shapes and gnarls the hands of committed players.[73] The toughnesss required to play was part and parcel of the status granted to the boys and men who became standout *rebote* players. It was a "man's game," in the sense that playing made one a man in the eyes of other men. Girls and women were often discouraged from playing, although some—like Amanda, who grew up in the neighborhood—would find a way to play both on the court itself and in the alley behind the court. Shaped by the adaptation of play to the balls and walls that were available, these informal games were played against different walls and thus had different bounces.

Michi and Tommy Nishiyama, a young Japanese-American couple, began running the grocery store attached to the court after their internment during World War II, eventually buying the property that included the court in the 1970s. Over the years, their store and the court became the center of the mostly Mexican American neighborhood. Michi extended credit and passed out sandwiches to people short of money. Inside the court, boys and men played handball, gambled, and drank. Women and girls would join for the dances and holiday celebrations that Michi organized. Amanda recalls, "There was a game going on all times of the week. You could hear the ball,

Figure 6.16
(Left) The Nishiyama family standing outside El Centro Grocery. (Right) Michi Nishiyama at the court, which is decorated for Christmas. Photographs courtesy of Amanda Perez.

pah-pow pah-pow pah-pow."[74] As Sweet relates it, on the sidewalk outside the court, people would deal marijuana and heroin. The criminalization of possession in 1956 led to more and more people from the neighborhood being imprisoned.[75] After the sheriff's department threatened the court with code violations as retribution for being refused entry during a police chase, Tommy Nishiyama, together with other mainstay players, organized to make it into an official men's club with dues-paying members and a lock on the door.[76] The fact that the court was privately owned by the Nishiyamas is what protected it when the Los Angeles City Council tore down most of the city's handball courts during the 1990s and 2000s on the grounds that they attracted gang activity.[77] After the Nishiyamas passed away (in 2006 and 2007, respectively), the property passed to a nephew who did not have a deep connection to the place. The grocery and the court stood empty and abandoned.

When she realized what had happened, Amanda started an effort to preserve the court and its history. Her commitment to preserving the court's history is shaped by the politicization that she, along with so many others, experienced as a Chicana student marching against the Vietnam War. In 1970—in response to the disproportionate number of Mexican Americans who were drafted, wounded, and killed in the war—the National Chicano Moratorium organized a protest march that drew over 20,000 people.[78] As Perez explains, the term "Chicano/a" was reclaimed from its origins as a racist and classist slur against low-income Mexicans and refigured in the

1940s as a politicized cultural identification. It is an identity born in and out of a contact zone. Sheila Marie Contreras puts it this way:

> To name oneself as "Chicana" or "Chicano" is to assert a gendered, racial, ethnic, class, and cultural identity in opposition to Anglo-American hegemony and state-sanctioned practices of representing people of Mexican descent in the United States. As it evokes the "radical" politics of cultural nationalism, "Chicano" stands against the institutionally normative "Hispanic," as well as the linguistically insistent "Latino." Chicano has a different set of complexities that attend the revival of the tradition.[79]

It is an identity that for many connects them to Indigenous heritage and traditions. For Amanda, whose Indigenous roots include Kikapu heritage on her mother's side and Yaki on her father's side, "Chicano is my culture, my belief, my origin."[80] The Maravilla Handball Court, along with the land that it is built on, likewise "is a sacred site. It's a place where you can say, this is us."[81]

Although her larger vision of purchasing the land and creating a community center was never realized, the Maravilla Historical Society, together with the Los Angeles Conservancy, were able to attain historical status for the court in 2012. When the court and adjoining house and store were sold in 2016 to someone from the neighborhood, the new owner was obligated

Figure 6.17
Amanda and Vergie putting up a banner to recruit people for the Maravilla Historic Society in 2008. Photograph courtesy of Amanda Perez.

to preserve the exterior walls, including the front and side walls of the court and the muraled back wall. As Amanda cleared out the space after the sale, she asked me to give some of the artifacts from the court a new home, including the wonderfully unruly rubber tree. I took it back to my apartment at the time, and it has moved with me ever since. A few years later, Amanda introduced me to her teacher—the Nahua elder, Sun dancer, master builder, and healer Piltzinkoyotl—who in turn introduced me to Raul and the San Fernando *ulama* team. And so I found myself, beginning in late 2022, spending long hours curled in a pink chair next to the tree that used to live behind the Maravilla Handball Court, following the chase lines of this cultural history of bounce. The disheveled basketballs, soccer balls, baseballs, softballs, volleyballs, tennis balls, golf balls, wiffle balls, handballs, beach balls, stress balls, kids' balls, dog balls, and others that I have collected from the streets of Los Angeles were piled high in the opposite corner, as I moved words to and fro, working to relay the ways that Mesoamerican ballgame and *rebote* echo through this city. Pok-ta-pok-ta-pok-ta-pok. Pah-pow, pah-pow, pah-pow. The vision of AJUPEME and the San Fernando City *ulama* team reaches into the future, as the tradition of playing *ulama* spreads across Los Angeles County, California, the United States, and beyond.

Figure 6.18
The Maravilla Handball Court rubber tree, in my writing corner, 2024.

Conclusion: Culture-Inside-Out

Off the Ball

Elijah makes a beeline for every boundary. As soon as he is able to worm his way across the floor, he's off to investigate where the carpet meets the floor, where the porch meets the stairs, where the bed ends, where the edge of the dresser begins. He bangs his hands against every surface within reach—bed, floor, body, face, toy, chair, changing table, highchair.

From one week to the next, I watched him learn that the floor is harder than the carpet. He began to be gentler with his floor *thawks*. He goes from clutching the quilted, highly grabbable ball that his grandmother gave him to launching himself toward the green exercise ball that is just a bit bigger than he is. Every ball presents a chance to grab, roll, and slap, making gloriously percussive sounds. Through one impact after another, after another, he is getting to know the boundaries of his body in and with the world, adjusting himself based on how much he does or does not rebound off any given surface. The day after he turned seven months old, he learned to clap. Immediately upon waking the next morning, he rolled on his side to begin bringing his hands together over and over, privately delighted. Feminist theorist and physicist Karen Barad writes: "When two hands touch, there is a sensuality of the flesh, an exchange of warmth, a feeling of pressure, of presence, a proximity of otherness that brings the other nearly as close as oneself. Perhaps closer. And if the two hands belong to one person, might this not enliven an uncanny sense of the otherness of the self, a literal holding oneself at a distance in the sensation of contact, the greeting of the stranger within?"[1] As Elijah repeated the gesture throughout the week, each time smiling with astonished joy, I delighted in his delight at the sensation of his hands making contact with each other—in his surprised elation in his own self and its otherness.

He loves to bounce and be bounced. Supported in a standing position, he can bounce himself up and down, up and down, and up and down, emitting a series of chortles and excited baby pterodactyl shrieks. In her book on how the brain develops during the first five years of life, neuroscientist Lise Elliot devotes an entire chapter to why babies love to be bounced. The answer, she explains, lies in a "highly developed vestibular system—a 'sixth' sense that allows us to perceive our body's movement and degree of balance. The vestibular senses are very old, in evolutionary terms, since all earthly organisms have had to orient themselves with respect to gravity and their own motion."[2] Anyone who has taken care of a baby is likely familiar with their love of bounce. I laughed out loud when a colleague who had a child at the same time as me wrote to say that he had caught himself bouncing a pack of paper towels in the supermarket. Many babies train their caretakers into movements that will achieve some vibrationary effect: walking on toes, rocking in chairs, bouncing in arms or on giant exercise balls, driving in circles or down long stretches of highway. Anything to achieve that up and down and up and down and up and down. A plethora of sleeping and playing devices on the market promise to relieve caretakers' tired arms by bouncing, rocking, jiggling, jumping, or otherwise soothing babies through mechanical motion for a spell.[3]

Extending bounce to how babies knock about, offers another angle on the idea that the identity produced by bouncing interactions is always a relational one. Elijah's slaps and knocks establish that his body and these other bodies—floor, bed, book, tub, mama's face, crocheted moon, wooden block, rubber teether—are similarly solid. On this basis, they interactively have enough in common to support each other in maintaining their current forms—different from the water that splashes and recedes in response to his eager slaps, different from the air that carries the wind that it took him (and the elastic tissue of his lungs) three months to develop the capacity to face and breathe. Bounce is one of the fundamental interactions through which humans learn and differentiate their own bodies and movements from other bodies, beings, densities, textures, and surfaces of the surrounding environments. These interactions are developed and elaborated into many different examples of deep play, the phrase of Jeremy Bentham's that Clifford Geertz made famous—"play in which the stakes are so high that it is, from [Bentham's] utilitarian standpoint, irrational for me to engage in it at all."[4] And precisely because of the irrationality of engagement, for Geertz, these forms of play are where societies catch sight of themselves.[5] Elaborated, bounce becomes a way to test for truth, communicate the science of the day, create characters, coordinate collective performances, and keep the universe in balance. The incredible ways that we bounce off of ourselves, and others, allow us to play with passing the *I* to and fro—what is me and not me, what and how we construct our *we's*.

Our (no longer just my) home is awash in rubber objects. I had already amassed a pile of scavenged playing balls in the living room, but they have been joined by a joyous jumble of rubber nipples, bowls, bibs, toys, and teethers in the shape of a small blue octopus, a mini-banana with helpful handles, a perfect simulacra of a radish, a multipiece, semiabstract avocado, and Sophie the Parisian giraffe. Elijah chews and gnaws on them as his teeth work their way through his gums. Like the balls, most of these objects are made of some kind of rubber—"natural rubber," "food-grade silicone." Some are wood. Some are plastic. Most have been produced with the aid of computers at multiple points in the process. Sometimes he bypasses them all in favor of a rock of just the right size.

As I watch him develop his early object relations and body techniques through these gestures, I wonder how much of the story of elastic materials and motion practices—told in this book through sports and games— might just as easily have been told from this prior moment of infant objects and actions. I consciously assist him in some ways—showing him specific techniques for latching to my breast, rolling over, pulling himself up, and finding his mouth with the small rubber spoon. We play peekaboo around the corner and in the mirror, hide-and-seek, chasing games—all iterations on *fort/da*, going away, hiding, running away, returning. I unconsciously transmit other ways of mirroring—he lights up at the sight of my iPhone, Ive's laptop, his grandmom's Apple watch. Leslie asks him, laughing, if he sees himself, as he stares a little extra intently whenever a picture of him is displayed on the Aura electronic picture frame that my mom gave me for Mother's Day. He is seven months old and already recognizes that screens— and their many mirrors—are part of the game.

Culture-Inside-Out

Anthropology, sociology, and cultural studies have a long tradition of describing how bodies come to hold cultural form. In 1936, Marcel Mauss proposed the term *body techniques* to describe "the ways in which, from society to society, men know how to use their bodies."[6] Body techniques are acquired through daily activities. Without realizing that this is what we are doing, we learn how to use our bodies, thereby naturalizing cultural gestures. Body techniques shape how we are in the world, how we are with others—and, how we hold ourselves and others in the world can make the world otherwise. Mauss illustrates his notion of "body techniques" with the example of his own outdated swimming technique: "The habit of swallowing water and spitting it out again has gone. In my day, swimmers thought of themselves as a kind of steamboat. It was stupid, but in fact I still do this: I cannot get rid of my technique. Here then we have a specific technique of the body, a gymnastic art perfected in our own day."[7] Mauss's example is a technique that is named and taught—a technique for how to go about a certain activity: a kind of leisure play refined into a "gymnastic art." Importantly, for all his anthropological specificity, Mauss often argues that one technique is objectively better than another.

This willingness to judge technique is less present in Pierre Bourdieu's term *habitus*, which describes the cultural milieu (language, landscape, cultural habits, etc.) that surrounds and fundamentally shapes an individual's speech, movement, and gesture early in life. This surround becomes "the basis of perception and appreciation of all subsequent experience."[8] Iris Marion Young extends and sharpens the point by naming the social and political consequences of the naturalization of cultural gestures in "Throwing Like a Girl: A Phenomenology of Feminine Body Comportment Motility and Spatiality." Young describes how children learn to occupy and move through space based on how those around them respond to their movements, and argues that children are responded to in different ways, depending on their perceived gender and on what that perceived gender means in their cultural context. Crucially, children learn gendered modes of occupying and moving through space long before their bodies are differentiated by puberty.[9] Young demonstrates that children learn to throw a ball differently—along a gendered axis—both explicitly (as a technique, à la Mauss) and implicitly (via habitus, à la Bourdieu). Young's analysis, first published in 1990, helps situate both the decades-long absenting of women from *FIFA* and the shaming of young boys for playing *ulama* "like a girl." While Young focuses on the ways that the phrase, "throwing like a girl," points to a set of pressures that shape the bodily comportment of girls, idiomatically, the phrase is also used to apply a different set of pressures to boys—indeed, it can apply pressure, in one direction or another, to any child.

Neither Young, nor Mauss or Bourdieu, explore how the kind of ball at play might affect how that ball is thrown—or, rather, what pressures that ball might put on the movement habits and practices. Ways of throwing gather around the balls that we are—or are not—given to throw: it matters who is given which balls to play with; who is encouraged and how; if those balls last for a day or a decade; if it takes nine or ninety thousand balls to play a professional tournament; if the gravity that balls are subject to is the same as our own bodies or is virtually manufactured. In short, the material and material conditions matter. Today, industrial objects are almost always also computational objects, made with the assistance of computer-aided design (CAD) and computer-aided production. Many are made for television or social media; their design explicitly tailored toward how they will appear on a television or a phone screen. We are trained into gender through gesture, yes, and also so much more. The kinds of balls at hand, and in play, shape the habits and inhabitation of their and our worlds.

There is an underlying logic at play here: we learn language, gesture, and manners through imitation, or more accurately, through mimesis. Making a similar argument in a different register, Michel Foucault defines "exercise" as "that technique by which one imposes on the body tasks that are both repetitive and different but always graduated . . . Having become an element in the political technology of the body and of duration"; exercise, he elaborates, "does not culminate in a beyond, but tends towards a subjection that has never reached its limit."[10] Exercise, for Foucault, does not have an end point. It is an asymptotic disciplining of the body. From an infant slapping at the floor to, an adult's pick up game at the park, to an athlete training to place a ball with reliable precision, repetition produces the ordered body. Both Foucault and Henri Lefebvre borrow the name of an elaborate equestrian sport to describe how humans are trained into culture: dressage. As Lefebvre describes it:

> One can and one must distinguish between education, learning and dressage or training . . . To enter into a society, group or nationality is to accept values (that are taught), to learn a trade by following the right channels, but also to bend oneself (to be bent) to its ways. Which means to say: dressage. Humans break themselves in like animals. They learn to hold themselves. Dressage can go a long way: as far as breathing, movements, sex. It bases itself on repetition. One

breaks-in another human being by making them repeat a certain act, a certain gesture, or movement . . . Liberty is born in a reserved space and time, sometimes wide, sometimes narrow; occasionally reduced by the results of dressage to an unoccupied lacuna.[11]

Lefebvre uses "dressage" to name how humans "bend" their bodies to specific *ways* of moving. He leaves liberty cordoned off and under siege.

Ranging across anthropology, sociology, and cultural studies, much of the twentieth- and twenty-first-century scholarship on sports and games fits comfortably within this paradigm of subject and cultural formation. The British school of sport studies, in particular, has produced a large body of work that situates narratives of embodied subject formation in the context of imperial power and educational ideology.[12] Scholars of game studies have considered how our conceptions of self—and the abiding transformations in the means of production—are materially linked to ideology through bodily performances.[13] In this paradigm, play is generally located on one end of a spectrum of performance that travels, bouncing from play to game to sport (or theater) and on to ritual. Performance theorist Richard Schechner phrases it this way: "Human performance is paradoxical, a practiced fixedness founded on pure contingency."[14] In all these tellings, cultural practice is what organizes and structures contingency and indeterminate play.

The key insight of these theories of culture formation is that the internalization and naturalization of cultural practices serve to cement power structures. The task, then, for the progressive intellectual, artist, activist, and scholar, is to reveal, denaturalize, and make visible the normally invisible workings of culture and power. Here, sport could be understood as a modern apparatus for capturing play.[15] A form of culture, sport trains and entrains variations of human behavior into rule-governed repetition. At its most blunt, play is captured by the act of extracting value from freedom, drawing back into use that which, by definition, exists apart from instrumentality. Yet instrumentality is paradoxical.[16] While treating ourselves, or worse, others as mere instruments easily tips into exploitation, not every end is to be regarded with the same degree of suspicion. In certain cases, a means to an end can be a means to a kind of autonomy—like a baby channeling their capacity into a technique for crawling forward. Theories of culture formation are essential for rendering the mechanics of culture and power visible. Still, they are inadequate tools for understanding the stakes of the microcultural motion practices of games and sport.

This is what bounce helps us understand. Sport is culture, but—and this is the crucial point—it is also culture-inside-out. Instead of training and entraining the variation of human behavior into rule-governed repetition—as argued by anthropologists and sociologists from Mauss to Bourdieu on forward—sport begins with rules and ends with variation. Playing sports re-create grounds for departure: calling forth a culminating moment that transcends the scope and authority of the rules guiding it. As the cases in this book demonstrate—from court tennis and modern tennis to *FIFA* and *ulama*—gestures and infrastructures in sport are less implicitly adopted through mimesis than the subject of explicit scrutiny by players, spectators, and officials. Each participant develops a keen awareness of the conditions of play and, in different moments, may support or challenge them. In the case of physical sports, extraordinary effort goes into creating a fixed ground with which to begin. In electronic games, moreover, additional effort is directed toward producing the possibility of variation. Both physical and digital games demonstrate how bodies and objects comport to infrastructure and how infrastructure is in turn assembled for imagined bodies. Culture-inside-out helps us see how ballgames—along with, and perhaps despite, their very real reinforcement of hegemonic values—are places to look (literally) for edges, limits, and what bounces extraordinarily beyond them. Animators search the archive of photographs of athletes mid-hurdle, or crossing the finish line, to learn how to portray the fullest range of expression. Games retain conservative vestiges that equalize chances, by way of handicapping, and thus warrant a second look.

When Lefebvre picks up dressage, he misses the point of his metaphor. In the actual sport of dressage, the rigorous training of horse and rider is what creates the conditions of possibility for the astonishing moment of variation. In sport, order produces the different body. It is from a fixed ground that we call forth the flash of luck, the incredibly timed leap, the brilliantly placed ball, the unimaginable retrieval. These moments do not actualize life by discipline so much as they provide us with a presentiment of the virtual upon which the actual is founded. These are creative gestures called forth from, and through, constraints and limits. This gives the sensation of "presence" or "epiphany."[17] Ball play is not merely a form of cultural and institutional entrainment and ideological mediation. Rather, ball play sets out frameworks and infrastructures—models of world-making—for extraordinary acts and feats, glitches and accidents, innovations and protests. What emerges is a collective cultural history created by all who play, watch, judge, organize, support, broadcast, and bet on the games.

On Fixed- and Unfixed-ness

The first chapter of the English edition of Álvaro Enrigue's novel *Sudden Death*, translated by Natasha Wimmer, concludes: "When speaking of someone's death in Mexico we say he 'hung up his tennis shoes,' that he 'went out tennis shoes first.' We are who we are, unfixable, fucked. We wear tennis shoes. We fly from good to evil, from happiness to responsibility, from jealousy to sex. Souls batted back and forth across the court. This is the serve."[18] *Sudden Death* braids together a set of storylines: one about Italian Baroque painter Caravaggio and Spanish poet Francisco de Quevedo facing off against each other in a court tennis match, using a ball stuffed with Anne Boleyn's hair; another about Malinalli (La Malinche) knitting a scapular made from Cuauhtemoc's hair and spinning tales about her daughters wending their ways through the fighting and fucking of conquest; and, another about the author himself, wrestling these histories into stories. The English edition opens with a description of a six-page letter that Enrigue's friend's father wrote to him after reading the original Spanish edition, detailing the many rules and descriptions of tennis that Enrigue had described in the book that were physical impossibilities.

I first read *Sudden Death* while in Sian Ka'an, a small biopreserve just south of Tulum, Mexico, in May 2016, a month after defending the dissertation that was the grounds for this book. I had traveled there after attending a new faculty orientation at Scripps College in Claremont, California, where I would begin teaching in the fall. As I read Enrigue's novel, over the course of that week, I had a sense of the universe winking at me. I walked the beach, accompanied by two dogs, collecting sun-beaten and waterlogged remnants of plastic balls scattered on the sand along with other plastic trash from North America that washes up on Mexico's beaches. I drove to see the Mesoamerican ballcourts at Chichén Itzá and Coba. To say, "He hung up his tennis shoes" (*colgar los tenis*) is to say that they are out of the game. It is also not to say, "He hung up his *faja*." This is a Mexico where tennis, not *ulama*, looms large enough in the cultural imaginary that it can be used to say what needs to be said about someone's death.

During that initial read, I kept catching myself thinking: "Oh, this was what I was trying to say." Not that I would or could have written *Sudden Death*, but in Enrigue's braid of past and present stories, I recognized the kind of complexity that I had been chasing. Everyday intimacies and antagonisms participate in the sweeping scale of history. Returning to it now, I enjoy the difference between his articulation and mine—and the thoughts that that distinction sparks. "We are unfixable . . ." One might read this—I initially read this line—as "We are broken." We cannot be repaired or made right. We cannot be fixed. Yes. And . . .

Language is unfixable, in every sense of the word. One can read, hear, and understand what it means to be unfixable in many ways. We are unfixable—yet changeable. We cannot be contained or held static or constant. Of course, we want to fix and be fixed. Recalling the smell of the fix bath in the black-and-white photo darkroom, the back of my nose and throat tingle. Photography upended all prior modes of representation, in part, because it appeared to fix a present moment—to pin down an instant, like some shiny beetle taxidermized for the future. But we are unfixable. Writing will not fix—neither repair nor hold constant—anything for very long. Nevertheless, we work to find words to say some true things, some things that will ring right and land true. A "rubber meets the road" kind of true—a true that lasts for some shorter or longer instant. This is what wrangling with the word "history" for over a decade—attempting to write sentences that I can stand behind—has left me understanding. Bourdieu writes, "The habitus, the product of history, produces individual and collective practices, and hence history, in accordance with the schemes engendered by history."[19] I take his elaborate and circular point to be a precise way of saying that the individual and collective practices of particular historical moments structure perception in a way that, in turn, produces history. A shift from one historical moment to the next is, in part, a reconfiguration of both practices and structures of perception.

For Sigmund Freud, it was "first and foremost, the body ego." Jacques Lacan's mirror stage figures the baby's recognition of themselves in the mirror (supported physically by a caretaker) as a bounded—or at least in some way fixed—being. This stages a drastic, irrecoverable moment of alienation. That is me, all neat and bounded! But all the sensations that I am experiencing, even as I am looking at this neat and bounded image, are not neat and bounded! That is me, and this is me! So this is not me? Or that is not me? What is to be made of the gap?

As Leslie Dick explained it to our class at California Institute of the Arts (CalArts): Think about watching the other students walking by, and as you see them, they are all obviously walking purposefully to their studios to make great art, while you are a flailing, floppy, dripping mess, with no idea what you are up to. And yet, in the initial moment, as Lacan describes it, "what demonstrates the phenomenon of recognition, implying subjectivity, are the signs of triumphant jubilation and the playful self-discovery that characterize the child's encounter with his mirror image."[20] This early scene of a baby's experience of the bizarre distances and differences between their bounded mirror baby self and their felt, embodied baby being is one that finds in the distance and difference an opportunity for triumphant, jubilant play. This is not yet the deep play that Geertz describes, where economic and emotional stakes create a mechanism for society to catch sight of itself. But mirrors (and all forms of mirroring) are another kind of deep play—a play that allows for catching sight of—even for just a momentary glimpse of—an "I" that continually surfaces and recedes through repeated bouncings and boundings between being, body, and boundary.

Norwegian-born, American Quaker sociologist and peace studies scholar Elise Boulding talked about a two-hundred-year present—one imagined by going back to the birth of the oldest person who touched you when you were a baby and then traversing forward to the death of the youngest person who will touch you before you die.[21] I first encountered the idea when listening to John Paul Lederach, interviewed by Krista Tippet for the *On Being* podcast. In his relaying of this idea, I liked that he did not presume or name the relations as necessarily familial, generating a more general image of responsibility—running across generations and enacted through small moments of contact.[22] This extended present situates the current moment, and those of us in it, in the middle of a long now instead of some never-ending frontier. This feels like one way out of pioneering, settler-colonial relations to time. There is a striking similarity between Boulding's idea and the Seventh Generation Principle, a principle common to many Indigenous people, and laid out in The Great Law of the Haudenosaunee Confederacy: "In our every deliberation, we must consider the impact of our decisions on the next seven generations."[23]

This principle guides decision-making about resources and relationships by foregrounding a vision of how decisions—made in given moments—ripple outward through the connective tissue of time. The language that explains the principle appeals to touch, although in a less literal way than Lederach's example: "The thickness of your skin shall be seven spans—which is to say that you shall be proof against anger, offensive actions and criticism."[24] Here, generations are figured as accumulated layers that create resilience against negative impact. When I finally sought out Boulding's own wording, it felt insistently numerical, conjuring up the image of a slide ruler. In her words, "The 200-year present began 100 years ago, with the year of birth of the people who reach their hundredth birthday today. The other boundary of the 200-year present, 100 years from now, is the 100th birthday of the babies born today. If you take that span, you and I have a lot of contact with a lot of people from different parts of that span."[25] I prefer other images—a gossamer web of connections. I am wondering: If the now is always an extended present, then what is the relationship between this long-now and the instance, the insistence, of this moment?

Throughout this book, as a way to bring forward a few of the many histories that accumulate into particular instants, I have explored instances of ball play. I wanted to surface things that pull at the edges of these instances and unsettle their smoothness. Supported by the arguments that it is possible to bounce off and through them, the bounded spectacle of ballgames show how histories can be pulled together in an instant. Accordingly, instances and moments are markers—milestones—even as they are always unfixed.

Bounce moves between object and surface, figure and ground—somehow a name for the moment of impact and the interval between impacts. I can still—just barely—feel it in my body. When I was at the top of my game, I would see my opponent hit a drop shot and respond just like that—arriving at the front of the court with time to spare, making space with my body, shaping my racket as the ball bounced on the floorboards, and tracking it as it ascended on its upward arc. For a moment, it would feel as if time were suspended—the ball hanging in the air, right at the top of the bounce as I waited to make contact—before I sent it in a new direction.

Figure C.1
Ball Ellipsis (Beached Balls, Si'an Kaan, Mexico), 2016. Photograph by the author.

Notes

Introduction

1. "David Hammons: Higher Goals"; and McGill, "ART PEOPLE: Hammons' Visual Music."

2. McGill, "ART PEOPLE."

3. "David Hammons: Higher Goals." The first iteration of *Higher Goals* was installed in a vacant lot at 121st Street and Frederick Douglas Boulevard in 1983. The Public Art Fund commission funded the more elaborate version of the piece that I am discussing. See also Lakin, "When Dawoud Bey Met David Hammons."

4. Allan's recommendation and his mentorship and friendship were gifts that I am grateful to have had. This book would not be what it is without him. I think of him often and miss him dearly.

5. James, *Beyond a Boundary*.

6. I have both apologized and thanked them many times since for spending weekend after weekend ferrying me to and from matches, sitting on hard benches watching me play, and sleeping in hotels that all too often had overactive fire alarm systems. They always say that they were happy to do it. It seems proper to thank them again here, along with my brother, Matthew Wing, who was brought along for the ride.

7. The sport's acceptance into the 2028 Summer Games in Los Angeles marks a long-awaited entrance onto the Olympic stage.

8. Guinness and Kent, *Contemporary Absurdities, Existential Crises, and Visual Art*, 1.

9. Warren, "Virtual Reality Will Save Games from Itself."

10. Other spectacles that make these kinds of lists include the *Apollo* moon landing; royal weddings, including those of Prince Charles and Lady Diana Spencer, Prince William and Catherine Middleton, and Prince Harry and Meghan Markle; Princess Diana's death; Muhammad Ali's funeral; the 9/11 attacks on the United States; and the Gulf War. "List of Most Watched Television Broadcasts."

11. Sportico. "Number of Broadcasts of the 100 Most Watched TV Broadcasts in the United States from 2020 to 2024, by Type of Content." Chart. January 3, 2025. *Statista*. Accessed May 23, 2025. https://www-statistsa-com.ccl.idm.oclc.org/statista/1445788 /broadcast-number-us-content-type

12. Hayes, "Computing Science: The Way the Ball Bounces," 331.

13. *Oxford English Dictionary*, s.v. "bounce (*v.*)," December 2024, https://doi.org/10 .1093/OED/8400796119.

14. *Oxford English Dictionary*, "bounce."

15. See Sylla, "Jacob Bernoulli and the Mathematics of Tennis," Beretta and Tosi, "Tennis and the Scientific Revolution." and other articles in *Nuncius*, volume 28, issue 1, which was devoted to the topic of "Tennis and the Scientific Revolution."

16. Burnett, "On the Ball," 68.

17. Hayes, "Computing Science," 335.

18. Hayes, "Computing Science," 331.

19. Gillmeister, *Tennis: A Cultural History*, 1.

20. When Gillmeister claims this as the first textual reference to tennis, he stretches the history of the game back to a moment far before it was called "tennis," and before it acquired its contemporary architecture and equipment. He does this by characterizing the ball as a special object and the gesture of hitting a ball back and forth as a common thread. I will address this history of the game of ball in chapter 1.

21. Freud, *Beyond the Pleasure Principle*, 11–16.

22. Freud, *Beyond the Pleasure Principle*, 14.

23. Winnicott relied on the work of Anna Freud and Melanie Klein to develop this object-based theory of communicative play.

24. Winnicott, Playing and Reality, 64.

25. Winnicott, *Playing and Reality*, 3, 73.

26. Peters, *Speaking into the Air*, 46.

27. Serres, *The Parasite*, 225.

28. Serres, *The Parasite*, 227.

29. Serres, *The Parasite*, 225.

30. Serres, *The Parasite*, 228. The passage continues: "Being is abolished for the relation. Collective ecstasy is the abandon of the 'I's' on the tissue of relations. This moment is an extremely dangerous one. Everyone is on the edge of his or her inexistence. But the 'I' as such is not suppressed. It still circulates, in and by

the quasi-object. This thing can be forgotten. It is on the ground, and the one who picks it up and keeps it becomes the only subject, the master, the despot, the god."

31. Serres, *The Parasite*, 228.

32. Gabrielle Hecht's definition of nuclearity inspired this formulation. See Hecht, *Being Nuclear: Africans and the Global Uranium Trade*.

33. In making the verb instead of the noun the departure point, I am in agreement with Karen Barad, who argues that the primary object referent or ontological unit should be phenomena rather than the object-world. Barad, *Meeting the Universe Halfway*, 139. However, nouns in turn produce new verbs, of course.

34. This is the recurring lyric in the Knife's song "Heartbeats," covered by Jose Gonzalez and licensed for the Sony BRAVIA ad.

35. Appadurai, "Mediants, Materiality, Normativity."

36. Peters, *The Marvelous Clouds: Toward a Philosophy of Elemental Media*, 27.

37. Peters, *The Marvelous Clouds*, 27.

38. Peters, *The Marvelous Clouds*, 17.

39. Young, "The KULTUR of Cultural Techniques: Conceptual Inertia and the Parasitic Materialities of Ontologization," 385.

40. Star and Griesemer, "Institutional Ecology, 'Translations' and Boundary Objects," Bowker and Star, *Sorting Things Out*.

41. See Shapin and Schaffer, *Leviathan*. Importantly, they turn to Ludwig Wittgenstein's idea of language games and thus deploy the metaphor of games for the purpose of making their argument throughout.

42. Gitelman, *Always Already New: Media, History, and the Data of Culture*.

43. Jhally, "The Spectacle of Accumulation."

44. See Taylor's *Raising the Stakes* and *Watch Me Play* (2018); Consalvo, Mitgutsch, and Stein's edited collection *Sports Videogames*; and Guins, Lowood, and Wing, *EA Sports FIFA: Feeling the Game*.

45. Scholars in sports studies who are thinking about the digitization of sports include Brookey and Oates, *Playing to Win*; and Miah, *Sports 2.0*.

46. McLuhan, *Understanding Media*, 237.

47. McLuhan, *Understanding Media*, 235, 237, 240.

48. Benjamin, "Work of Art," 22. Spectatorship has perhaps been best theorized by the traditions of art history and cinema studies. If we go along with Benjamin's

comparison of sport and cinema here, we could ask why cinema studies have developed such a strong tradition within academia, while sports studies have been, for the most part, sidelined.

49. Csíkszentmihályi, *Flow*, 4. Appadurai has indicated to me that Csíkszentmihályi's work on this was deeply influenced by John MacAloon, who was his graduate student at the time.

50. Recall Winnicott's definition of magic here. I understand magic as the hyperuse of a technique intended to sustain the precarious interplay between psychic reality and the experience of control of objects.

51. Perinbanayagam, *Games and Sport*, 111.

52. One look across the headlines of the daily tabloids during the infamous 2015 "Deflategate" scandal makes this point all too clearly. New England Patriots quarterback Tom Brady is "BALL BUSTED!" "NFL probe has Brady BY THE BALLS," says another *New York Post* headline (in the *Daily News*, it is Brady's coach, Bill Belichick, whom the NFL has "BY THE BALLS!" Brady's wife, the supermodel Gisele Bündchen, is called a "BALL BUSTER" for saying, "My husband cannot f***ing throw the ball and catch the ball at the same time," and ventriloquized via a speech bubble asking, "Tommy, why did your balls go soft?" *New York Post*, May 12, 2015; *Daily News*, January 21, 2015, and *New York Post*, February 7, 2012; *New York Post*, January 2, 2015, respectively. In contrast to the case of a hard or soft penis, the hardness or softness of a man's balls is not a reference to sexual prowess, but instead to the capacity to endure extreme vulnerability and physical pain.

53. Whether knowingly or not, the documentary's title recalls Tennyson's 1959 article, "They Taught the World to Play." The documentary is based on a book titled *The Ball: The Object of the Game* by John Fox, which I reviewed for the *American Journal of Play* (2014, 6:2). The book and the film are different in a number of ways, but both naturalize the problem of gender and miss an opportunity to talk explicitly about masculinity.

54. Early versions of the following paragraphs were published in Wing, "Hitting Walls (v.XXVIII): Captured Play."

55. Doyle, "Dirt off Her Shoulders," 419, 421.

56. Doyle, "Dirt off Her Shoulders," 421. This indicator of how deeply modern sport is invested in the measuring of capacity is another reason that games of the "measured self" deserve such close attention.

57. Doyle, "Dirt off Her Shoulders," 426.

58. Barthes, "The Grain of the Voice," 276.

Chapter 1

1. I will use "court tennis," along with "the game of ball," to refer to the earlier enclosed game, and "modern tennis" to refer to the game that is common today.

2. The experience of Grandma Selma not coming to watch me play lodged itself in my twelve-year old self as part of my slow coming into an as-yet unarticulated but already intuited understanding of whiteness. Growing up as the daughter of a Jewish mother and white atheist Episcopalian father in New York City, I became accustomed to being read as white, by which I mean, among other things, that I move through public spaces, interact with the police, and enter spaces such as private clubs and campuses with the ease of whiteness. I felt the echoes of my time in those spaces as I entered the halls of R&T Club.

3. Serres, "Theory of the Quasi-Object."

4. I take the name "game of ball" from P. A. Negretti's translation of Scaino's original title *Trattato del Givoco Della Palla*.

5. Stuchtey, "Colonialism and Imperialism, 1450–1950," 1.

6. Beretta and Tosi, "Tennis and the Scientific Revolution," 1. Beretta and Tosi cite the source of this claim as an ambassador from the Doge's court in Venice.

7. Beretta and Tosi, "Tennis and the Scientific Revolution."

8. Turkle, *Evocative Objects*. 305.

9. For an account of the end of the tournament era, see Huizinga, *Autumn of the Middle Ages*.

10. For a discussion of the sounds of this time period, see Smith, "Soundscapes."

11. Greenblatt, *The Swerve*.

12. I use the term "get a handle on" here to refer to a method of bringing knowledge into the body by way of practices that depend on eye-hand coordination.

13. For more on the role of measurement in court tennis, see Perinbanayagam, "The Measured Self."

14. Latour, *Pasteurization of France*, 87.

15. Edwards, *The Closed World: Computers and the Politics of Discourse in Cold War America*, 12–13.

16. There is a repeated assertion in the literature that the word "tennis" comes from the French word *tenez*, meaning "to hold," and this was called out by the server to mark the start of play, but this has never been verified by the leading historians in the field. Both Gillmeister and Morgan consider the origin of the English word "tennis" an

open question. The German names for the game, *kateschen* and *kaatsen*, came from the Flemish game *cache* (Gillmeister, *Tennis: A Cultural History*, 103). In Italy, the enclosed version of the game was called *gioco della palla corda* (game of the ball and cord). It was one of a group of games that included *pallone, scanno, palla da mano, palla da rachetta,* and *pallacorda*, which were played with different balls ("solid" versus "wind") and different instruments ("open hand," "clenched fist," and "fist with instrument") but with similar rules.

17. Gillmeister, *Tennis: A Cultural History*, 2.

18. Gillmeister, *Tennis: A Cultural History*, 39, 105. He dates the first documented usage of the phrase to a verse of a medieval play written in 1300, in which the queen of Scotland, adrift on the ocean, laments the way that the sea plays with her, as with a tennis ball, casting her about here and there: "Car souvent la mer par mainte onde / Jouoit de moy comme à la bonde / Et me jettoit puis ça puis là." [For the sea with its innumerable waves often / Played with me as with a tennis ball, / And cast me first here and there.] The analogy of a person in the world to a ball in a game appears repeatedly in early literature. Part of its metaphorical power comes from the fact that a ball is an object that humans themselves can easily handle, hit, and otherwise cast about. Thus, our inability to control how the world affects us is itself bounded through the use of this graspable metaphor.

19. Gillmeister, *Tennis: A Cultural History*, 39, 105.

20. Gillmeister, *Tennis: A Cultural History*, 105.

21. Gillmeister, *Tennis: A Cultural History*, 77, 133, 168.

22. *Oxford English Dictionary,* s.v. "bound (n.1)," June 2024, https://doi.org/10.1093/OED/4754105080.

23. I mean by this something similar to what Karen Barad means when she uses the term "cut" to describe how any act of observation produces its object by drawing a line between what is included and excluded from consideration. See Barad, *Meeting the Universe Halfway*.

24. We know that a version of the chase rule was already in place in *jeu de bonde* as early as 1316, from its appearance in *Roman du Comte d'Anjou*, a story of three burgesses' sons from Orleans meeting on a fine summer day to play tennis: "Le giue de bonde commencierent / L'un fiert l'estuef, l'autre rachace, / Chascun pour faire bonne chace."[They began with "the bounce game": / One hits the ball, the other chases it back, / Each in order to make a good chase.] Gillmeister, *Tennis: A Cultural History*, 39.

25. Morgan, *Tennis*, 100.

26. The French and the English chase lines differ slightly because they employ different units of measurement.

27. In fact, it is more complicated than this because when the ball bounces between chase lines, the chase that is set can be half of the area between the lines; thus, the size of the chase area itself varies.

28. *Oxford English Dictionary*, s.v. "ricochet (n.)," December 2024, https://doi.org/10.1093/OED/2792414713.

29. *Oxford English Dictionary*, s.v. "ricochet (v.)." July 2023, https://doi.org/10.1093/OED/9003803765.

30. William Shakespeare, *Henry V*, Act 1, Scene 2, lines 270–277.

31. Gillmeister, *Tennis: A Cultural History*, 35.

32. Scaino, *Treatise*, 25.

33. Scaino, *Treatise*, 11.

34. Scaino, *Treatise*, 60. As "percussion of the air" implies, Scaino is a plenist—he believes that the air is made up of small materials and argues against the existence of a vacuum in nature.

35. Beretta, "Training Tennis Players," 26. We shall see this analogy of sound as a projectile reappear in the twentieth century with radar and computing.

36. Beretta, "Training Tennis Players," 32. Beretta cites Brunet as evidence against the rarity of the *Treatise*.

37. Beretta and Tosi, "Tennis and the Scientific Revolution," 2.

38. Iatromechanics is the medical application of physics. Beretta and Tosi, "Tennis and the Scientific Revolution," 4.

39. Scaino, *Treatise*, 117 (emphasis added).

40. Scaino, *Treatise*, 3.

41. For a discussion of the conceptions of enclosure that laid the foundation for the moment that I am discussing, see Howie, *Claustrophilia*.

42. Lake, "Real Tennis."

43. Perinbanayagam, "The Measured Self,"123.

44. Lake, "Real Tennis," 559.

45. Scaino, *Treatise*, 3.

46. The limited seating and standing that were available for spectators in the galleries that ran under the penthouses put the viewer directly in the path of the ball, which is presumably why these windows were eventually covered with netting.) Some courts also had walkways around the very top, allowing viewing from above.

47. Scaino, *Treatise*, 117.

48. Lake, "Real Tennis," 564.

49. Lake, "Real Tennis," 25 (emphasis added).

50. Geertz, *The Interpretation of Cultures*, 432 (emphasis added).

51. Geertz, *The Interpretation of Cultures*, 433.

52. Geertz, *The Interpretation of Cultures*, 434.

53. Scaino, *Treatise*, 38.

54. Scaino, *Treatise*, 17.

55. Scaino, *Treatise*, 38.

56. In this game, you can bet as a player—as part of the players' coalition—or you can bet on a particular player or a particular rally. This is a side bet.

57. Several types of point advantages were available. Players could start with some number of points added or subtracted from their score, or the weaker player could be granted *bisques*—points that could be taken at any time during the game. A player was allowed to take a *bisque* only during a chase, not during a normal exchange (called a "rest"). Players could also agree in advance that a certain number of faults would be forgiven for one player. Another less common type of handicap, called "cramped odds," involved banning the use of certain strokes or types of serves for one player or both players.

58. The current classification system used to ensure fair competition at the Paralympics offers a good example of this practice today. Play theorists now often call for us to play with limits, to play within the rules, because they see a capacity to do so as a demonstration of true agency or power or freedom. Court tennis includes a built-in system for playing within and modifying its own rules and limits.

59. For the player on the service side, this was the winning gallery and the grille; for the player on the hazard end, it was the dedans.

60. The 40 was originally 45, presumably shortened once the aura of breaking the unit of 60 into quarters had faded. Gillmeister and Morgan both discuss the fuzzy origins of the scoring at length. They speculate that it had something to do with the importance of the number 60 in astronomy and watches and clocks. Gillmeister, *Cultural History;* and Morgan, *Tennis*.

61. There are sports such as high jump or javelin, where contestants are given multiple tries, but every contestant gets the same number of tries and the top score for each person is then recorded.

62. In enclosed courts, they aim for the "nick," where the wall meets the floor, knowing that the possible ways that the balls can react are far harder to predict and

the nick offers a good set of possibilities for holding onto the ball and earning them the point. This kind of chance is a product of interaction, and there is more of it in enclosed games of to-and-fro than in races or timed events.

63. Scaino, *Treatise*, 30.

64. Shapin and Schaffer, *Leviathan*, 225.

65. Shapin and Schaffer, *Leviathan*, 32 and 38.

66. Shapin and Schaffer, *Leviathan*, 56. Boyle makes his case by citing, among other things, the fact that one witness at a trial is not considered proof, while two credible witnesses are enough to count a person guilty.

67. MacIntosh, Anstey, and Jones, "Robert Boyle," *Stanford Encyclopedia of Philosophy*, https://plato.stanford.edu/entries/boyle/#BoylLaw.

68. "On 12/22 January 1671 Sir Robert Moray reported to the Royal Society that 'the King has laid a wager of fifty pounds to five for the compressions of air by water; and that it was acknowledge that his majesty had won the wager'." Birch, *History*, vol. II, p. 463; Shapin and Schaffer, *Levithan*, 134.

69. See Shapin and Schaffer, *Leviathan*, 3, 15, 46, 72, 110, 174, 192, 298, 310, 336, 339.

70. Shapin and Schaffer, *Leviathan*, 109.

71. Shapin and Schaffer, *Leviathan*, 174.

72. There are sports today where place matters and the difference between play spaces is not just an unfortunate inevitability. These sports directly engage with natural or built environments: golf, skateboarding, parkour, surfing, and skiing, for instance. It is easy to imagine the early street contests of this game of ball and even the codified and partially standardized cord game as something like a kind of parkour for tennis.

73. Latour, *We Have Never Been Modern*, 24. He paraphrases Shapin and Schaffer: "If science is based not on ideas but on a practice, if it is located not outside but inside the transparent chamber of the air pump, and if it takes place within the private space of the experimental community, then how does it reach 'everywhere'? How does it become as universal as 'Boyle's laws' or 'Newton's laws'? The answer is that it never becomes universal—not at least in the epistemologists' terms! Its network is extended and stabilized."

74. Latour, *Pasteurization of France*, 87.

75. "Robert Boyle," *Stanford Encyclopedia of Philosophy*.

76. Even today, no two courts are alike. The rules of the International Real Tennis Professional Association affirm the prerogative of the home court in Rule 1.2: "From time to time the Association, host club or the organisers of an event may lay down

special rules relating to a Court, a match or to a tournament. In such an event, the Laws of Tennis shall prevail except where they are inconsistent with those rules."

77. Home court rules also took priority over the published code, as they still do.

78. The consequences of pursuing standard enough environments will become apparent in the discussion of modern tennis in chapter 2.

79. Hacking, *Emergence of Probability*, 72. For further work on the history of probability during this period, see Daston, *Classical Probability*, and Hald, *History of Probability*.

80. Huygens published *On Reckoning in Games of Chance* in 1657.

81. Sylla, "Jacob Bernoulli," 154. Sylla writes: "At least three factors were involved in Bernoulli's development of a mathematics for calculating the parts in tennis. First, he needed to have a definition of the concept of *partie* or part, share, lot, or expectation. Second, he had to have a notion of a player's strength or agility, which was not a single entity or property existing in the outside world, but was rather a concept that captured multiple significant causes(s) of the effects observed. Third—in what Bernoulli came to believe was his most significant accomplishment—he eventually had to demonstrate his fundamental theorem that if sufficiently many observations are made, one can increase as far as one wishes the probability that one has discovered, at least approximately, the underlying ratios of possible cases or causal factors." Sylla, "Jacob Bernoulli," 154–155.

82. Bernoulli, *Art of Conjecturing*, 363.

83. Bernoulli, *Art of Conjecturing*, 363.

84. Bernoulli, *Art of Conjecturing*, 98.

85. Bernoulli, *Art of Conjecturing*, 363–364.

86. Daston, *Classical Probability*, 21.

87. Sylla, "Jacob Bernoulli," 146. Sylla also writes that "according to both Huygens and Bernoulli, calculations in games of chance should be founded not on relative frequencies, but on equity: what people pay to play a game of chance and what they receive if the game is broken off before the planned end should be proportional to their expectations at those points" (Sylla, "Jacob Bernoulli," 146).

Chapter 2

1. I find these discarded, bleached-out, chewed-up objects to be completely beautiful, blindingly charismatic. I imagine that this is not true for most people. And so I wonder at what has shaped my sight that I experience them this way. In some ways, it is quite simple—an irrational attraction and attachment that still feel unarticulated, even after all these words.

2. See chapter 1. Lawn tennis came to dominate its parent sport to such an extent that it became known simply as "tennis," while the older version was forced to take on various adjectives to distinguish it from the upstart game—*real* tennis, *court* tennis, and *royal* tennis, among others. Enthusiasts today still insist on calling the older game "tennis" and refer to the modern sport as "lawn tennis."

3. Litman (2014).

4. Today, tennis ball cores are made by joining two dumbbell-shaped pieces of vulcanized rubber treated with chemicals that produce a gas resulting in approximately twelve pounds more pressure per square inch inside the core of the sphere once the pieces are joined than the pressure per square inch of the external ambient air.

5. Lehrer, "The Physics of Grass, Clay, and Cement."

6. "Made for TV" is a 1960s advertising slogan describing films made for distribution on television networks rather than to be shown in cinemas; it was aimed at encouraging people to stay home and watch more television.

7. Patent No 685, A.D. 1874, reprinted in Alexander, *Wingfield, Edwardian Gentleman*, 199.

8. Alexander, *Wingfield, Edwardian Gentleman*, 91.

9. Alexander, *Wingfield, Edwardian Gentleman*, 201.

10. Wingfield, *The Game of Sphairistike; or Lawn Tennis*, 1876. By Royal Letters Patent, French & Co, London SW, 18, reprinted in Alexander, *Wingfield, Edwardian Gentleman*, 234.

11. Gillmeister, *Tennis: A Cultural History*, 177.

12. Gillmeister, *Tennis: A Cultural History*, 177.

13. This is Arjun Appadurai's term for a form that resists structural change through "a set of links between value, meaning, and embodied practice," and Appadurai identifies sports like cricket as examples of this, in that it "changes those who are socialized into it more readily than it itself is changed." *Modernity at Large*, 90.

14. Loadman, *Tears of the Tree: The Story of Rubber—a Modern Marvel*, 26.

15. A game that is being played after the final result has already been determined is known as a "dead rubber."

16. *Oxford English Dictionary*, s.v. "rubber (n.1)," December 2024, https://doi.org/10.1093/OED/9023324788.

17. As I describe in "True Bounce: Stories of Dunlop and Vulcanized Play," the concept of sport, as used here, is tied to the concept of modernity. The English word "sport" was derived from the French word *desport*, meaning both entertainment and comportment.

These two meanings point toward the way that forms of play and sport shape people as embodied subjects. "Sport" took on a new meaning of physical competition in the middle of the nineteenth century with the rise of organized sports such as football, rugby, cricket, and athletics. In the process, the concept became associated on the one hand with a set of ideals around athleticism, masculinity, and imperialism that were fostered in the British public schools; on the other hand, it became the name for newly professional kinds of competitions: spectacles that sold seats and merchandise to an emerging middle class equipped with new kinds of leisure. Modern sport is a form of institutionalized play, and as such it contains the shape and form and material conditions of society. For more, see Mangan (1981) and Mangan and McKenzie (2010).

18. I call this an era of vulcanized *play* rather than vulcanized *sport* because it is not just our institutionalized forms of play that rest on rubber.

19. *Patents for Inventions*.

20. For a discussion of the role of visuality in the plantation complex, see Mirzoeff, *Right to Look*. Mirzoeff locates the key moment for the plantation complex just before rubber becomes a plantation crop, but he also notes that even as that form is superseded by the imperial complex as the dominant authority relation, it of course persists.

21. For example, it was not until 1910 that the Penang Sugar Estates, which Sidney Mintz writes about in *On Sweetness and Power*, officially changed its name to the Penang Rubber Estates. Tully, *The Devil's Milk*, 189.

22. Loadman, *Tears of the Tree*, 151.

23. At these sites, Sigmund Freud's theory of the fetish seems to emerge as a distorted echo of itself, with the fear of castration and amputation being all too justified. Casement (1994) details cases of Belgian soldiers cutting off and smoking hands and penises to send with the shipment of rubber as evidentiary explanation for why the load was light. Casement, an Irishman by birth, was later hanged for treason. To create support for the hanging, the British leaked parts of his journal to the press, which detailed his homosexual fantasies about and relations with native men. Casement is a figure in which Marx's and Freud's respective forms of fetishism can be thought together very productively, as has been done to a certain extent by Michael Taussig. Speaking about Casement, Taussig writes, "And if Casement slept with his fetishes 'colored like the very tree-trunks they flitted among like spirits of the woods'" (Taussig, *Shamanism, Colonialism, and the Wild Man*, 128–130). Others who played a significant role in witnessing, documenting, and writing about the massacres in Congo include Edmund Morel, Joseph Conrad, and George Washington Williams (Hochschild, *King Leopold's Ghost*). In *King Leopold's Ghost*, Hochschild describes how in Congo, when people failed to meet rubber quotas, Belgian "Force Publique soldiers or rubber company "sentries" often killed everyone they could find." See Hochschild, *King Leopold's Ghost;* and also Loadman, *Tears of the Tree*, 128.

24. Dunlop is commonly identified as the inventor of the pneumatic tire because it was his invention that took off, and the company bearing his name went on to become one of the largest British multinationals. However, he was not even the first Scottish man to come up with the idea. Robert William Thompson patented a design for a pneumatic tire in 1845, but in that moment, without the nascent automobile industry to drive demand, no momentum accrued around the invention. The first US car race was held in Chicago just six years after Dunlop's first pneumatic tire experiments. In 1908, the first Model T Ford rolled off the assembly line and demand for tires skyrocketed over the next decades.

25. Dunlop, *The History of the Pneumatic Tyre*, 13 and 15. See also Cooke, *John Boyd Dunlop*, 15.

26. The name "Dunlop" stayed constant through multiple other changes for just under a century. Today, the name persists for former subsidiaries that are now independent companies but continue to use the name for brand recognition.

27. The company saw its role in sport as bringing "the advantages of scientific design" to existing games. The first Dunlop sports product were golf balls, with the "Orange Spot" golf ball debuting in 1910, followed in 1912 by the Dunlop "V," which was "the outcome of a close study of ballistics and flight." Dunlop began producing tennis balls in 1922 and created the Dunlop Sports subsidiary in 1928. See Dunlop Rubber Company, *The Story of Dunlop through the Reigns*, 12.

28. I have argued this point at greater length elsewhere. This chapter incorporates pieces of this argument. See Wing, "True Bounce: Stories of Dunlop and the Rise of Vulcanized Play."

29. McMillan, *The Dunlop Story: The Life, Death, and Re-Birth of a Multi-National*, 37.

30. Dunlop Rubber Company, *Story of Dunlop*, 10.

31. "Police Massacre of Striking Dunlop Rubber Plantation Workers in Malaya."

32. As I argue in "True Bounce,": Dunlop Sports and the Dunlop Sport brand were eventually sold off after the acquisition of the company by British Tyre and Rubber (BTR) in 1985. The remaining entity was renamed Invensys as part of a transformation into a conglomerate of companies that focus on software, industrial automation, energy controls, and appliances, a brand that in turn was acquired by Schnieder Electric. But the brand name Dunlop, which was first printed onto dirt roads in 1891, and the Flying D logo, which was launched in 1960, persist. Today, they are used by companies around the world that sell Dunlop tires, conveyor belts, tennis and golf balls, tennis rackets, mattresses, and footwear. In acquiring the right to use the Dunlop brand, these disparate companies acquired a common lineage while having no other necessary relation to each other. Their wielding of the brand can in turn render the lineages of the current

companies (Goodyear, Sumitomo Rubber, the Ruia Group, Pacific Brands, Sports Direct) harder to see. One can think of this game of hide-and-seek as another kind of vulcanized play.

33. Mangan, *The Games Ethic and Imperialism: Aspects of the Diffusion of an Ideal*, 18.

34. Mangan, *The Games Ethic and Imperialism.*18.

35. Tennyson, "They Taught the World," 211. Also see the many works of J. A. Mangan and his students. Mangan's works include *The Cultural Bond, Athleticism,* and *The Games Ethic.*

36. Gillmeister, *Tennis: A Cultural History,*186.

37. See du Cros, *Wheels of Fortune: A Salute to Pioneers; On the First Manufacturers of Pneumatic Tyres,* 220–221.

38. Miles, *Racquets, Tennis, and Squash,* 269.

39. "No Cause for Worry: European War Will Have No Effect on American Sporting Implements," *The New York Times*, August 21, 1914.

40. "Lawn Tennis," *The Times* (London), July 25, 1913, 15.

41. "Lawn Tennis," *The Times* (London).

42. Dunlop Rubber Company, *The Story of Dunlop through the Reigns,* 58–59.

43. The USTA uses the ITF specifications for all its testing. These specifications used by the ITF and USTA are actually broader than the ones manufacturers set for themselves. Interview with Suresh Ponnusamy, October 17, 2014.

44. The lab was set up to test three types of balls: the Type 2 balls used in competition, as well as the faster Type 1 and slower Type 3. All are made with what is called a "natural rubber core." Most manufacturers have several production lines, one for producing these three types of top-quality balls—balls with true bounce that will be used in official competitions and sold in pro shops—and one or more additional production lines for producing more imperfect objects.

45. Wilson Sporting Goods is an American company originally started by a meatpacking company in 1913 to produce goods such as tennis racket strings from animal by-products. It was named after its new president, Thomas E. Wilson, in 1915. It was purchased by the Finnish multinational Amer Sporting Company in 1989 and in turn was purchased by the Chinese company Anta Sports in 2019.

46. Interview with Suresh Ponnusamy, October 17, 2014.

47. See Whannel, "Television and the Transformation of Sport"; Rowe, *Sport, Culture and the Media: The Unruly Trinity*; Miah, *Sports 2.0.*

48. "Centre Court," *The Listener*. (With thanks to Hannah Zeavin, who pointed me to television scholar Fanny Cradock's tweet of this gem.)

49. See "Centre Court," *The Listener*. The piece goes on to discuss how television forces a change to commentary from play by play, which is always behind the action, to a quieter commentator, who reminded the viewer only from time to time which player was on which side of the net and what the score was.

50. For a fantastic collection of pieces on the status of sport in the early twenty-first century see Kashmere and Suparak, *INCITE: Journal of Experimental Media*.

51. Flusser, *Into the Universe of Technical Images*, 55.

52. Murray, *Bright Signals: A History of Color Television*, 126.

53. Attenborough, "David Attenborough Reveals How He Brought Colour Television to Wimbledon."

54. Murray, *Bright Signals: A History of Color Television*, 116.

55. Murray, *Bright Signals: A History of Color Television*, 120.

56. Attenborough, "David Attenborough Reveals How He Brought Colour Television to Wimbledon."

57. Attenborough, "David Attenborough Reveals How He Brought Colour Television to Wimbledon." Meanwhile, Wimbledon, the footage of which had spurred Attenborough's suggestion, continued to use the traditional white ball until 1986.

58. *2004 U.S. Open*, CBS, August 30, 2004–September 12, 2004. In these more recent iterations, the ball also acquired a cometlike tail as it travels with no discernible spin toward its point of impact.

59. In this, it becomes part of the history of instant replay, which starts with action replays, slow motion, and assistive graphics in the 1950s. Hanson, "The Instant Replay: Time and Time Again," 51.

60. Hanson, "The Instant Replay: Time and Time Again."

61. Hanson, "The Instant Replay: Time and Time Again," 52.

62. Collins, Evans, and Higgins, *Bad Call*, 18. They go on to argue that "part of specialist umpiring and refereeing skills comprises "somatic tacit knowledge"—the kind of skills we build up in our bodies like the ability to drive, or type fluently . . . [and that this enables] . . . a match official to appraise a rapidly unfolding situation in an instant without a conscious calculation. But with TV replays this skill becomes redundant as time is effectively slowed." Collins et al., *Bad Call*, 18.

63. Hanson, "The Instant Replay: Time and Time Again," 57. (emphasis in original)

64. Collins et al., *Bad Call*, 7–8.

65. Collins et al., *Bad Call*, 16. False transparency is in contrast to presumptive justice (justice presumed to be done, with no need to see it in person), transparent justice (justice witnessed), and transparent injustice (injustice witnessed).

66. Collins et al., *Bad Call*, 6–7.

67. Berry, *A People's History of Tennis*, 2.

68. Berry, *A People's History of Tennis*, 139. Also see the filmmaker Darius Clark Monroe's series of shorts based on stories from *Racquet* magazine: "South Oxford," in which Brooklyn houses a unique history for Richard and Ann Northern as the site of their former tennis club; "All Iowa Lawn Tennis Club," in which a father's home-made tennis court holds special significance after a family tragedy; and "Serve," a meditation on finding strength and peace in the ritual of the serve.

69. Rankine, *Citizen: An American Lyric*.

70. Serena Williams, "Serena Williams: I'm Going Back to Indian Wells."

71. Serena Williams, "Serena Williams: I'm Going Back to Indian Wells."

72. Sheppard, *Sporting Blackness*.

73. I chose Rankine's *Citizen: An American Lyric* as the text for my Core 1 lecture to Scripps students in 2018 because I wanted them to read good writing that addressed their contemporary moment and immediate context. She wrote much of the book while teaching at Pomona, one of the other Claremont Colleges drawing on her experience as a Black faculty member at a predominantly white institution. *Citizen* presents a host of decisive moments—moments when racism appears suddenly, shockingly, or even when expected—sometimes in intimate interactions among friends, sometimes in public, and sometimes during live broadcast media spectacles. To present these stories, Rankine deploys the "I" and the "you" in a way that asks readers to confront and inhabit their relationship to the history of whiteness and anti-Black racism in the United States, and to the ways that this history shapes available forms of living and dying in the present day.

74. Rankine, *Citizen: An American Lyric*, 25 and 29.

75. Hazlitt, "The Death of John Cavanagh."

76. Rankine, *Citizen*, 30.

77. Rankine, *Citizen*, 25 and 30.

78. Rankine, "The Meaning of Serena Williams."

79. Rankine, "The Meaning of Serena Williams."

80. Reverend angel Kyodo williams's guidelines for the "no big deal" sit that she initiated are: "Come as you are," "Leave as you must," and "Mind your business." While attending that sit online during the COVID-19 pandemic, it became clear to me that discerning what is and is not our business is a crucial part of the practice of individual and collective liberation.

81. Rankine, *Citizen*, 27.

82. Television extended who needed to see the ball. Historically, most ballgames have depended on sightedness, with compelling exceptions, such as goalball, developed in the wake of World War II, which requires players to wear blinders; and Beep Baseball, which uses sound to orient players. There are also audio description services and other assistive technologies that enable blind people to watch sports, a phrase that in this context is shown to have always been figurative rather than merely literal.

Chapter 3

1. Wells, *Animation, Sport and Culture*, 37. The lamp in question was modeled after a Luxo lamp designed by Scandanavian designer Jac Jacobsen, which was sitting on director John Lasseter's desk when he decided to make the film.

2. The director of *Luxo Jr.*, John Lasseter, was a former Disney animator. To Lasseter's mind, 3D computer animation at the time was bad because those working in the field were not familiar with the principles of good animation developed by 2D animators at Disney over the course of the twentieth century. He addresses this using the principles developed at Disney in 3D animation in his article "Principles of Traditional Animation Applied to 3D Animation," 35.

3. Hayes, "Computing Science: The Way the Ball Bounces," 331.

4. Manovich, *The Language of New Media*, 7.

5. What is colloquially called the "spinning beach ball of death" began as a black-and-white quartered circle and eventually became a colorful pinwheel. It has been updated routinely as part of Apple's operating system updates. My thanks to J. D. Connor for directing my attention to the bounce scroll, which relies on an equation for exponential decay.

6. Manovich, *The Language of New Media*, 7.

7. Of course, they are also different from the bounce that we see in computer games and other computing contexts today.

8. Needham, *Science and Civilization in China*, vol. 4, *Physics and Physical Technology*, part 1, *Physics*, 123–124. Some sources claim that Ding Huan invented a device similar to the nineteenth-century zoetrope, but this is disputed, and I cannot find any good verification. In any case, applying the name "zoetrope" to Ding Huan would be a back-formation that obfuscates what this historical thread might have to tell us.

9. Wells, *Animation, Sport and Culture*, 37.

10. Wells, *Animation, Sport and Culture*, 7.

11. Canales, "Desired Machines: Cinema and the World in Its Own Image."

12. There is some dispute about this point, as is so often the case. William Henry Fitton was known to have performed a trick with a coin demonstrating the principle

behind the thaumatrope, and Ayrton Paris may very well have gotten the idea from Fitton.

13. Paris, *Philosophy in Sport*, 20 (emphasis in original).

14. Ayrton Paris presented the thaumatrope to the Royal College of Physicians in 1824 to prove his claim that "the eye has also its sources of fallacy." Paris, *Philosophy in Sport*, 337.

15. Paris, *Philosophy in Sport*, 40.

16. Paris, *Philosophy in Sport*, 46, 47, 94, 106, 140, 150.

17. Paris, *Philosophy in Sport*, 10, note to the reader.

18. Rousseau, quoted in Paris, *Philosophy in Sport*, vii. Here, we can begin to think about the questions of gender representation in the natural sciences that persist today. We need to look at the metaphors that we rely on to understand and communicate about these fields and then ask ourselves if those metaphors are drawn from fields that welcome women and girls. While girls are now encouraged to play sports, I wonder if the metaphors have shifted accordingly to realms that they are still not wholly welcomed into.

19. The book employs a tone and style that make it read like a precursor to the schoolboy novel, which would become wildly popular in England a few decades down the road.

20. Wells, *Animation, Sport and Culture*, 34.

21. Wells, *Animation, Sport and Culture*, 34.

22. Gunning, "Hand and Eye," 495–497.

23. Gunning, "Hand and Eye," 500.

24. With thanks to Tricia Wang for shared conversations as I thought through these histories of the relationship between elasticity and character construction and she thought through ideas of an elastic self as a name for identity in the context of social media.

25. Canales, "Desired Machines,"329.

26. Marey, *Movement*, 1.

27. Cartwright, *Screening the Body*, 36.

28. Cartwright, *Screening the Body*, 36.

29. Wells, *Animation, Sport and Culture*, 36.

30. Marey, *Movement*, 50, 85.

31. Marey, *Movement*, 51.

32. This is also notable because of the presence of a man (possibly Marey himself) and a chronometric dial in the image to mark the method of measurement

33. Marey, *Le Movement*, 51.

34. Cartwright, *Screening the Body*, 36. She quotes Steve Neale on Marey's opposition to the pictorial codes of photographic representation:

> Marey was even more insistent [than Eadweard Muybridge] that research and analysis entailed not only breaking with age-old conventions of representation and perception, but also breaking with the perception of photography itself, the very ideology in which it was caught and which it hitherto had been used to support. . . . In order to trace the shapes and patterns of movement, he did everything possible to break down the sensuous density of the photograph, posing his subjects in black against a carefully constructed black surface, with only white abstract shapes on various portions of the limbs to mark stages of movement, replacing the seamless analog of the conventional photographic image with posed, constructed and specifically arbitrary signs.

Neale, *Cinema and Technology*, quoted in Cartwright, *Screening the Body: Tracing Medicine's Visual Culture*, 34–35.

35. Cartwright, *Screening the Body*, 36.

36. Marey, *Movement*, 161.

37. Marey, *Movement*, 161–162.

38. Another (counter) set of balls appears throughout the book, not recorded but used to initiate the recording. These objects are hollow India rubber balls, squeezed by hand or stepped on by foot or horse hoof to press air through a narrow tube, the pressure of which released a not-too-distant shutter and initiated recording. These balls are at what we would call today the site of interaction, the interface.

39. For an extended discussion of the relationship between Marey and Bergson see Martha Blassnigg, *Time, Memory, Consciousness, and the Cinema Experience*.

40. Bergson, *Laughter: An Essay on the Meaning of the Comic*, 17.

41. "The victim, then, of a practical joke is in a position similar to that of a runner who falls—he is comic for the same reason. The laughable element in both cases consists of a certain mechanical inelasticity, just where one would expect to find the wide-awake adaptability and the living pliableness of a human being." Bergson, *Laughter: An Essay on the Meaning of the Comic*, 14.

42. Bergson, *Laughter*, 38 (emphasis added). The phrase in the French original is "l'illusion de la vie." Bergson, *Le rire: Essai sur la signification du comique*, 35.

43. Bergson, *Laughter*, 33–34.

44. Bergson, cited in Canales, "Desired Machines," 352.

45. Bergson, *Matter and Memory*, 277.

46. Canales, "Desired Machines," 329.

47. Canales, "Desired Machines," 352.

48. Bergson, *Laughter*, 33–34.

49. *Balle rebondissante, étude de la trajectoire* traces a bounce trajectory in a manner similar to a number of images that would later be produced at the Massachusetts Institute of Technology (MIT): oscilloscope displays of bounce programs made by early programmers of the Whirlwind computer at MIT (see chapter 5); photographs made by Harold Edgerton, a professor of electrical engineering at MIT from 1934 until 1977, best known for his work developing techniques for high-speed photography; and photographs by Berenice Abbott, who spent several years at MIT in the late 1950s creating a series of images for a new physics curriculum. One important difference between the images that Edgerton and Abbott produced and Marey's work is the direction of the ball's bounce. The ball in Marey's photograph goes from right to left, while later ball trajectories (unnecessarily) follow the direction of Latin languages and are displayed bouncing from left to right.

50. Feynman, *The Feynman Lectures on Physics*.

51. Crafton, *Before Mickey: The Animated Film 1898–1928*, 7.

52. Working in stacks also allowed them to test sequences in the manner of a flip-book. The pages were held in place by something called a "peg bar," whose pegs fit the holes punched in the cels and held a stack of pages in place. The celluloid's transparency eventually enabled the development of an assembly-line mode of production, wherein the majority of effort could be directed toward drawing many images to represent a character's movements while drawing only one for those parts of the sequence that remained constant (such as the background). Eventually, more experienced animators would draw only "key frames," drawings that defined the starting and ending points of a movement, leaving lower-level animators known as "inbe-tweeners" to fill in the remaining frames.

53. Barrier, *The Animated Man: A Life of Walt Disney*, 26. Barrier's source here is an August 17, 1937, conversation between Walt Disney and Irene Gentry from the Walt Disney Archives.

54. Wells argues that for Disney and other twentieth-century animation studios, "sport was a vehicle by which animation as an evolving medium could readily test itself technically, aesthetically, and socioculturally." Wells, *Animation, Sport and Culture*, 58.

55. Walt Disney as a person and Disney as a company both played a role in this history.

56. Crafton, *Before Mickey*, 7.

57. Raj Kottamasu, conversation with author, January 25, 2015.

58. Thomas and Johnston, *Disney Animation: The Illusion of Life*, 9.

59. Thomas and Johnston, *Disney Animation*, 12. The stages were Squash and Stretch, Anticipation, Staging, Straight Ahead Action and Pose to Post, Follow Through and Overlapping Action, Slow In and Slow Out, Arcs, Secondary Action, Timing, Exaggeration, Solid Drawing, and Appeal.

60. Thomas and Johnston, *Disney Animation*, 50. "When depicting a ball or character using squash and stretch, "the squashed position can depict the form either flattened out by great pressure or bunched up and pushed together. The stretched position always shows the same form in a very extended condition." Thomas and Johnston, *Disney Animation*, 48.

61. Thomas and Johnston, *Disney Animation*, 50–51 (emphasis added). The quote continues:

> Some men added distinction by starting with a big bounce, followed by shorter and shorter ones as the ball gradually lost its spring. Some put the action in perspective to show how well they could figure a complicated assignment, or they added a stripe around the ball to show how much it turned during the whole action. These men were grabbed quickly by the Effects Department, which specialized in a mechanical type of animation. Those more interested in a livelier type of entertainment preferred surprise endings: the ball exploding on contact, or crashing like a broken egg on the second bounce, or sprouting wings and flying off.

62. There is an added layer to this aside, which depicts Mexico as the home of unpredictable, unruly hoppings-about, even though Mesoamerican civilizations developed the first technologies of reliable rubber bounce that Europeans later adapted and industrialized to produce the very object that the animators were working with in Disney's studios. Indeed, Thomas and Johnston's description of the circular forms on the page taking off "as if they had a life of their own" echoes uncannily the language used by the Dominican chroniclers who first encountered the rubber balls used by the Aztec for their spectacular ballgames.

63. Thomas and Johnston, *Disney Animation*, 50.

64. Wells, *Animation, Sport and Culture*, 58.

65. Wells, *Animation, Sport and Culture*, 89.

66. Wells, *Animation, Sport and Culture*, 39.

67. Shale, *Donald Duck Joins Up*, 4–5.

68. With the help of Dr. Lee de Forest.

69. Thomas and Johnston, *Disney Animation*, 287.

70. Thomas and Johnston, *Disney Animation*.

71. Thomas and Johnston, *Disney Animation*, 290.

72. Disney, "Method of and Means for Scoring Motion Pictures," 34–77. Patent 1,913,048 was granted on June 6, 1933.

73. Disney, "Method," lines 5–11.

74. Thomas and Johnston, *Disney Animation*, 290. Milton Erwin Kahl was an animator during the early days of Disney Studios.

75. Turner, *The Democratic Surround: Multimedia and American Liberalism from World War II to the Psychedelic Sixties*, 33.

76. Adorno, "On Popular Music," 460.

77. Wilfred Jackson, interview by Michael Barrier, Milton Gray, and Bob Clampett, 1973. http://www.michaelbarrier.com/cgi-sys/suspendedpage.cgi. I accessed the interview on a site that has since gone down, but the interview has been published in: Didier Ghez, *Walt's People: Talking Disney with the Artists Who Knew Him*. United States: Theme Park Press, 2017.

78. My younger brother drove the purchase of games and consoles. I followed his lead and reaped the rewards.

79. Many thanks to Matt Whitt for that gift and the support that he provided in that time. I also learned how to infuse vodka with beets to create a bright pink hue that seemed in keeping with the game's vivid palette.

80. *Sonic the Hedgehog*, 2020.

81. *Sonic the Hedgehog*.

82. Bowman, "Sonic the Hedgehog," *100 Greatest Videogame Characters*, 182.

83. Quoted in *From Airline Reservations to Sonic the Hedgehog*, 98

84. *From Airline Reservations to Sonic the Hedgehog*, 101.

85. *Service Games: The Rise and Fall of SEGA*.

Chapter 4

1. In the original paper, the images are presented on sequential pages rather than adjacent to each other, as I have placed them here.

2. Licklider, "The Computer as a Communication Device," 34–35.

3. Rowland Wilson drew the cartoons for Licklider's article at the very start of his career. He went on to draw cartoon illustrations regularly for *Playboy*, *TV Guide*, and *The New Yorker*, among others. Later in his career, he moved to animation,

eventually landing at Disney Studios in the 1980s. Wilson described his approach to drawing illustrations this way: "The picture is not the event. It is a telling of the event by an artist who was there. It should always look like his drawing and never the real event. The story is a system within a system." Wilson, *Rowland B. Wilson's Trade Secrets: Notes on Cartooning and Animation*, 4.

4. Licklider, "The Computer as a Communication Device," 23.

5. The visual pun on the phrase "under the table" places the blame for the communication failure squarely on the player and the ball machine, which break the rules both by ignoring the proper way of sending messages (refusing to keep the messages neatly within the proper frame of the table tennis court) and by sending message after message without waiting for a reply (ignoring the pattern and rhythm of good interaction).

6. The celluloid material of the balls winks at the material underpinnings of cinema and cel animation.

7. Hayes, "Computing Science: The Way the Ball Bounces," 331.

8. "Boing Bouncing Ball Demo Amiga/Atari/Commodore 64/Sinclair QL Comparison | Nostalgia Nerd," uploaded by B:\Nostalgia Nerd, March 11, 2016, https://www.youtube.com/watch?v=fSwwqt3ue2M.

9. See Katlin Clifton Forcier for a discussion of the way that screensavers, unlike other moving-image genres, are *boundless* because they continue for indefinite periods of time and "articulate a logic of endlessness that is associated with the emergence of networked digital culture," which offers *After Dark*'s screensavers as a key case study for understanding the way that screensavers echo the reconfiguration of labor and time under a 24/7 logic of late capitalism. Forcier, "Never Idle: The Animated Screensaver and the Culture of Always-On Computing," 79.

10. "Never Idle," 335.

11. "Never Idle," 331.

12. "Never Idle."

13. Sobchak, *Carnal Thoughts: Embodiment and Moving Image Culture*, 156.

14. See *Oxford English Dictionary*, s.v. "ping (int. & n.)," December 2024, https://doi.org/10.1093/OED/7249873519; and *Oxford English Dictionary*, s.v. "ping (v.2)," December 2023, https://doi.org/10.1093/OED/3427078489.

15. *Oxford English Dictionary*, s.v. "ping (int. & n.)." Depending on the response—on what does or does not bounce back—it becomes possible to learn different things about the networked device: Is the connection working properly? How long does it take for the packet to be received and returned?

16. Almeter, "I'll Ping You."

17. Gere, "Genealogy of a Computer Screen," 151. Gere's point is that the images appearing on radar screens are physical traces (indexes) of the objects that they are representing. The conditions through which the presence and nonpresence of a surface are established (the conditions of observation) are based both on the material properties of whatever does the bouncing, be it a lead weight tied to a rope, a sound wave, or an electromagnetic wave, and on the techniques of observation (the techniques that make these material properties meaningful).

18. Shiga, "Ping and the Material Meanings of Ocean Sound," 15.

19. Shiga, "Ping," 15. (The projection of sound waves is called "sonar," initially the acronym SONAR, for SOund NAvigation and Ranging, which eventually became a word. Sonar is more effective underwater than radar.)

20. Shiga, "Ping," 17.

21. Gere, "Genealogy," 144.

22. This conceptualization was fundamental to the development of the information theory of communication. The same year that the Whirlwind was first hooked up to an oscilloscope, Warren Weaver famously defined communication as "all the procedures by which one mind may affect another . . . all procedures by means of which one mechanism (say automatic equipment to track an airplane and to compute its probable future positions) affects another mechanism (say a guided missile chasing this airplane.)" This is a theory of communication that imagines communication as arcing trajectories followed by impact. Weaver, "Some Recent Contributions to the Mathematical Theory of Communication," 2.

23. Everett, quoted in Metropolis, Howlett, and Rota, *A History of Computing in the Twentieth Century: A Collection of Essays*, 365. The Whirlwind had had visual displays of one kind or another from the start. But the oscilloscope was the first progressive display (i.e., the first that updated continuously in real time).

24. This continued the tradition of MIT's close connection to the military. After World War I, the school began serving as the home of the first Army and Navy ROTC programs. The Servomechanism Lab was an offshoot of the Radiation Laboratory at MIT, which had played such a key role in the war effort. The word "servomechanism" is used to describe a self-correcting machine or organism.

25. Metropolis et al., *History of Computing*, 366.

26. Edwards explains: "In the 1940s, flight simulators were servo-operated, electromechanical devices that mimicked an airplane's attitudinal changes in response to movements of its controls" (*The Closed World: Computers and the Politics of Discourse in Cold War America*, 76). As the end of World War II bled into the beginning of the Cold War, the computing research being funded by the US military became more explicitly concerned with building machines capable of real-time interactions with human

users. The Whirlwind project facilitated a number of key firsts in computing, including Forrester's invention of random-access, coincident-current magnetic storage, which became the standard memory device for digital computers, replacing electrostatic tubes.

27. Metropolis et al., *History of Computing*, 366. Kent C. Redmond and Thomas M. Smith quote Forrester's triumphant declaration in a 1949 report: "The computing section of WWI [Whirlwind I], has just passed a most significant milestone: solving an equation and displaying its solution" on an oscilloscope, showing values for x, x2, and x3. Previous test problems had called for only 'single-point' solutions," whereas the progressive display required by this problem, "no matter how simple, can result only when all the basic parts of the computer act in harmony." Redmond and Smith, *Project Whirlwind: The History of a Pioneer Computer*, 180. Forrester describes the display not simply as a conveyor of a desired result, such as the answer to a problem, but as *a result in and of itself.* The display was in turn a requirement of the kind of problems that were being solved, which were differential equations. A *differential equation* is a kind of equation that contains a derivative (an expression of a rate of change) of one or more of the variables involved, such as velocity, position, and mass. Differential equations were part of the invention of calculus. There are several kinds, but in this context, we are most interested in those used in classical mechanics.

28. Redmond and Smith, *Project Whirlwind*, 180.

29. Redmond and Smith, *Project Whirlwind*, 331.

30. I will not address the special case of hyperelastic materials here.

31. Weisberg, "Computer-Aided Design's Strong Roots at MIT," 3–5.

32. Adams, "Small Problems on Large Computers," 101.

33. Burnham, *Supercade: A Visual History of the Videogame Age 1971–1984*, 44. Burnham's criteria for a good computer toy are that it should show off as many of the computer's resources as possible and tax those resources to the limit; that it should be interesting, within a consistent framework; and that it should make the viewer a participant—in other words, it should be interactive (Burnham, *Supercade*, 45).

34. Electronic Computer Division Staff, "Bi-weekly Report, Project 6673, February 2, 1951," 5. When Saxenlan refers to "the new men," he is giving an accurate description of the makeup of the student body. At the time, MIT only occasionally admitted women and students of color under the category of "special students," so the students, with few exceptions, would have been white men.

35. Digital Computer Laboratory Administrative Staff, "Biweekly Report, March 27, 1953," 34 (emphasis added); Electronic Computer Division Staff, "Biweekly Report, March 13, 1953," 30.

36. Scientific and Engineering Computation Group, "Biweekly Report, June 1, 1953," 7–8.

37. Scientific and Engineering Computation Group, "Biweekly Report, June 1, 1953," 7–8.

38. Scientific and Engineering Computation Group, "Biweekly Report, July 13, 1953," 6. The relevant passage reads: "All four demonstration tapes have been tested with the new scope decoders,and are working properly. They are: Bouncing Ball, Tape #2690; Polynomial, Tape #2691; Number Display, Tape #2692; RLC Display, Tape #2693."

39. Nyitray, "William Alfred Higinbotham," 98–99. I am grateful to Raiford Guins and Laine Nooney for pointing me toward Higinbotham's game. The documentary *When Games Went Click*, which Guins and Nooney wrote the script for, provides a more detailed walk-through of the cultural history of the game.

40. Higinbotham, "Brookhaven TV-Tennis Game," 2.

41. Higinbotham, "Brookhaven TV-Tennis Game," 4.

42. Higinbotham, "Brookhaven TV-Tennis Game," 1.

43. Higinbotham, "Deposition," 2.

44. Higinbotham, "Deposition," 1. They had in fact made controls for velocity but decided that players would have trouble operating an additional control, and programming the analog computer required separating the commands for the action of the hit and for the degree of force.

45. *When Games Went Click*, 9:32.

46. Brookhaven National Laboratory. "The First Video Game?"

47. This phrasing comes from Devon Tannahill's unpublished master's thesis "Rise of the Machine: The Making of the Video Game Industry and Military Simulation."

48. *Pong* appears in most video game histories. For an excellent walk-through of the history of the game, see Henry Lowood's "Videogames in Computer Space: The Complex History of Pong." And for a compelling reading of the canonical game with an equally canonical text from queer theory that locates resonances between queerness and games at the start of video game history, see Bo Ruberg's chapter "Between Paddles: *Pong, Between Men*, and Queer Intimacy in Video Games," in their book, *Video Games Have Always Been Queer*.

49. Magnavox Company, "Odyssey Installation and Game Rules," Strong Museum of Play, Electronic and Video Game Collection.

50. Baer, "Television Gaming and Training Apparatus," 1: 41–44

51. Baer, "Television Gaming and Training Apparatus," 1: 9.

52. Baer, "Television Gaming and Training Apparatus," 1: 9.

53. The result of this definition of "action," which ties the concept to visibility or measurability, is a world in which only those things that can be counted in the numerical sense end up counting in the qualitative sense.

54. Baer, quote from oral history interview, on the website of the National Museum of American History, Washington, D.C.

55. Van Burnham, *Supercade*, 55.

56. Baer, "Television Gaming and Training Apparatus," [57].

57. The name was shortened simply to "Games" at the turn of the century.

58. Baer, "Television Gaming and Training Apparatus," 1: 9.

59. The entire gaming apparatus (controller, receiver, and television screen) are understood as a kind of body. The patent uses the word "embodiment" repeatedly and includes an entire section laying out preferred embodiments of the gaming system. Baer, "Television Gaming and Training Apparatus."

60. Magnavox Company, "Odyssey Installation and Game Rules."

61. Baer, "Television Gaming and Training Apparatus," [57].

62. The first prototype was installed in 1972 at a local restaurant, Andy Capp's Tavern, in Sunnyvale, California.

63. For example, see Postigo, "From Pong to Planet Quake: Post-industrial Transitions from Leisure to Work," and Wolf, *The Video Game Explosion: A History from Pong to Playstation and Beyond*.

64. Tran, "Man with Brain Implant Challenges Neuralink's Monkey to 'Pong' Game."

65. Lowood, "Videogames in Computer Space: The Complex History of Pong," 13.

66. Lowood, "Videogames in Computer Space."

67. Lowood, "Videogames in Computer Space."

68. I first had an opportunity to think through the squareness of the ball for a short piece on the *Pong* ball that I was invited to write for *100 Greatest Video Game Characters*, edited by Jamie Banks, Robert Meija, and Aubrie Adams.

69. Burnham describes how, following Bushnell's instructions, Alcorn engineered and built the first prototype: using solid-state circuitry and transistor-transistor logic (TTL)—all wired up to a standard television display. As in its predecessor, *Computer Space*, there were separate circuit boards for each of the game's functions—one for the paddles, one for the ball, and one for the scoring. Burnham, *Supercade*, 87.

70. Sito, *Moving Innovation: A History of Computer Animation*, 110.

71. Burnham, *Supercade*, 87.

72. Ole Capriani, quoted in Kanitsakis, "An Audio Research on the Gameplay Sounds of Atarti's 'Pong' and the silence of Magnavox Odyssey's "Tennis." Capriani notes that the distance between the paddle and wall sounds is a pleasing octave, while the interval between these and the point sound is slightly more than a semitone, and thus strikingly disharmonic.

73. Burnham, *Supercade*, 110.

74. *Pong*, Atari, 1972.

75. Sito, *Moving Innovation*, 110. This innovation made Pong a field-changing breakthrough for computer graphics in eyes.

76. Lowood, "Videogames in Computer Space," 5.

77. Lowood, "Videogames in Computer Space," 10.

78. Lowood, "Videogames in Computer Space,"10. Lowood quotes Larry Kerecman, "Computer Space," here: "The brilliance of these machines was that Nolan Bushnell and company took what was computer programming (in *Spacewar*) and translated it into a simpler version of the game (no gravity) using hard-wired logical circuits." Kerecman, "Computer Space," quoted in Lowood, "Videogames in Computer Space," 10.

79. Guins, *Game After: A Cultural Study of Video Game Afterlife*, 60.

80. Burnham, *Supercade*.

81. Anderson, "Who Really Invented the Video Game?" 8.

82. Higinbotham, "The Brookhaven TV-Tennis Game." As a technician, Higinbotham makes it clear that he is not impressed by the early Magnavox games. In his 1983 article in *Creative Computing* that situated Higinbotham as the inventor of the first video game, John Anderson observes that Higenbotham's "implementation [of TV-Tennis] was very much more sophisticated than the first 'Pong' games." Anderson, "Who Really Invented the Video Game?" 8.

83. Lowood, "Videogames in Computer Space," 17.

84. Montfort and Bogost, *Racing the Beam: The Atari Video Computer System*, 10.

85. ACM SIGGRAPH History: Information and Artifacts, "Audience Participation, by Carpenter."

86. For a firsthand account of the evening, see Denslow, "A Report on the Recent Siggraph '91 Conference: Fun and Games in Las Vegas."

87. Director Adam Curtis uses this footage to open and close "Love and Power," the first segment of his three-part series *All Watched over by Machines of Loving Grace*

(2011). Curtis is famous for combining many kinds of visual and sonic material, building arguments by way of juxtaposition, sequencing, and strategic cuts.

88. "Audience Participation, by Carpenter."

89. Acevedo-Ryker, Solomon, and Briseno, *306090 06: Shifting Infrastructures,* 14.

90. By the winter of 1991, Loren had filed a patent claim on this "method and apparatus for audience participation by electronic imaging," in which "the audience controls the screen image for purposes of voting or playing a game" and Rachel had become chief executive officer and a producer for Cinematrix Interactive Entertainment Systems, a company that the couple cofounded that would eventually incorporate in 1993.

91. Carpenter and Carpenter, "Cinematrix Founders."

92. Curtis, "Love and Power," 11:17.

93. Michael Scroggins, personal communication, October 5, 2022.

94. Curtis, "Love and Power," 58:17.

95. Latour, "Visualisation and Cognition: Drawing Things Together," 30. There are, of course, many nondigital traditions of animation and huge swaths of computing that are not primarily concerned with visual display. But the predominant forms of both live at their convergence.

96. Licklider, "The Computer as a Communication Device," 22

97. Licklider, "The Computer as a Communication Device," 22. Licklider was building on the communication theory of information laid out by Norbert Weiner, Claude Shannon, and Warren Weaver, one of the main tenets of which is "that information be measured by entropy," where "information, in communication theory, is associated with the amount of freedom of choice we have in constructing messages." Shannon and Weaver, "A Mathematical Theory of Communication," 6.

98. Licklider, quoted in Pierce, Shannon, Rosenblith, and Bush, "What Computers Should Be Doing," 318–319.

99. Licklider, quoted in Pierce et al., "What Computers Should Be Doing."

100. Hayes, "Computing Science," 331.

Chapter 5

1. This chapter is a revised and extended version of my chapter published in the collection *EA Sports FIFA: Feeling the Game,* which I coedited with Raiford Guins and Henry Lowood. My thinking for this book, and particularly this chapter, has been shaped by Iris Marion Young's foundational essay "Throwing Like a Girl: A

Phenomenology of Feminine Body Comportment Motility and Spatiality," which has provided rich grounds for my thinking, writing, and teaching about embodiment, movement, and gender.

2. I first fell for this adjective in the context of Ungainly Sculpture, a course taught by Martin Kersels and Leslie Dick at California Institute of the Arts (CalArts), which I never took but enjoyed being in the vicinity of.

3. Released annually since 1993, sales of *FIFA* generally bring in between 10 percent and 15 percent of EA's annual net revenue, which was $5.53 billion in the 2019–2020 fiscal year and up to 7.562 billion in 2023–2024 fiscal year. In addition to the substantial chunk of revenue coming from game sales, sales of extra content within the Ultimate Team mode, which allows players of EA's sports games "to collect current and former professional players in order to build, and compete as, a personalized team," represents between 20 percent and 30 percent of annual net revenue, most of which is derived from the *FIFA Ultimate Team* (FUT). (Electronic Arts Inc. Annual Report 2020, accessed on March 7, 2020 https://sec.report/Document/0000712515 -20-000019/ and Electronic Arts Inc. Notice of 2014 Annual Meeting and Proxy Statement.) For an earlier version of this chapter and a more extensive treatment of the game series, see Guins, Lowood, and Wing, *EA Sports FIFA: Feeling the Game.*

4. Steam, "Steam Charts."

5. As my collaborator, John Dieterich, and I have written, "You can find soccer fields in almost every corner of the world—in stadiums and parks, prisons and schools, on military bases, attached to factory complexes, medical facilities, corporate campuses, private clubs, and youth detention centers. More than this, being a global game means being a global media spectacle. You can find soccer matches broadcast on televisions in living rooms and bars and bus riders' smartphones; dorms, office break rooms, and arcades outfitted with fusball tables." See "Any Given Moment," in *Interactive Storytelling for the Screen.*

6. Guins, Lowood, and Wing, "Pre-match Commentary," 1.

7. The games range from AAA and independent game company releases to individuals and teams releasing their side hustle or class project on Itch.io or Steam. A starting list of published games would include *Pro Evolution Soccer (PES), Rocket League, Football Masters, Actua Soccer Virtua Striker, New Star Soccer, Kick Off 2, Head-On Soccer/ Fever Pitch Soccer, World Cup 98, Sensible Soccer, Glen Hoddle Soccer, UEFA Champions League, Emlyn Hughes International Soccer, Gazza's Super Soccer, Three Lions, Kidz Sports International Soccer, RedCard 20–03, Freestyle Street Soccer, Pure Futbol, David Beckham Soccer, Kris Kamara's Street Soccer, Codemasters Club Football, Sega Soccer Slam, Lego Soccer Mania, Pele II: World Tournament Soccer, Pro Evolution Soccer Management, Brian Clough's Football Fortunes, Dino Dini's Soccer, O'Leary Manager,* and *Gravity Soccer!* There are more than one hundred soccer games on Itch.io and more than eight hundred ballgames with "Football" in their name and six hundred with "Soccer" in their name on Steam.

8. This is the official tagline of *FIFA '15*. For an extended address of *FIFA*, see *EA Sports FIFA: Feeling the Game*.

9. *EA Sports FC '25*, https://www.ea.com/games/ea-sports-fc.

10. This is not to say that I never play games. I spent a streak of months truly captivated by playing *Angry Birds* on the New York City subways, and there was one retro summer focused on beating *Sonic the Hedgehog*. But I write about video games not from the position of an avid player, but from the position of someone compelled by ball play, by forms of movement, and by their representations.

11. Abe Stein, "Playing the Game on Television," in *Sports Video Games*, 117.

12. While the vast majority of game controllers are designed for hands, there is a robust history of alternative interfaces designed to accommodate those wanting or needing to voice, mouth, limbs, or feet to communicate commands to a computer. One of the most dynamic parts of the annual Game Designer Conference is always the alt.ctrl. showcase, which presents alternative controllers created by independent developers.

13. This would normally be a phenomenal feat. As my coeditor, Henry Lowood, points out, there are many videos online documenting top players aiming for the crossbar as a demonstration of their virtuosity.

14. Matt Bilby, in Kay Hill (producer), *Megafactories Documentary: The Making of EA FIFA '12*.

15. For example, see the following threads: https://answers.ea.com/t5/Other-FIFA -Games/Invisible-ball/td-p/4183316; https://www.reddit.com/r/FIFA/comments/3rxc 8c/there_is_no_ball_visible_when_playing_fifa_16/; https://answers.ea.com/t5/FIFA -17/Invisible-ball-Fifa-17/m-p/5710541#:~:text=Hi%2C,it%20visible%20again%20 while%20playing;and https://www.reddit.com/r/FIFA/comments/ji5fyi/floating_ball _glitch/.

16. Henry Lowood, "Where There Is Smoke, There Is Fire . . .": The FIFA Engine and Its Discontents," *EA Sports FIFA: Feeling the Game* (London: Bloomsbury 2022).

17. To date, discussions of possible *FIFA* NFTs focus on the card packs, not the in-game balls.

18. Gaboury, *Image Objects*, 32.

19. Gaboury, *Image Objects*, 32.

20. There are configurations that alter this, such as the Player Relative setting for Right Stick Switching, which I will not have a chance to discuss here, but the ball is the default; and even when a player switches from Ball Relative to Player Relative, the rest of the game's avatars are still pegged to the ball's position.

21. Swink, *Game Feel*.

22. Swink, *Game Feel*, xiii.

23. Wardrip-Fruin, "Gravity in Computer Space."

24. See Hayes, "Computing Science: The Way the Ball Bounces," for an excellent walk-through of the different approaches to programming bounce.

25. Geertz, *Local Knowledge: Further Essays in Interpretive Anthropology*, 75.

26. Maher, *The Future Was Here: The Commodore Amiga*, 20.

27. Jules Burt, interview by Ross Silifant, *Atari Compendium*, 2016.

28. Maher, *The Future Was Here*, 30.

29. MCV Staff, "Rejection, Tragedy and Billions of Dollars—The Story of FIFA."

30. Parkin, "Fifa, the Video Game That Changed Football."

31. Parkin, "Fifa, the Video Game That Changed Football." This, of course, raises the question: What does it mean to understand the activity of playing a screen-based game as *feeling* like soccer, and to whom is that feeling available?

32. Parkin, "Fifa, the Video Game That Changed Football." For example, the movement in Amiga Soccer, which launched in 1988, feels almost like watching a game of foosball where the plastic players have escaped their metal rods.

33. Deshbandhu, "Toward a Monopoly: Examining FIFA's Dominance in Simulated Football"; or see the PES versus FIFA discussion on r/FIFA, https://www.reddit.com/r/FIFA/comments/czfsic/pes_vs_fifa/.

34. Chiaet, "FIFA Physics."

35. Chiaet, "FIFA Physics."

36. As Henry Lowood relates, this aspect is particularly strange in *FIFA '21*, with the balls flying off heads and feet halfway or more down the pitch, and passes inexplicably zooming downfield.

37. Parkin, "Fifa, the Video Game That Changed Football."

38. See Stanfill and Salter, "Avatar Bodies That Matter: The Work of 'Realism' in Gendered Representation," for further discussion of *FIFA*'s long-standing association with realism in the context of gendered representation.

39. Newer versions of the game also include a smaller, less bold, off-white or yellow arrow indicating which is the next-closest player to the ball.

40. In 2019, Right Stick Switching was introduced to allow advanced players to actively choose which avatar they control at any moment, based on proximity to either the current avatar player or the ball.

41. The collapse into one-on-one play is, at least in part, an artifact from the moment when the game was created when soccer was thought to be too complex a sport to simulate. Jan Tian has said of programming the first version of the game that programming the positioning, distance between players, and player relationship to the ball was the most difficult thing. Parkin, "Basement Idea to Blockbuster: The Story of Fifa, the Video Game." This was an underlying principle of the series for decades, and it is something that has been shifting in recent years, with different kinds of cooperative play being added to different areas of the game. But the most elite, competitive versions remain one-on-one.

42. Serres, *The Parasite*, 228.

43. In soccer, this phrase is usually understood to refer to players with possession of the ball, but being "off the ball" is understood to mean movements that are done away from where the ball is, so I am thinking here of defenders who are specifically challenging for possession as being "on the ball" as well.

44. Bateson, *Steps to an Ecology of Mind*, 321.

45. Tucker, "Physiology of Football: Profile of the Game."

46. As motion capture acquisition specialist Nigel Nunn describes it: "We have the balls on specific parts, the joints, parts that move, and a red light bounces off the balls and goes back into the cameras—so the camera only sees those reflections. It's kind of like [global positioning system], it tracks the movement, and watches how players move. It's like connect the dots, you just see all these markers, and someone goes in and turns it into a stick man moving about—we use that to drive the 3D model of that person." Price, *FIFA Football: The Story behind the Video Game Sensation*, part 6, location 1867.

47. Consalvo, "Women, Sports, and Video Games."

48. At that time, there were no transgender or nonbinary professional soccer players who had come out publicly. In the past few years, a number of athletes who play on professional women's league and national teams have come out as transgender or nonbinary, making it clear that these are the teams to look at for progressive reimaginings of sport beyond gender segregation.

49. Price, *FIFA Football*.

50. As Geoff Harrower explained to me, a set of dance cards is called a "movement set," and a full locomotion set for an avatar is made of several movement sets that differ from each other aesthetically (e.g., near, far, tired).

51. Phillips, *Gamer Trouble*.

52. As Brett Kashmere and Astria Suparak put it, "In the information age . . . athletes are objectified as data, becoming sets of statistical profiles and avatars." Kashmere and Suparak, "Introduction: A Non-Zero-Sum Game," 6.

53. Frye, "The Athlete's Two Bodies,"

54. In late 2021, EA and FIFA announced that they had failed to come to an agreement to renew the license, throwing the future of the series up in the air. Raiford Guins, Henry Lowood, and I address this briefly in our introduction to *EA Sports FIFA: Feeling the Game*.

55. See Pennington, "Ritualised Exclusion," as well as Elsey and Nadel, *Futbolera: A History of Women and Sports in Latin America*.

56. Elsey and Nadel, *Futbolera*, 231.

57. More recent examples include a number of players complaining about being forced to play their World Cup matches on artificial turf during the 2015 FIFA Women's World Cup and the organization's infamously dishonest president, Sepp Blatter, threatening to retaliate against the complaining players; and FIFA introducing fundamental rule changes just months before the 2019 FIFA Women's World Cup tournament, making the women's biggest event a kind of test ground for the new rules. Adding insult to injury, two other major men's tournaments were scheduled at the same time as the Women's World Cup that year.

58. Heckmann and Furini, "The Introduction of Women's Teams in FIFA 16 and How Brazilian Women Reacted to It," 257.

59. Orland, "The Tech That's Putting Women in EA's *FIFA* Games for the First Time."

60. Ahmed, "A Phenomenology of Whiteness," 163.

61. See also Stanfill and Salter, "Avatar Bodies That Matter," which uses Judith Butler's work on bodies that matter to address the addition of playable women to *FIFA* and the question of gendered bodies in *FIFA* and other video games.

62. Price, *FIFA Football*, location 2956.

63. Takahashi, "How Females in FIFA Led to a Diversity Movement at EA."

64. Blake, "FIFA 16 Introduces Female Footballers for the First Time."

65. Takahashi, "How Females in FIFA Led to a Diversity Movement at EA."

66. Price, *FIFA Football*, location 2956.

67. Shade, "Top Ten Players in Every FIFA (94–21)."

68. Cox and Thompson, "Multiple Bodies: Sportswomen, Soccer and Sexuality"; and Kolnes, "Heterosexuality as an Organizing Principle in Women's Sport."

69. Doyle, "Dirt off Her Shoulders."

70. See Elsey and Nadel, *Futbolera*, for an excellent discussion of these intersections in the context of the history of women's football in Latin America.

71. Doyle, "Dirt off Her Shoulders," 420.

72. As Doug Goodwin put it to me, "Isn't this game playing out Newtonian physics plus Joseph Campbell's universal hero one more time? As the crowd goes wild."

73. Orland, "The Tech That's Putting Women in EA's *FIFA* Games for the First Time."

74. Blake, "FIFA 16 Introduces Female Footballers for the First Time."

75. I take the term "soccerwomen" from Clarke's *Soccerwomen: The Icons, Rebels, and Trailblazers Who Transformed the Beautiful Game.*

76. A wealth of scholars have written about the politics and practices of Black hair and addressed the construction of ideals of "good" and "bad" hair in the context of the history of the enslavement of Black people and the construction of whiteness as the measure of beauty in the United States. For example, see Mercer. "Black Hair/Style Politics"; and Lester, "Nappy Edges, Goldy Locks: African-American Daughters and the Politics of Hair."

77. *EA Forums*, https://fifaforums.easports.com/en/discussion/83866/why-85-to-unlock-afros-lower-that-please and https://fifaforums.easports.com/en/discussion/185417/pro-clubs-is-racist.

78. Chancy319, October 2, 2016. https://fifaforums.easports.com/en/discussion/185417/pro-clubs-is-racist.

79. "Frostbite: Full Hair Tech Demo 2019," https://www.youtube.com/watch?v=8wlRCiIjbSs.

80. See Lowood, "Joga Bonito," and Stein, "Playing the Game on Television," for extended discussions of *FUT*.

81. See Emma Witkowski and Rune Nielson on the game as a third object, and Raiford Guins on the game as a medium, in *EA Sports FIFA: Feeling the Game.*

82. Russell, *Glitch Feminism: A Manifesto*, 24.

Chapter 6

1. California AJUPEME Delegation, "CA.AJUPEME.USA 🌐🎽⚽🔟🎽 WWW.AJUPEME.-USA.COM," Instagram, July 27, 2022. https://www.instagram.com/reel/CghZ-uAlKvP/?igshid=MzRlODBiNWFlZA==.

2. Bede and all other names in this chapter are pseudonyms used to protect the identities of the participants. Real names are used for those people who have given me consent and hold public positions or are otherwise named in previously published texts or films. I have used the real names of those who hold public positions in AJUPEME and those involved with the Maravilla handball court. These are Armando Uscanga and Reyna Puc, the founders of AJUPEME; Antonio Chavez, president of AJUPEME-USA; Raul Herrera, vice-president of AJUPEME-USA; Aztlan

Tenochtitlan; Miguel Duran, the National Ulama Trainer; Amanda Perez, founder of the Maravilla Historical Society; and Piltzinkoyotl.

3. The practice of asking permission is something that I have learned from and through this project from scholars like Robin Wall Kimmerer, as well as from dear friends like Amanda Perez.

4. "Reviving a 3,000-Year-Old Ancient Ballgame," 3:20, https://ajupeme-usa.com/ (translated from spoken Spanish).

5. It is not known what the Olmec called themselves. Only one example of their writing has survived, and it remains undeciphered. Western scholarship uses the name that the Nahuatl-speaking peoples of central Mexico, collectively known as the Aztec, gave to this much earlier civilization, which has been translated as "rubber people." "Aztec" is itself a contested designation as is Mesoamerica. Occidental historiography uses "Aztec," both as an umbrella term for all Nahuatl-speaking societies (even though they themselves had, and continue to have, distinct names) and the name for the most prominent of these societies encountered by the Europeans in the sixteenth century (who called themselves the "Mexica" and had their capital at Tenochtitlan, now Mexico City). These confusions are the result of the imposition of European concepts of "nation" and "empire" onto the societal structures of the pre-Columbian Americas. In her entry "Americas" in *Keywords for Latina/o Studies*, Alexandra T. Vasquez describes Mesoamerica:

> As a catch-all, it supports a lazy and willful forgetting of Tawantinsuyu (Quechua for the four regions of the Inca Empire) and Anáhuac (Nahuatl for the Aztec's "land by the waters"). Why these names don't roll off all our tongues suggests the unfamiliar and unrelenting consonants roiling under our collective surface. This imposition of the Old World and all its diseased baggage atop the New, the actual and discursive annihilation of what and who was here before is just one method in the genocidal repertoire enabled by what Walter Mignolo calls the "two entangled concepts" of "modernity and coloniality" (Mignolo 2005, 2011; Quijano 2007).

Vasquez, in *Keywords for Latina/o Studies*, 10.

6. While it was previously thought that the Olmec invented *ulama*, new research including the identification of ball court mounds dating to the Archaic period now supports the argument that the Olmec were not the originators of the game, although they are still credited with being the first to leave significant evidence of ballgame practices being central to culture. Taube, *Olmec Art at Dumberton Oaks*, 2004; Scarborough and Wilcox, *The Mesoamerican Ballgame*, 1991; Diehl, *The Olmecs: America's First Civilzation*," 2004; also cited in Ramos, "Women Playing a Man's Game: Reconstructing Ceremonial and Ritual History of the Mesoamerican Ballgame."

7. Personal communication with Manuel Aguilar-Moreno, 2020. When I first wrote this sentence, there were over 1,700. See Taladoire, "The Architectural Background of the Pre-Hispanic Ballgames," 2001; Santley et al., "The Politicization of the Meso-american Ballgame and Its Implications for the Interpretation of the Distribution

of Ballcourts in Central Mexico," 1991). When I interviewed art historian Manuel Aguilar-Moreno, who has been studying surviving versions of the ballgame for decades, he said that it was closer to 3,000.

8. The game played such a central role that for several centuries, a pyramid and a ballcourt served as the public architecture marking the center of cities. The game was especially prominent in the Aztec and Maya empires, where gambling was a central feature, as both players and spectators bet heavily on the outcomes of matches. Members of the Aztec nobility both played the game themselves and organized professional teams.

9. Players are allowed to use only one hand to support themselves when they drop to the ground, but they can use the other to get up after they have touched the ball.

10. Ramos, "Women Playing a Man's Game: Reconstructing Ceremonial and Ritual History of the Mesoamerican Ballgame."

11. Ramos, "Women Playing a Man's Game."

12. For an excellent collection of essays on the ballgames, see Whittington, *The Sport of Life and Death: The Mesoamerican Ballgame.*

13. Ashmore, "Mesoamerican Cosmologies, Worldviews, and Sacred Landscapes." 2007; also cited in Ramos, "Women Playing a Man's Game," 141.

14. Loadman, *Tears of the Tree*, 26.

15. Varieties of Nahuatl are spoken by around 1.7 million people in Mexico and in smaller populations in the United States as of 2023.

16. *Online Nahuatl Dictionary.*

17. Aguilar-Moreno and Brady, "The Ulama Ballgame: Past, Present, and Future," 42.

18. Fox, *The Ball: Discovering the Object of the Game*, 123.

19. Fox, *The Ball*, 123.

20. Tarkanian and Hosler, "America's First Polymer Scientists: Rubber Processing, Use and Transport in Mesoamerica," 473.

21. Tarkanian and Hosler, "America's First Polymer Scientists." While I see the attraction of the choice that Tarkanian and Hosler make to call the Olmecs and following cultures the "first polymer scientists," I wonder about the way that back-formation of the terms "polymer" and "scientist" risk eliding the differences in understandings, knowledges, and relationships to materials.

22. Kelly, *Notes on a West Coast Survival of the Ancient Mexican Ballgame*; Rochín (dir.), *Ulama, el juego de la vida y la muerte*; Nadal, "Rubber and Rubber Balls in Mesoamerica"; Coggins and Ladd, "Copal and Rubber Offerings"; and Tarkanian, "3,500 Years Before

Goodyear: Rubber Processing in Ancient Mesoamerica"; and "Prehistoric Polymer Engineering: A Study of Rubber Technology in the Americas," cited in Tarkanian and Hosler, "America's First Polymer Scientists," 479.

23. Mártir de Angelería, *Décades del Nuevo Mundo*, in Tarkanian and Hosler, "An Ancient Tradition Continued: Modern Rubber Processing in Mexico," 118.

24. Sahagún, *The Florentine Codex, Book 10*. 1961, 87, quoted in Miller, "The Maya Ballgame: Rebirth in the Court of Life and Death," 79.

25. Durán, *Book of the Gods and Rites and the Ancient Calendar*, 316, also cited in Nadal, "Rubber and Rubber Balls in Mesoamerica," 23.

26. Cline, "Cortes and the Aztec Indians in Spain," 70.

27. Cline, "Cortes and the Aztec Indians in Spain," 87–88.

28. I take this translation of the caption in the *Trachtenbuch des Christoph Weiditz*, 1530–1540, from Cline, "Hernando Cortés," 75.

29. Schiebinger, *Plants and Empire: Colonial Bioprospecting in the Atlantic World*, 3 (emphasis in original).

30. Schiebinger, *Plants and Empire*, 3 (emphasis in original).

31. Tarkanian and Hosler, "Ancient Tradition Continued," 117.

32. As well as using it to make balls for their ritual games, Aztec societies burned rubber as incense and used it for waterproofing and hafting weapons, and, in its liquid form, to mark the bodies of those about to be sacrificed to the gods.

33. A century or so later, rubber did begin to circulate around Europe, mostly as a scientific and technical curiosity. The syringe function noted by French explorer Charles Marie de La Condamine led to experiments using rubber to coat thin wire catheters. The most popular initial rubber commodity was a waterproof shoe. In 1755, King Joseph I of Portugal sent his boots to Pará to be waterproofed. In 1769, a French chemist named Pierre Joseph Macquer made a pair of waterproof boots for Frederick the Great.

34. The arc of history of Anáhuac swerved sharply with the arrival of Europeans and their project of colonization, something that Walter D. Mignolo dubs "the darker side of the Renaissance." This powerful new colonial project, which served as the underwriter of modernity, rested intellectually on constructing a moral and semiotic hierarchy with European Christian people and alphabetic Latin languages at the top, and materially on the extraction of gold, silver, labor, and anything else identified as a resource from what Amerigo Vespucci coined "Mundus Novus" (the New World). The name made sense to the people who used it, even as it encapsulated their mistaken belief that the world was called into existence with their arrival and its resources rightly belonged to them. The world was new only to

them. It was not new to the 100 million or so people who lived on the continental mass at the time, nor to the millions more who had lived there over the prior thousands of years.

35. Personal interview with Manuel Aguilar-Moreno, 2020.

36. See Arrizón, "Mestizaje," and Doremus, "Indigenism, Mestizaje, and National Identity in Mexico during the 1940s and the 1950s." *Mexican Studies/Estudios Mexicanos,* 375–402.

37. Arrizón, "Mestizaje."

38. Arrizón, "Mestizaje." She concludes by introducing Gloria Anzaldúa's remaking of the term by proposing a "new" mestizaje "as a "method" in Chicana feminism, supporting a transcultural form of consciousness, constantly traveling back and forth between race, gender, sexuality, language, and nations."

39. Moreno, "Ulama: The Pre-Columbian Ballgame Survives Today."

40. Aguilar Moreno, personal communication, 2020.

41. Rochín, *Ulama: The Game of Life and Death,* 67.

42. Only one of the existing Indigenous recipes for stabilizing sap into a variously elastic and durable material, as well as the methods for producing different rubber objects, were recorded by the European colonizers, and none were transferred across the ocean. The history of vulcanization, industrialization, and platationization of rubber that I tell in chapter 2 will happen three hundred years later.

43. Leyenaar, *Ulama: The Perpetuation in Mexico of the Pre-Spanish Ballgame Ullamaliztli,* 59–60.

44. Over two decades later, over the course of the Ulama Project, Aguilar-Moreno came to a similar conclusion and began leading ball-making workshops in Sinaloa and Los Angeles.

45. A new ball court was built in Xcaret. Because the performances here are primary for tourists, the game is played with a Western cumulative scoring system. Moreno, "Ulama: The Pre-Columbian Ballgame Survives Today."

46. Sanchez, "Ancient Ballgame: Indigenous Game Ulama Is Being Practiced Again."

47. As Raul put it to me in 2021, "People started doing other things, people were mourning, people just left."

48. Official competitions currently are divided into men's and women's teams.

49. AJUPEME-USA, https://ajupeme-usa.com/.

50. I attended a handful of the team's demonstrations, including Mundo Maya Festival at Levitt Pavillion in MacArthur Park in downtown Los Angeles; the Día

de Los Muertos Festival in a neighborhood park in the San Fernando Valley; and Mt. San Antonio College, a community college that is federally designated a Hispanic Serving Institution and Asian American and Native American Pacific Islander Serving Institution.

51. This is the same kind of wrap that is used by women during and after pregnancy to help support the belly and hips. Piltzinkoyotl gave me a red *faja* for this purpose to support me in my seventh month of pregnancy right as I was finishing this manuscript.

52. The founders of AJUPEME learned the techniques at ball-making workshops that Aguilar-Moreno ran as part of the Ulama Project.

53. Personal communication with Antonio Chavez, president of AJUPEME USA, January 23, 2022. I have been thinking about this as I write. Thinking about the energy that I am giving these words and sentences as I move my fingers across the keys of a QWERTY keyboard and scrawl colored ink on printed drafts. Over the decade that I have spent working on this book, it has brought out and been through many energetic moments, and my techniques for shaping thoughts and sentences have shifted so many times that it has been hard to keep up.

54. This rule is a site of contention. AJUPEME is not the only group finding ways to connect to and extend the tradition of the Mesoamerican ballgame. In Los Angeles, Chavez tells me, there is a "ghost team" that had its ball taken away when they did not go along with the rules of the association. And at the ceremony to inaugurate a stretch of 6th Avenue in Los Angeles as the Mayan Mile, the San Fernando team was joined by another team that is not part of AJUPEME.

55. Fox, *The Ball*, 123.

56. *Ulama* balls are usually kept hung in a cloth from the ceiling to preserve their round shape, although Daniel told me that he tested this by leaving a ball on a flat surface for several months and found that it did not lose its shape.

57. Personal communication with Miguel, June 6, 2021.

58. Personal communication with Daniel, September 13, 2021.

59. Personal communication with Daniel, September 13, 2021.

60. Personal communication with Daniel, September 13, 2021.

61. See Burstyn, *The Rites of Men: Manhood, Politics, and the Culture of Sport*; Pennington, "Ritualised Exclusion"; and Elsey and Nadel, *Futbolera: A History of Women and Sports in Latin America*. Trans athletes, nonbinary athletes, and indeed any athlete who does not neatly fit into the binary to begin with face additional degrees of exclusion and sometimes are subjected to verbal and physical violence from sports players, fans, and institutional bodies. Many scholars have addressed this point in

sports studies. For example, see Anderson and Travers (eds.), *Transgender Athletes in Competitive Sports*; and Greey and Lenskyj (eds.), *Justice for Trans Athletes: Challenges and Struggles.*

62. Elsey and Nadel, *Futbolera,* 55.

63. Ramos, "Women Playing a Man's Game," 253–255. On machismo, she writes, "While Americo Paredes contends that it began during the conquest period when Cortés and his soldiers arrived in New Spain and raped Aztec women (1967: 65), Patricia Fernández-Kelly asserts that the behaviors identified today as *macho* existed since pre-colonial times (2005: 78). Although this is still a matter of debate in scholarship, there is a general consensus that *machismo* became popularized in the forties and fifties with the figure of the revolutionary warrior personified in the *charro* (Mexican cowboy)" (Paredes, "Estados Unidos, Mexico y el Machismo"; and Fernández-Kelly, "Reforming Gender: The Effects of Economic Change on Masculinity and Femininity in Mexico and the U.S"; both also cited in Ramos, "Women Playing a Man's Game," 256).

64. Ramos, "Women Playing a Man's Game," 251.

65. This rubber tree is the second one that I have lived with. It has moved from place to place with me ever since it came to me in 2016. I have taken to writing next to it. I imagine that I am writing from it. I also sleep next to another, smaller rubber plant that my friend K-Sue gave me when she moved from Los Angeles to D.C. in 2019. I brought home my first rubber plant in 2010 from Natty Garden, a nursery a few blocks from my gargoyle-adorned former-seminary-turned-apartment building on the corner of Atlantic Avenue and Washington Avenue in Brooklyn. I watched it grow as I read piles of books about the history of rubber. This pursuit of rubber and its bounce is how I first met Amanda. In preparation for the workshop, I began learning about the history of handball in the city. Becerra, "Extending a Hand to a Faded East L.A. Handball Court."

66. Personal communication with Amanda Perez.

67. To see some of what came out of this effort, see Huerta (dir.), *Maravilla Handball Court: A Place That Matters"*; Sweet, "501 N. Mednik," *All Night Menu, Vol. 3*; Monroe, *Racquet: Long Play.*

68. Sweet, "Remembering El Rebote," 36. The variable origins of the game are visible in the variation in the number of walls that a court can have—ranging from one to four—and in the size and hardness of the balls used to play on them. The Irish game was first brought to Brooklyn in the 1800s. The Mexican game is a version of games that have been played on this continent for thousands of years inflected with games brought by the Spanish.

69. Sweet, "Remembering El Rebote," 36.

70. Standardization in sport, as we saw in chapter 2, supports a particular kind of capacity to measure and compare players against each other. Taken to an extreme, it produces greater and greater attunement to minute differences and less and less tolerance for particularity, variation, and chance. Good enough standardization allows a game to be recognizably played in different places while allowing the pleasure and challenge of differences among those places.

71. Mesoamerican cultures have played handball games with rubber balls for thousands of years. Early versions of handball in what is now Europe were played with balls made from other materials, mostly cloth or leather stuffed with various things. Those games converted to using rubber balls with the industrialization of rubber.

72. Sweet, "501 N. Mednik," 36.

73. I first witnessed this when I met longtime Los Angeles handball player, coach, and muralist Ernesto de la Loza in 2011. He lived around the corner from the Machine Project off Alvardo and Sunset, and we sat and talked a few times while I was on a residency at Machine planning the first ball-making workshop. He had met the flying bounce of handballs with his hands for over four decades, leaving them looking like tree roots, swollen and twisted about in surprising directions. The years of playing on cement—a surface with no bounce that forces all motion into the players' joints—also left him with fluid in his knee that he drained himself weekly in order to be able to continue to play.

74. Amanda Perez, in Monroe, *Racquet Long Play*, 26:37.

75. Sweet, "501 N. Mednik," 36.

76. Sweet, "501 N. Mednik," 36. The Maravilla Handball Club started in 1968, was joined by the Maravilla Racquetball Club in the 1980s, when that game found some popularity, and ran until the deaths of Michi and Tommy in 2006. The painted signs for the Maravilla Handball Club 1968–2000 and the Maravilla Racquetball Club 1980–2004 that list the members' names lived under the covered porch of my house for a year or so, visible to anyone who approached, before Amanda found a permanent home for them. The names are almost all identifiably male and Chicano, with the exception of Tommy Nishiyama.

77. Sweet, "501 N. Mednik," 38.

78. The march remains a touchstone in the history of the city of Los Angeles and the Chicano movement, because of both the massive scale of participation and the violent response of the sheriff's department. The sheriff's deputies brutally attacked the crowd, killing three, including the *Los Angeles Times* reporter and KMEX-TV news director Ruben Salazar.

79. Contreras, "Chicana, Chicano, Chican@, Chicanx."

80. Perez, in *Racquet: Long Play*, 27:45.

81. Perez, in *Racquet: Long Play*, 27:45.

Conclusion

1. Barad, "On Touching—The Inhuman That Therefore I Am."

2. Elliot, *What's Going on in There: How the Brain and Mind Develop in the First Five Years of Life.*

3. To follow this thought further, see Hannah Zeavin's excellent history of American techno-parenting, *Mother Media: Hot and Cool Parenting in the 20th Century.*

4. Geertz, *Interpretation of Cultures*, 432–433.

5. Geertz, *Interpretation of Cultures.*

6. Mauss, "Techniques of the Body," 455.

7. Quoted in Mauss, "Techniques of the Body," 456.

8. Bourdieu, *Logic of Practice*, 78.

9. Young, "Throwing Like a Girl: A Phenomenology of Feminine Body Comportment Motility and Spatiality."

10. Foucault, *Discipline and Punish: The Birth of the Prison*, 161.

11. Lefebvre, *Rhythmanalysis: Space, Time, and Everyday Life*, 39.

12. Three of the touchstone texts that serve as a prehistory for the field of sport studies, described earlier in this book, are all also in some way about the British educational system. Thomas Hughes's 1857 semiautobiographical book *Tom Brown's School Days* is an account of a nineteenth-century British boy being educated in the spirit of the Rugby School through sport. In his 1959 article "They Taught the World to Play," Charles Tennyson declares Victorian England "the world's games-master." And C. L. R. James's 1963 autobiographical *Beyond a Boundary* is an account of the formative power, deep pleasure, and embedded politics of the play of cricket in Trinidad.

13. Many scholars working on sport and play have written about this, while also showing how conceptions of self are stitched to social orders via bodily performance and affective attachment. See Geertz, *The Interpretation of Cultures*; Doyle, "Dirt off Her Shoulders"; Huizinga, *Homo Ludens*; and MacAloon, *Rite, Drama, Festival, Spectacle: Rehearsals towards a Theory of Cultural Performance.*

14. Richard Schechner, quoted in Turner, *Anthropology of Performance*, 10. Turner meanwhile defines "play" as "categorically uncategorizable" and "a liminal or liminoid mode, essentially interstitial, betwixt and between all standard taxonomic

nodes," with "liminality" defined as a "suspension of quotidian reality." *Anthropology of Performance*, 17, 102.

15. Giorgio Agamben, taking up and extending Foucault's term "apparatus," defines it as "anything that has in some way the capacity to capture, orient, determine, intercept, model, control, or secure the gestures, behaviors, opinions, or discourses of living beings." Agamben, *What Is an Apparatus?* 14. For Agamben, apparatuses capture or otherwise shape living beings (substances), with subjects emerging as the product of the struggle between living beings and apparatuses.

16. Jonathan Sterne articulated this paradox of instrumentality during his Leboff seminar "Instruments and Instrumentality" at the Department of Media, Culture, and Communication, New York University, 2013. This book would not be what it is without Jonathan's remarkable work and the stunning example he set for us all of how to think, write, teach, and mentor others in ways that deeply enrich the world. He is dearly missed.

17. It is not surprising then, that in his discussions of epiphany, Hans Ulrich Gumbrecht often turns to sporting metaphors. See Gumbrecht, *Production of Presence: What Meaning Cannot Convey.*

18. Enrigue, *Sudden Death*, 5.

19. Bourdieu, *Logic of Practice*, 8.

20. Lacan, *Écrits: A Selection*, 18.

21. I first wrote about this in an essay titled "Swerve and Return," for *Tech/Know/Future: From Slang to Structure,* a catalog edited by Charlotte Kent for an exhibition curated by Tom Leeser that was repeatedly postponed due to the pandemic.

22. However, in his book *On This Moment,* Lederach does use the familial example of touching his grandmother's hands to illustrate Boulding's idea.

23. 7 Generation Foundation, http://7genfoundation.org/7th-generation. I also hear an echo here with Resmaa Menakem, trauma therapist and author of *My Grandmother's Hands: Racialized Trauma and the Pathway to Healing Our Hearts and Bodies,* who says that he thinks it will take nine generations to achieve the somatic abolition of white supremacy. "Resmaa Menakem on Trauma & White Body Supremacy," Episode 315, The Relationship School Podcast, accessed December 24, 2020, https://relationshipschool.com/podcast/resmaa-menakem-on-trauma-white-body-supremacy-315/).

24. Parker, *The Constitution of the Five Nations, or The Iroquois Book of the Great Law,* 38–39.

> The thickness of your skin shall be seven spans—which is to say that you shall be proof against anger, offensive actions and criticism. Your heart shall be filled with peace and good will and your mind filled with a yearning for the welfare of the people of the Confederacy. With

endless patience you shall carry out your duty and your firmness shall be tempered with tenderness for your people. Neither anger nor fury shall find lodgement in your mind and all your words and actions shall be marked with calm deliberation. In all of your deliberations in the Confederate Council, in your efforts at law making, in all your official acts, self interest shall be cast into oblivion. Cast not over your shoulder behind you the warnings of the nephews and nieces should they chide you for any error or wrong you may do, but return to the way of the Great Law which is just and right. Look and listen for the welfare of the whole people and have always in view not only the present but also the coming generations, even those whose faces are yet beneath the surface of the ground—the unborn of the future Nation.

This passage from the Constitution of the Iroquois Nation (The Great Binding Law), put in writing after generations of being communicated through oral tradition, is contextualized on the 7 Generation Foundation website with this explanation:

Today, The Seventh Generation Principle usually applies to decisions about the energy we use, water and natural resources, and ensuring those decisions are sustainable for seven generations in the future. We should apply the Seventh Generation Principle to relationships—so that every decision we make results in sustainable relationships that last at least seven generations into the future. In particular relationships between Indigenous and non-Indigenous peoples should be forged with the Seventh Generation Principle in mind, so that future relationships will be positive for many generations to come.

25. Elise Boulding, interview with Juan Portilla, *Beyond Intractability*, 2003. Accessed at www.beyondintractability.org/audiodisplay/boulding-e.

Bibliography

Abbeel, Pieter. "Geoff Hinton, the Godfather of AI." *The Robot Brains Podcast*, May 10, 2023. https://covariant.ai/insights/the-robot-brains-podcast-geoff-hinton -the-godfather-of-ai/.

Acevedo-Ryker, Patricia, Jonathan D. Solomon, and Alexander Briseno. *306090 06: Shifting Infrastructures*. Princeton, NJ: Princeton Architectural Press, 2004.

ACM SIGGRAPH History. Information and Artifacts. "Audience Participation, by Carpenter," 1991, https://history.siggraph.org/animation-video-pod/audience-partici pation-by-carpenter/.

Adams, Charles W. "Small Problems on Large Computers." In *Proceedings of the 1952 ACM National Meeting (Pittsburgh)*, 99–102. New York: Association for Computing Machinery, 1952.

Adorno, Theodor W. "On Popular Music." In *Essays on Music*, edited by Richard Leppert and translated by Susan H. Gillespie, 437–469. Berkeley: University of California Press, 2002.

Agamben, Giorgio. *What Is an Apparatus? And Other Essays*. Translated by David Kishik and Stefan Pedatella. Stanford, CA: Stanford University Press, 2009.

Aguilar-Moreno, Manuel. "Ulama: The Pre-Columbian Ballgame Survives Today." *American Indian* 17, no. 2 (Summer 2016). https://www.americanindianmagazine .org/story/ulama-pre-columbian-ballgame-survives-today.

Aguilar-Moreno, Manuel. "We Have Come Only to Dream." *Calliope* 16, no. 4 (December 2005).

Aguilar-Moreno, Manuel. Personal interview by Carlin Wing via phone. Friday, August 14, 2020. Los Angeles.

Aguilar-Moreno, Manuel, and James F. Brady. "The Ulama Ballgame: Past, Present, and Future." In *Material Intelligence*, edited by Glenn Adamson. Milwaukee: Chipstone Foundation, 2002. https://www.materialintelligencemag.org/rubber/?fb3d-page=44/.

Ahmed, Sara. "A Phenomenology of Whiteness." *Feminist Theory* 8, issue 2 (2007). https://doi.org/10.1177/1464700107078139.

Ahmed, Sara. *Queer Phenomenology: Orientations, Objects, Others*. Durham, NC: Duke University Press, 2006.

Alexander, George. *Wingfield, Edwardian Gentleman*. Portsmouth, NH: Peter Randall Publishing, 1986.

Almeter, Gary M. "I'll Ping You." *McSweeney's*, March 26, 2018. https://www .mcsweeneys.net/articles/ill-ping-you.

Anderson, Eric, and Ann Travers (eds.). *Transgender Athletes in Competitive Sport*. New York: Routledge, 2017.

Anderson, John. "Who Really Invented the Video Game?" *Creative Computing Video & Arcade Games* 1, no. 1 (Spring 1983): 8. http://www.atarimagazines.com/cva/v1n1 /inventedgames.php.

Appadurai, Arjun. "Mediants, Materiality, Normativity." *Public Culture* 27, no. 2 (Spring 2015): 221–237.

Appadurai, Arjun. *Modernity at Large: Cultural Dimensions of Globalization*. Minneapolis: University of Minnesota Press, 1996.

Arrizón, Alicia, "Mestizaje." In *Keywords for Latina/o Studies*, edited by Deborah R. Vargas, Nancy Raquel Mirabal, and Lawrence La Fountain-Stokes, 134. New York: NYU Press, 2017.

Ashmore, Wendy. "Mesoamerican Cosmologies, Worldviews, and Sacred Landscapes." Paper read at Cosmology and Society in the Ancient Amerindian World, 2007.

Attenborough, David. "David Attenborough Reveals How He Brought Colour Television to Wimbledon." *Radio Times*. July 3, 2017. https://www.radiotimes.com /tv/sport/tennis/david-attenborough-reveals-how-he-brought-colour-television-to -wimbledon/.

Ayrton Paris, John. *Philosophy in Sport Made Science in Earnest: Being an Attempt to Illustrate the First Principles of Natural Philosophy by the Aid of Popular Toys and Sports*. London: Longman, Rees, Orme, Brown, and Green, 1827.

B:\Nostalgia Nerd. "Boing Bouncing Ball Demo Amiga/Atari/Commodore 64/Sinclair QL Comparison | Nostalgia Nerd Extra," March 11, 2016. https://www.youtube .com/watch?v=fSwwqt3ue2M.

Baer, Ralph. Oral history interview. Quoted in "The Brown Box, 1967–1968," on the website of the National Museum of American History, Washington, DC. http:// americanhistory.si.edu/collections/search/object/nmah_1301997.

Baer, Ralph. "Television Gaming and Training Apparatus." US Patent 3,728,480, filed March 22, 1971, and issued April 17, 1973.

Banks, Jamie, Mejia, Robert, and Adams, Aubrie, eds. *100 Greatest Video Game Characters*. Washington, DC: Rowman & Littlefield, 2017.

Banks, Jamie. Robert Meja, and Aubrie Abrams. *100 Greatest Videogame Characters*. Lanham: Rowman & Littlefield, 2017.

Barad, Karen. *Meeting the Universe Halfway: Quantum Physics and the Entanglement of Matter and Meaning*. Durham, NC: Duke University Press, 2007.

Barad, Karen. "On Touching—The Inhuman That Therefore I Am." In *differences: A Journal of Feminist Cultural Studies*, ed. Sophia Roosth and Astrid Schrader, 23, no. 3 (2012): 206–223. https://doi.org/10.1215/10407391-1892943.

Barrier, Michael. *The Animated Man: A Life of Walt Disney*. Berkeley: University of California Press, 2007.

Barthes, Roland. "The Grain of the Voice." In *The Responsibility of Forms*, translated by Richard Howard, 267–277. New York: Hill and Wang, 1985.

Bateson, Gregory. *Steps to an Ecology of Mind*. Northdale, NJ: Jason Aronson, 1972.

Becerra, Hector. "Extending a Hand to a Faded East L.A. Handball Court." *Los Angeles Times*, February 14, 2010.https://www.latimes.com/archives/la-xpm-2010-feb-14-la-me-handball14-2010feb14-story.html.

Benjamin, Walter. *Illuminations*. Translated by Harry Zohn. New York: Schocken Books, 1985.

Benjamin, Walter. *The Work of Art in the Age of Mechanical Reproduction*. Translated by J. A. Underwood. London: Penguin, 2008.

Beretta, Marco. "Training Tennis Players through Natural Philosophy: From Scaino's *Trattato* to Garsault's *Art Du Paumier*." *Nuncius* 28, no. 1 (January 1, 2013): 19–42. https://doi.org/10.1163/18253911-02801003.

Beretta, Marco, and Alessandro Tosi. "Tennis and the Scientific Revolution." *Nuncius* 28, no. 1 (January 1, 2013): 1–4. https://doi.org/10.1163/18253911-02801001.

Bergson, Henri. *Laughter: An Essay on the Meaning of the Comic*. Translated by Cloudesley Brereton and Fred Rothwell. New York: MacMillan, 1911.

Bergson, Henri. *Matter and Memory*. Translated by Nancy Margaret Paul and W. Scott Palmer. New York: MacMillan, 1913.

Bergson, Henri. *Le rire: Essai sur la signification du comique*. Paris: Éditions Alcan, 1924.

Bernoulli, Jacob. *The Art of Conjecturing, Together with Letter to a Friend on Sets in Court Tennis*. Translated by Edith Dudley Sylla. Baltimore: Johns Hopkins University Press, 2006.

Berry, David. *A People's History of Tennis*. London: Pluto Press, 2020.

Best, David, and Brian Rich. *Disturb'd with Chaces: Tennis Courts, Celebrities and Scandals of Yesteryear*. Oxford, UK: Ronaldson Publications, 2009.

Bijker, W. E. *Of Bicycles, Bakelites, and Bulbs: Toward a Theory of Sociotechnical Change*. Cambridge, MA: MIT Press, 1997.

Birch, Thomas. *The History of the Royal Society of London, for Improving of Natural Knowledge, from its First Rise*. Vol. 2. London: A. Millar, 1756.

Blake, Vikki. "FIFA 16 Introduces Female Footballers for the First Time," *IGN*, September 6, 2016. https://www.ign.com/articles/2015/05/28/fifa-16-introduces-female -footballers-for-the-first-time.

Blassnigg, Martha. *Time, Memory, Consciousness and the Cinema Experience*. Leiden, NL: Brill, 2010

Boulding, Elise. "Interview." Interviewed by Juan Portilla. *Beyond Intractability*, 2003. www.beyondintractability.org/audiodisplay/boulding-e.

Bourdieu, Pierre. *Logic of Practice*. Stanford, CA: Stanford University Press, 1990.

Bowker, Geoffrey C., and Susan Leigh Star. *Sorting Things Out: Classification and Its Consequences*. Cambridge, MA: MIT Press, 1999.

Bowman, Nicholas David. "Sonic the Hedgehog," In *100 Greatest Videogame Characters*. Edited by Jamie Banks, Robert Meja, and Aubrie Abrams. Lanham, MD: Rowman & Littlefield, 2017: 80–81.

Brockedon, William. *Italy, Classical, Historical, and Picturesque*. London: Duncan & Malcolm, 1842.

Brookey, Robert Alan, and Thomas P. Oates. *Playing to Win: Sports, Video Games, and the Culture of Play*. Bloomington: Indiana University Press, 2015.

Brookhaven National Laboratory. "The First Video Game?" Brookhaven National Laboratory. https://www.bnl.gov/about/history/firstvideo.php.

Brown, Harry. *Golf Ball*. New York: Bloomsbury, 2015.

Burnett, D. Graham. "On the Ball." *Cabinet* 26 (Winter 2014–2015): 64–72.

Burnham, Van. *Supercade: A Visual History of the Videogame Age 1971–1984*. Cambridge, MA: MIT Press, 2001.

Burstyn, Varda. *The Rites of Men: Manhood, Politics, and the Culture of Sport*. Toronto: University of Toronto Press, 1999.

Burt, Jules. Interview by Ross Silifant. *Atari Compendium*. 2016. https://www .ataricompendium.com/archives/interviews/jules_burt/interview_jules_burt.html.

California AJUPEME Delegation. "CA.AJUPEME.USA 🖐 ▓ ⊕ ⊙ ▓ WWW.AJUPEME .-USA.COM," Instagram, July 27, 2022. https://www.instagram.com/reel/CghZ -uAlKvP/?igshid=MzRlODBiNWFlZA==.

Campbell-Kelly, Martin. *From Airline Reservations to Sonic the Hedgehog: A History of the Software Industry*. Cambridge, MA: MIT Press, 2004.

Canales, Jimena. "Desired Machines: Cinema and the World in Its Own Image." *Science in Context* 24, no.3 (2011): 329–359.

Canton Papers. MS/597, 1738–1772. Royal Society of London.

Caprani, Ole. "The PONG Game." Aarhus University—Department of Computer Science, Sound as Media, between Signal and Music. Introduction to Digital Audio, 2014. http://web.archive.org/web/20180309023632/http://cs.au.dk/~dsound/DigitalAudio .dir/Greenfoot/Pong.dir/Pong.html.

Carey, James W. *Communication as Culture: Essays on Media and Society*. New York: Routledge, 1992.

Carpenter, Loren C. "Method and Apparatus for Audience Participation by Electronic Imaging." US Patent 5,210,604, filed December 10, 1991, and issued May 11, 1993.

Carpenter, Loren. "SIGGRAPH 98 Interactive Experience" by Cinematrix. https:// history.siggraph.org/animation-video-pod/siggraph-98-interactive-experience-by -cinematrix/.

Carpenter, Loren, and Rachel Carpenter. "Cinematrix Founders." *Cinematrix Interactive Entertainment Systems*. http://www.cinematrix.com/founders.html.

Cartwright, Lisa. *Screening the Body: Tracing Medicine's Visual Culture*. Minneapolis: University of Minnesota Press, 1995.

Casement, Roger. *Correspondence and Report from His Majesty's Consul at Boma Respecting the Administration of the Independent State of Congo*. 1904. https://archive .org/stream/CasementReport/CasementReportSmall_djvu.txt.

Cavendish, Margaret. *Observations upon Experimental Philosophy*. Edited by Eileen O'Neill. Cambridge: Cambridge University Press, 2001.

Cavendish, Margaret. *Sociable Letters*. Edited by James Fitzmaurice. Orchard Park, NY: Broadview Press, 2004.

Chiaet, Julianne. "FIFA Physics." *Scientific American* 309, no.6, 19 (December 2013). doi:10.1038/scientificamerican1213–1219.

Ciesielski, Andrew. *Introduction to Rubber Technology*. Shawbury, UK: Smithers Rapra, 2001.

Clark Monroe, Darius (dir.). *Racquet Shorts: Long Play*. First Look Media, 2020.

Clarke, Gemma. *Soccerwomen: The Icons, Rebels, and Trailblazers Who Transformed the Beautiful Game*. New York: Bold Type Books, 2019.

Cline, Howard F. "Hernando Cortés and the Aztec Indians in Spain." *Quarterly Journal of the Library of Congress* 26, no. 2 (April 1969): 70–90.

Coggins, Clemency Chase, and John M. Ladd. "Copal and Rubber Offerings." In *Artifacts from the Cenote of Sacrifice, Chichén Itzá, Yucatán,* edited by Clemency Chase Coggins, 345–357. Cambridge, MA: Harvard University Press, 1992.

Collins, Harry, Robert Evans, and Christopher Higgins. *Bad Call: Technology's Attack on Referees and Umpires and How to Fix It*. Cambridge, MA: MIT Press, 2017.

Consalvo, Mia. "Women, Sports, and Video Games." In *Sports Video Games*, edited by Mia Consalvo, Konstantin Mitgutsch, and Abe Stein, 87–112. Cambridge, UK: Routledge, 2013.

Consalvo, Mia, Konstantin Mitgutsch, and Abe Stein (eds.). *Sports Video Games*. Cambridge, UK: Routledge, 2013.

Contreras, Sheila Marie. "Chicana, Chicano, Chican@, Chicanx." In *Keywords for Latina/o Studies*, edited by Deborah R. Vargas, Nancy Raquel Mirabal, and Lawrence La Fountain-Stokes, 32. New York: NYU Press, 2017.

Cooke, J. *John Boyd Dunlop*. Dublin: Dreoilín Publications, 2000.

Cox, Barbara, and Shona Thompson. "Multiple Bodies: Sportswomen, Soccer and Sexuality." *International Review for the Sociology of Sport*, 35, no. 1 (2000), https://doi .org/10.1177/101269000035001001.

Crafton, Donald. *Before Mickey: The Animated Film 1898–1928*. Chicago: University of Chicago Press, 1993.

Crary, Jonathan, and Sanford Kwinter. *Incorporations*. New York: Zone, 1992.

Csíkszentmihályi, Mihály. *Flow: The Psychology of Optimal Experience*. New York: Harper Perennial Modern Classics, 1990.

Cunning, David. "David Hammons: Higher Goals." Public Art Fund, May 8, 2024 .https://www.publicartfund.org/exhibitions/view/higher-goals/.

Daston, Lorraine. *Classical Probability in the Enlightenment*. Princeton, NJ: Princeton University Press, 1988.

David, Nicol. Personal interview by Carlin Wing, New York. October 2, 2013.

Day, Jane Stevenson. "Performing on the Court." In *The Sport of Life and Death: The Mesoamerican Ballgame*, edited by E. Michael Whittington, 65–77. New York: Thames & Hudson, 2001.

Denslow, Phillip Kelly. "A Report on the Recent Siggraph '91 Conference: Fun and Games in Las Vegas." *Animation Magazine*, September/October 1991. Available on Denslow's personal website, http://www.denslow.com/articles/sig91.html.

Deshbandhu, Aditya. "Toward a Monopoly: Examining FIFA's Dominance in Simulated Football." *gameviroments* (July 31, 2020), 49–76.https://doi.org/10.26092/elib/178.

Diehl, Richard A. "The Precolumbian Cultures of the Gulf Coast." In *The Cambridge History of the Native Peoples of the Americas*. Vol. II, Part I, edited by Richard E. W. Adams and Murdo J. Macleod, 156–196. Cambridge: Cambridge University Press, 2000.

Dieterich, John, and Carlin Wing. "Any Given Moment." In *Interactive Storytelling for the Screen* (PERFORM), edited by Sylke Meyer and Gustavo Aldano. London: Routledge, 2021.

Digital Computer Laboratory Administrative Staff. "Biweekly Report, March 13, 1953." Project Whirlwind Reports. Dome. MIT Libraries. http://hdl.handle.net/1721.3/39683.

Digital Computer Laboratory Administrative Staff. "Biweekly Report, March 27, 1953." Project Whirlwind Reports. Dome. MIT Libraries. http://dome.mit.edu/handle/1721.3/39663.

Disney, Roy O. "Method of and Means for Scoring Motion Pictures." US Patent 1,913,048, filed October 16, 1928, and issued June 6, 1933.

Disney, Walt. Disney-Seversky interview in the Disney archives. Quoted in Richard Shale, *Donald Duck Joins Up: The Walt Disney Studio during World War II*, UMI Research Press, 1982.

Doremus, Anne. "Indigenism, Mestizaje, and National Identity in Mexico during the 1940s and the 1950s," *Mexican Studies/Estudios Mexicanos* 17, no. 2 (Summer 2001): 375–302.

Doyle, Jennifer. "Dirt off Her Shoulders." *GLQ: A Journal of Lesbian and Gay Studies* 19, no. 4 (2013): 419–433. https://doi.org/10.1215/10642684-2279897.

du Cros, Sir Arthur Philip. *Wheels of Fortune: A Salute to Pioneers; On the First Manufacturers of Pneumatic Tyres*. London: Chapman & Hall, 1938.

Dunlop Rubber Company. *Dunlop in Malaya* [35mm film]. 1955.

Dunlop, John Boyd. *The History of the Pneumatic Tyre*. Dublin: A. Thom & Company, 1925.

Dunlop Rubber Company. *The Story of Dunlop through the Reigns*. London: The Firm, 1953.

Dunlop, J. B. *10,607*. London: Patent Office, 1888.

Durán, Diego. *Book of the Gods and Rites and the Ancient Calendar*. 2nd ed. Oklahoma City: University of Oklahoma Press, 1971.

Dyer-Witheford, Nick, and Grieg de Peuter. *Games of Empire: Global Capitalism and Video Games*. Minneapolis: University of Minnesota Press, 2009.

Edgerton, David. *The Shock of the Old: Technology and Global History since 1900*. Oxford: Oxford University Press, 2007.

Edwards, Paul N. *The Closed World: Computers and the Politics of Discourse in Cold War America*. Cambridge, MA: MIT Press, 1996.

Electronic Arts Inc. *Annual Report 2020*, accessed on March 7, 2020. https://sec.report /Document/0000712515-20-000019/.

Electronic Arts Inc. *EA Sports FC*. Developers EA Vancouver, EA Romania, EA Mobile, EA Spearhead, KLab Games. Microsoft Windows, Nintendo Switch, PlayStation 4, PlayStation 5, Xbox One, Xbox Series X/S. 2023–2024.

Electronic Arts Inc. *FIFA*. Developers Extended Play Productions (1993–1997), EA Vancouver (1997–2022), EA Romania (2016–2022). Sega Mega Drive/Genesis, Amiga, DOS, N-Gage, 32X, Mega-CD/Sega CD, Master System, Game Gear, Sega Saturn, GameCube, Gizmondo, PlayStation, PlayStation 2, PlayStation 3, PlayStation 4, PlayStation 5, PlayStation Portable, PlayStation Vita, Super NES, Nintendo 64, Nintendo DS, Nintendo 3DS, Wii, Wii U, Nintendo Switch, Game Boy, Game Boy Color, Game Boy Advance, 3DO Interactive Multiplayer, Microsoft Windows, iOS, Java Platform, Micro Edition, Android, Xperia Play, Xbox 360, Xbox, Xbox One, Xbox Series X/S, Windows Phone, macOS, Zeebo, Stadia. 1993–2022.

Electronic Arts Inc. Notice of 2024 Annual Meeting and Proxy Statement, accessed Mar 18, 2025. https://s204.q4cdn.com/701424631/files/doc_financials/2024/ar/electronic -arts-2024-proxy.pdf.

Electronic Computer Division Staff. "Bi-weekly Report, Project 6673, February 2, 1951." Project Whirlwind Reports. Dome. MIT Libraries. http://hdl.handle.net/1721 .3/39690.

Elliot, Lise. *What's Going on in There: How the Brain and Mind Develop in the First Five Years of Life*. New York: Bantam Books, 2000.

Elsey, Brenda, and Joshua H. Nadel. *Futbolera: A History of Women and Sports in Latin America*. Austin: University of Texas Press, 2019.

Enrigue, Alvaro. *Sudden Death*. New York: Penguin, 2017.

Fernández-Kelly, Patricia. "Reforming Gender: The Effects of Economic Change on Masculinity and Femininity in Mexico and the U.S." In *Women's Studies Review*, (2005): 69–101.

Fitzmaurice, James. *Margaret Cavendish: Sociable Letters*. New York: Taylor and Francis, 2012.

Flusser, Vilém. *Into the Universe of Technical Images*. Translated by Nancy Ann Roth. Twin Cities: University of Minnesota Press, 2011.

Forcier, Kaitlin Clifton. "Never Idle: The Animated Screensaver and the Culture of Always-on Computing." *After Image* 48, no. 2 (2021). 79–93.

Foucault, Michel. *Discipline and Punish: The Birth of the Prison*. New York: Vintage Books, 1995.

Fox, John. *The Ball: Discovering the Object of the Game*. New York: Harper Perennial, 2012.

Freud, Sigmund. *Beyond the Pleasure Principle*. Translated by C. J. M. Hubback. London: International Pyscho-Analytical Press. 1911.

Frye, Brian. "The Athlete's Two Bodies." In *INCITE Journal of Experimental Media: Sports*, edited by Brett Kashmere and Astria Suparak. Oakland, CA: INCITE Journal of Experimental Media (2016–2017): 50–65.

Gaboury, Jacob. *Image Objects: An Archaeology of Computer Graphics*. Cambridge, MA: MIT Press, 2021.

Gadys, Jason. "Resmaa Menakem on Trauma & White Body Supremacy." Episode 315, The Relationship School Podcast, accessed December 24, 2020, https://relationshipschool.com/podcast/resmaa-menakem-on-trauma-white-body-supremacy-315/.

Galloway, Alexander R. *The Interface Effect*. Cambridge, UK: Polity, 2012.

Geertz, Clifford. *The Interpretation of Cultures: Selected Essays*. New York: Basic Books, 1973.

Geertz, Clifford. *Local Knowledge: Further Essays in Interpretive Anthropology*. New York: Basic Books, 1983.

Gere, Charlie. "Genealogy of the Computer Screen." *Visual Communication* 5, no. 2 (June 1, 2006): 141–152. https://doi.org/10.1177/1470357206065306.

Ghez, Didier. *Walt's People: Talking Disney with the Artists Who Knew Him*. United States: Theme Park Press, 2017.

Gillmeister, Heiner. *Tennis: A Cultural History*. New York: New York University Press, 1998.

Gitelman, Lisa. *Always Already New: Media, History, and the Data of Culture*. Cambridge, MA: MIT Press, 2008.

Great Big Story, Jan 2, 2018. "Reviving a 3,000-Year-Old Ancient Ballgame." [Video]. YouTube. https://www.youtube.com/watch?v=oYJxng6i4NQ&t=1s.

Greenblatt, Stephen. *The Swerve: How the World Became Modern.* New York: Norton, 2011.

Greey, Ali Durham, and Helen Jefferson Lenskyj (eds.). *Justice for Trans Athletes: Challenges and Struggles.* Bingley, UK: Emergal Publishing, 2023.

Guinness, Katherine, and Charlotte Kent. *Contemporary Absurdities, Existential Crises, and Visual Art.* Chicago: Chicago University Press, 2024.

Guins, Raiford. *Game After: A Cultural Study of Video Game Afterlife.* Cambridge, MA: MIT Press, 2014.

Guins, Raiford. "Playing Games with My Feelings." In *EA Sports FIFA: Feeling the Game,* 213–230. New York: Bloomsbury, 2022.

Guins, Raiford, Henry Lowood, and Carlin Wing (eds.). *EA Sports FIFA: Feeling the Game.* New York: Bloomsbury, 2022.

Guins, Raiford, Henry Lowood, and Carlin Wing. "Pre-match Commentary." In *EA Sports FIFA: Feeling the Game,* 1–25. New York: Bloomsbury, 2022.

Gumbrecht, Hans Ulrich. *Production of Presence: What Meaning Cannot Convey.* Stanford, CA: Stanford University Press, 2004.

Gunning, Tom. "Hand and Eye: Excavating a New Technology of the Image in the Victorian Era." *Victorian Studies* 54, no. 3 (April 1, 2012): 495–516. https://doi.org/10.2979/victorianstudies.54.3.495.

Hacking, Ian. *The Emergence of Probability: A Philosophical Study of Early Ideas about Probability, Induction and Statistical Inference.* Cambridge: Cambridge University Press, 1975.

Hacking, Ian. "Experimentation and Scientific Realism." *Philosophical Topics* 13, no. 1 (Fall 1982): 71–87. https://doi.org/10.5840/philtopics19821314.

Hald, Anders. *A History of Probability and Statistics and Their Applications before 1750.* Hoboken, NJ: John Wiley & Sons, 2003.

Hancock, T. *Personal Narrative of the Origin and Progress of the Caohtchouc or India-Rubber Manufacture in England.* London: Longman, Brown, Green, Longmans, & Roberts, 1857.

Hanson, Christopher. "The Instant Replay: Time and Time Again." *Spectator,* 28, no. 2 (Fall 2008): 51–60.

Harman, Mike. "Police Massacre of Striking Dunlop Rubber Plantation Workers in Malaya." Libcom.org. March 23, 2018.https://libcom.org/article/police-massacre-striking-dunlop-rubber-plantation-workers-malaya.

Hayes, Brian. "Computing Science: The Way the Ball Bounces." *American Scientist* 84, no. 4, (July–August 1996): 331–335.

Hazlitt, William. "The Death of John Cavanagh." *The Examiner,* February 7, 1817.

Hecht, Gabrielle. *Being Nuclear Africans and the Global Uranium Trade*. Cambridge, MA: MIT Press, 2012.

Heckmann, Pedro, and Liana Furini. "The Introduction of Women's Teams in FIFA 16 and How Brazilian Women Reacted to It." *Estudos em Comunicação* no. 26, vol. 1 (May 2018), 247–260.

Hermosillo, Carmen (humdog). "Pandora's Vox: On Community in Cyberspace." May 5, 1994. http://alphavilleherald.com/2004/05/introducing_hum.html.

Higinbotham, William A. "The Brookhaven TV-Tennis Game." William A. Higinbotham Game Studies Collection. Stony Brook University.

Higinbotham, William A. "Deposition." April 29, 1976. Available on the website of the Brooklyn National Laboratory. https://www.bnl.gov/about/docs/Higinbotham_Deposition.pdf.

Hobbes, Thomas. *Leviathan: Or, The Matter, Forme & Power of a Commonwealth, Ecclesiasticall and Civill*. Edited by A. R. Waller. Cambridge: Cambridge University Press, 1904. Available on Google Books. https://books.google.com/books?id=2oc6AAAAMAAJ.

Hochschild, Adam. *King Leopold's Ghost: A Story of Greed, Terror, and Heroism in Colonial Africa*. Boston: Houghton Mifflin, 1998.

Hookway, Branden. "Cockpit." In *Cold War Hothouses: Inventing Postwar Culture, from Cockpit to Playboy*, edited by Beatriz Colomina, Annmarie Brennan, and Jeannie Kim, 22–54. New York: Princeton Architectural Press, 2004.

How to Play Baseball. Directed by Jack Kinney. Los Angeles: Walt Disney Productions/RKO Radio Pictures, 1942.

Howie, Cary. *Claustrophilia: The Erotics of Enclosure in Medieval Literature*. Basingstoke, UK: Palgrave Macmillan, 2007.

Huerta, Manuel (dir.). *Maravilla Handball Court: A Place That Matters* (8 min). 2010.

Hughes, Thomas. *Tom Brown's School Days*. A revised edition with illustrations by Arthur Hughes and Sydney Prior Hall. New York: Macmillan, 1882(orig. 1857).

Huizinga, Johan. *The Autumn of the Middle Ages*. Chicago: University of Chicago Press, 1996.

Huizinga, Johan. *Homo Ludens*. London: Routledge and Kegan Paul, 1949.

Hutton, Sarah. "In Dialogue with Thomas Hobbes: Margaret Cavendish's Natural Philosophy." *Women's Writing* 4, no. 3 (1997): 421–432. https://doi.org/10.1080/09699089700200024.

Intellectual Property Office. "Class 132: Games." *British Patent Index*. 1909–1915.

Intellectual Property Office. "Class 132: Toys, Games, and Exercises." *British Patent Index*. 1855–1866.

International Real Tennis Professional Association. "Real Tennis Rules." IRTPA. http://www.irtpa.com/real-tennis-rules/.

Jackson, Wilfred. Interviewed by Michael Barrier, Milton Gray, and Bob Clampett, 1973. http://www.michaelbarrier.com/cgi-sys/suspendedpage.cgi.

James, C. L. R. *Beyond a Boundary*. London: Yellow Jersey, 2005.

Jhally, Sut. "The Spectacle of Accumulation: Material and Cultural Factors in the Evolution of the Sports/Media Complex." *Insurgent Sociologist*, 12, no. 3, 1984: 41–57.

Juhasz, Alexandra, and Anne Balsamo. "An Idea Whose Time Is Here: FemTechNet—A Distributed Online Collaborative Course (DOCC)." *Ada: A Journal of Gender, New Media, and Technology* 1 (2012). http://adanewmedia.org/2012/11/issue1-juhasz/.

Kanitsakis, Stelios. "An Audio Research on the Gameplay Sounds of Atarti's 'Pong' and the Silence of Magnavox Odyssey's 'Tennis'." *Medium*, March 26, 2020. https://stelioskanitsakis.medium.com/an-audio-comparison-between-the-sounds-of-ataris-pong-and-the-silence-of-magnavox-odyssey-s-83e6fac56653.

Kashmere, Brett, and Astria Suparak (eds.). *INCITE: Journal of Experimental Media*, no. 7–8 (2016–2017).

Kelly, Isabel. *Notes on a West Coast Survival of the Ancient Mexican Ballgame*. Notes on Middle American Archaeology and Ethnology, Vol. I, No. 26. Cambridge, MA: Carnegie Institution of Washington, Division of Historical Research, 1943.

Kerecman, Larry. "Computer Space." Arcade History Database. http://www.arcade-history.com/?n=computer-space&page=detail&id=3388.

Khan, Latasha. Personal interview. New York, October 2, 2013.

Kolnes, Liv-Jorunn. "Heterosexuality as an Organizing Principle in Women's Sport." *International Review for the Sociology of Sport*, 30, no. 1 (1995).

Lacan, Jacques. *Écrits: A Selection*. Translated by Alan Sheridan. New York: W. W. Norton, 1977.

Lake, Robert J. "Real Tennis and the Civilising Process." *Sport in History* 29, no. 4 (2009): 553–576.

Lake, Robert J. "Social Exclusion in British Tennis: A History of Privilege and Prejudice." PhD diss., Brunel University of London, 2008.

Lakin, Max. "When Dawoud Bey Met David Hammons." *The New York Times*, May 1, 2019.

Lasseter, John. "Principles of Traditional Animation Applied to 3D Animation." *Computer Graphics* 21, no. 4 (July 1987): 35–44.

Latour, Bruno. *The Pasteurization of France*. Cambridge, MA: Harvard University Press, 1988.

Latour, Bruno. "Visualisation and Cognition: Drawing Things Together." In *Knowledge and Society: Studies in the Sociology of Culture Past and Present*. Vol. 6, edited by Henrika Kuklick, 1–40. Jai Press, 1986.

Latour, Bruno. *We Have Never Been Modern*. Cambridge, MA: Harvard University Press, 1993.

"Lawn Tennis: The Davis Cup: Challenge Round." *The Times*, July 25, 1913, 15.

Lefebvre, Henri. *Rhythmanalysis: Space, Time, and Everyday Life*. London: Continuum, 2004.

Lehrer, Jonah. "The Physics of Grass, Clay, and Cement." *Grantland*, September 8, 2011.

Lester, Neal. "Nappy Edges, Goldilocks: African-American Daughters and the Politics of Hair." *The Lion and the Unicorn* 24, no. 2 (April 2000). 201–224.

Leyenaar, Ted J. J. *Ulama: The Perpetuation in Mexico of the Pre-Spanish Ballgame Ullamaliztli*. Leiden, Netherlands: E. J. Brill, 1978.

Licklider, J. C. R. "The Computer as a Communication Device." *Science and Technology* 17 (April 1968): 20–41.

"List of Most Watched Television Broadcasts." *Wikipedia*, accessed May 10, 2024. https://en.wikipedia.org/wiki/List_of_most-watched_television_broadcasts#.

Litman, L. "Alex Morgan on Why Artificial Turf Is Tough for Players." *USA TODAY*, October 15, 2014. https://www.usatoday.com/story/sports/soccer/2014/10/15/alex -morgan-us-women-artificial-turf-world-cup/17295011/.

Loadman, John. *Tears of the Tree: The Story of Rubber—a Modern Marvel*. Oxford: Oxford University Press, 2005.

"Love and Power." Part of the series *All Watched Over by Machines of Loving Grace*. Directed by Adam Curtis. BBC, 2011.

Lowood, Henry. "Joga Bonito: Beautiful Play, Sports, and Digital Games." In *Sports Videogames*, edited by Mia Consalvo, Konstantin Mitgutsch, and Abe Stein. New York: Routledge, 2013: 67–86.

Lowood, Henry. "Videogames in Computer Space: The Complex History of Pong." *IEEE Annals of the History of Computing* 31, no. 3 (July–September 2009): 5–19.

Lowood, Henry. "Where There Is Smoke, There Is Fire . . .": The FIFA Engine and Its Discontents." In *EA Sports FIFA: Feeling the Game*, 105–124. London: Bloomsbury 2022.

MacAloon, John J. *This Great Symbol: Pierre de Coubertin and the Origins of the Modern Olympic Games*. Chicago: University of Chicago Press, 1981.

MacAloon, John J., ed. *Rite, Drama, Festival, Spectacle: Rehearsals towards a Theory of Cultural Performance*. Philadelphia: Institute for the Study of Human Issues, 1984.

MacIntosh, J. J., Peter Anstey, and Jan-Erik Jones. "Robert Boyle." In *The Stanford Encyclopedia of Philosophy* (Winter 2022 edition), edited by Edward N. Zalta and Uri Nodelman. https://plato.stanford.edu/archives/win2022/entries/boyle/.

Magnavox Company. "Odyssey Installation and Game Rules." The Magnavox Company, 1972. Electronic and Video Game Collection, Strong Museum of Play.

Maher, Jimmy. *The Future Was Here: The Commodore Amiga*. Cambridge, MA: MIT Press, 2012.

"The Making of FIFA '12." *Megafactories*. National Geographic Channel. Season 6, Episode 1. 2012.

Malaysian Rubber Board. "Global Consumption of Natural and Synthetic Rubber from 1990 to 2015 (in 1,000 Metric Tons)." *Statista*, accessed March 26, 2016 .http://www.statista.com/statistics/275399/world-consumption-of-natural-and -synthetic-caoutchouc/.

Mangan, J. A. *Athleticism in the Victorian and Edwardian Public School: The Emergence and Consolidation of an Educational Ideology*. Cambridge: Cambridge University Press, 1981.

Mangan, J. A. *The Cultural Bond: Sport, Empire, Society*. London: F. Cass, 1992.

Mangan, J. A. *The Games Ethic and Imperialism: Aspects of the Diffusion of an Ideal*. New York: Viking, 1986.

Mangan, J. A., and Callum C. McKenzie. *Militarism, Hunting, Imperialism: Blooding the Martial Male*. London: Routledge, 2010.

Manovich, Lev. *The Language of New Media*. Cambridge, MA: MIT Press, 2002.

Marey, Étienne-Jules. *Movement*. Translated by Eric Pritchard. London: William Heinemann, 1895.

Mártir de Anglería, Pedro. *Décadas del Nuevo Mundo*. Sociedad Dominicana de Bibliofilos. Santo Domingo, Dominican Republic, 1989.

Mauss, Marcel. "Techniques of the Body." In *Incorporations*, edited by Jonathan Crary and Sanford Kwinter, 455–477. New York: Zone Books, 1992.

Maynes-Aminzade, Dan, Randy Pausch, and Steve Seitz. "Techniques for Interactive Audience Participation," *ICMI '02 Proceedings of the 4th IEEE International Conference on Multimodal Interfaces*, October 2022. https://dl.acm.org/doi/abs/10.1109/ICMI.2002 .1166962.

McGill, Douglas C. "ART PEOPLE: Hammons' Visual Music." *The New York Times*, July 18, 1986.

McLuhan, Marshall. *Understanding Media: The Extensions of Man*. Cambridge, MA: MIT Press, 1994 (orig. 1964).

McMillan, James. *The Dunlop Story: The Life, Death, and Re-birth of a Multi-national*. London: Weidenfeld and Nicolson, 1989.

MCV Staff. "Rejection, Tragedy and Billions of Dollars—The Story of FIFA." *MCV/DEVELOP*, August 16, 2013. https://www.mcvuk.com/business-news/publishing /rejection-tragedy-and-billions-of-dollars-the-story-of-fifa/.

Menakem, Resmaa. *My Grandmother's Hands: Racialized Trauma and the Pathway to Healing Our Hearts and Bodies*. Las Vegas: Central Recovery Press, 2017.

Mercer, Kobena. "Black Hair/Style Politics." *new formations: A Journal of Culture, Theory, & Politics*, no. 3 (Winter 1987), 33–54.

Metropolis, N., J. Howlett, and Gian-Carlo Rota (eds.). *A History of Computing in the Twentieth Century: A Collection of Essays*. New York: Academic Press, 1980.

Miah, Andy. *Sports 2.0: Transforming Sports for a Digital World*. Cambridge, MA: MIT Press, 2017.

Mignolo, Walter D. *The Darker Side of the Renaissance: Literacy, Territoriality, and Colonization*. 2nd ed. Ann Arbor: University of Michigan Press.

Miles, Eustace. *Racquets, Tennis, and Squash*. London: Ward, Lock & Co, 1902.

Miller, Mary. "The Maya Ballgame: Rebirth in the Court of Life and Death." In *The Sport of Life and Death: The Mesoamerican Ballgame*, edited by E. Michael Whittington, 78–87. New York: Thames & Hudson, 2001.

Mintz, Sidney W. *Sweetness and Power: The Place of Sugar in Modern History*. New York: Penguin Books, 1986.

Mirzoeff, Nicholas. *The Right to Look: A Counterhistory of Visuality*. Durham, NC: Duke University Press, 2011.

Montfort, Nick, and Ian Bogost. *Racing the Beam: The Atari Video Computer System*. Cambridge, MA: MIT Press, 2009.

Morgan, Roger. *Tennis: The Development of the European Ballgame*. Oxford, UK: Ronaldson, 1995.

Morgan, Roger. *Tudor Tennis: A Miscellany*. Oxford, UK: Ironbark Ronaldson, 2001.

Murray, Sue. *Bright Signals: A History of Color Television*. Durham, NC: Duke University Press, 2018.

Nadal, Laura Filloy. "Rubber and Rubber Balls in Mesoamerica." In *The Sport of Life and Death: The Mesoamerican Ballgame,* edited by E. Michael Whittington, 20–31. New York: Thames & Hudson, 2001.

Neale, Steve. *Cinema and Technology.* Quoted in Lisa Cartwright, *Screening the Body: Tracing Medicine's Visual Culture,* 34–35. Minneapolis: University of Minnesota Press, 1995.

Needham, Joseph. *Science and Civilization in China,* vol. 4, *Physics and Physical Technology,* part 1, *Physics.* Cambridge: Cambridge University Press, 1962.

Newbolt, Henry. "Vitaï Lampada." In *Admirals All and Other Verses.* New York: John Lane, 1898.

Newton, Paula. "Pakistan's World Cup Stitch-Up." *CNN,* June 10, 2010.

"No Cause for Worry: European War Will Have No Effect on American Sporting Implements." *The New York Times,* August 21, 1914.

Nyitray, Kristen J. "William Alfred Higinbotham: Scientist, Activist, and Computer Game Pioneer." *IEEE Annals of the History of Computing* 33, no. 2 (April–June 2011): 96–101.

Online Nahuatl Dictionary, Stephanie Wood, ed. (Eugene, OR: Wired Humanities Projects, College of Education, University of Oregon, 2000–present) https://nahuatl.uoregon.edu/.

Orland, Kyle. "The Tech That's Putting Women in EA's *FIFA* Games for the First Time." *Ars Technica,* May 28, 2015. https://arstechnica.com/gaming/2015/05/the-tech-thats-putting-women-in-eas-fifa-games-for-the-first-time/.

Paredes, Americo. "Estados Unidos, Mexico y el Machismo." *Journal of Inter-American Studies* 9(1), 1967: 65–84.

Parker, Arthur C. *The Constitution of the Five Nations, or The Iroquois Book of the Great Law.* Ohsweken, Canada: Iroqrafts, 1991.

Parkin, Simon. "Basement Idea to Blockbuster: The Story of Fifa, the Video Game." *Irish Times,* December 22, 2016. www.irishtimes.com/sport/soccer/basement-idea-to-blockbuster-the-story-of-fifa-the-video-game-1.2915567.

Parkin, Simon. "Fifa, the Video Game That Changed Football." *The Guardian,* December 21, 2016. https://www.theguardian.com/technology/2016/dec/21/fifa-video-game-changed-football.

Patents for Inventions: Abridgements of Specifications, Class 132, Games, 1909–1915. London: Love and Malcomson, 1915.

Pennington, Michael. "Ritualised Exclusion." In *EA Sports FIFA: Feeling the Game,* edited by Raiford Guins, Henry Lowood, and Carlin Wing. London: Bloomsbury 2022.

Pepys, Samuel. *The Diary of Samuel Pepys, Vol. 5, 1664*. Edited by Robert Latham and William G Mathews. Los Angeles: University of California Press, 2000.

Perinbanayagam, Robert S. *Games and Sport in Everyday Life: Dialogues and Narratives of the Self*. Boulder, CO: Paradigm Publishers, 2006.

Perinbanayagam, Robert S. "The Measured Self: Tennis." In *Games and Sport: Dialogues and Narratives of the Self*, 111–123. Boulder, CO: Paradigm Publishers, 2006.

Peters, John Durham. *The Marvelous Clouds: Toward a Philosophy of Elemental Media*. Chicago: University of Chicago Press, 2015.

Peters, John Durham. *Speaking into the Air: A History of the Idea of Communication*. Chicago: University of Chicago Press, 1999.

Pettus, Sam. *Service Games: The Rise and Fall of Sega*. CreateSpace Independent Publishing Platform, 2018.

Phillips, Amanda. *Gamer Trouble*. New York: NYU Press, 2020.

Pierce, John R., Claude E. Shannon, Walter A. Rosenblith, and Vannevar Bush. "What Computers Should Be Doing." In *Management and the Computer of the Future*, edited by Martin Greenberger, 290–324. New York: MIT Press/John Wiley & Sons, 1962.

Pong (game and arcade cabinet). Atari/Syzygy, 1972.

Ponnusamy, Suresh. Personal interview, October 17, 2014.

Postigo, Hector. "From Pong to Planet Quake: Post-industrial Transitions from Leisure to Work." *Information, Communication & Society* 6, no. 4 (2003): 593–607.

Price, Lee. *FIFA Football: The Story behind the Video Game Sensation*. Oakamoor, UK: Bennion Kearny Ltd., 2015.

PricewaterhouseCoopers. "Video Games." *Global Entertainment and Media Outlook 2015–2019*. http://www.pwc.com/gx/en/industries/entertainment-media/outlook/segment-insights/video-games.html.

Ramos, Isabel. "Women Playing a Man's Game: Reconstructing Ceremonial and Ritual History of the Mesoamerican Ballgame." Unpublished dissertation, University of California San Diego, 2012.

Rankine, Claudia, *Citizen: An American Lyric*. Minneapolis: Graywolf Press, 2014.

Rankine, Claudia. "The Meaning of Serena Williams." *The New York Times*, August 30, 2015. https://www.nytimes.com/2015/08/30/magazine/the-meaning-of-serena-williams.html.

r/EASportsFC. "PES vs FIFA." *Reddit.com*. 2019. https://www.reddit.com/r/EASportsFC/comments/czfsic/pes_vs_fifa/.

Redmond, Kent C., and Thomas M. Smith. *Project Whirlwind: The History of a Pioneer Computer*. Bedford, MA: Digital Press, 1980.

Rochín, Roberto. *Ulama: The Game of Life and Death*. Culiacàn, Mexico: Universidad Autónoma de Sinalo, 2010, 67.

Rowe, David. *Sport, Culture and the Media: The Unruly Trinity*. Buckingham, UK: Open University Press, 1999.

Ruberg, Bo. *Sex Dolls at Sea: Imagined Histories of Sexual Technologies*. Cambridge, MA: MIT Press, 2022.

Ruberg, Bo. *Video Games Have Always Been Queer*. New York: NYU Press, 2019.

Russell, Legacy. *Glitch Feminism: A Manifesto*. New York, Verso, 2020.

Sahagún, Bernardino de. *The Florentine Codex, Book 10*. Translated by Arthur J. O. Anderson and Charles E. Dibble. Santa Fe: School of American Research and the University of Utah, 1961.

Sanchez, Arturo, interviewed for TRT World. "Ancient Ballgame: Indigenous Game Ulama Is Being Practiced Again." YouTube, June 1, 2019, 0:36. https://www.youtube.com/watch?v=Jp0709-CnkY.

Santley, Robert S., Michael J. Berman, and Rani T. Alexander. "The Politicization of the Mesoamerican Ballgame and Its Implications for the Interpretation of the Distribution of Ballcourts in Central Mexico." In *The Mesoamerican Ballgame*, edited by Vernon L. Scarborough and David R. Wilcox, 3–24. Tucson: University of Arizona Press, 1991.

Scaino, Antonio. *Treatise on the Game of Ball*. Translated by P. A. Negretti. London: Raquetier Productions, 1984.

A bilingual edition printed side-by-side with a facsimile of the original Italian (1555).

Scarborough, Vernon, and David Wilcox. *The Mesoamerican Ballgame*. Tucson: University of Arizona Press, 1991.

Schiebinger, Londa L. *Plants and Empire: Colonial Bioprospecting in the Atlantic World*. Cambridge, MA: Harvard University Press, 2004.

Scientific and Engineering Computation Group. "Biweekly Report, June 1, 1953." Project Whirlwind Reports. Dome. MIT Libraries. http://hdl.handle.net/1721.3/39661.

Scientific and Engineering Computation Group. "Biweekly Report, July 13, 1953." Project Whirlwind Reports. Dome. MIT Libraries. http://hdl.handle.net/1721.3/39681.

Serres, Michel. *The Parasite*. Translated by Lawrence R. Schehr. Minneapolis: University of Minnesota Press, 2008.

7th Generation Foundation. "7th Generation Principle." 7genfoundation.org. Accessed on March 19, 2025.

Shade. "Top Ten Players in Every FIFA (94–21)." YouTube. October 11, 2020 .https://www.youtube.com/watch?v=yVYKSu0kiGs.

Shakespeare, William. *Henry V*. Folger Shakespeare Library edition. Edited by Barbara A. Mowat and Paul Werstine. New York: Washington Square Press, 1995.

Shale, Richard. *Donald Duck Joins Up: The Walt Disney Studio during World War II*. Ann Arbor, MI: UMI Research Press, 1982.

Shapin, Steven, and Simon Schaffer. *Leviathan and the Air Pump: Hobbes, Boyle, and the Experimental Life*. Princeton, NJ: Princeton University Press, 1985.

Sheppard, Samantha, *Sporting Blackness: Race, Embodiment, and Critical Muscle Memory on Screen*. Oakland: University of California Press, 2020.

Shiga, John. "Ping and the Material Meanings of Ocean Sound." In *Sustainable Media*, edited by Nicole Starosielski and Janet Walker, 128–145. New York: Routledge, 2016.

Siegert, Bernhard. *Cultural Techniques: Grids, Filters, Doors, and Other Articulations of the Real*. Translated by Geoffrey Winthrop-Young. New York: Fordham University Press, 2015.

Sito, Tom. *Moving Innovation: A History of Computer Animation*. Cambridge, MA: MIT Press, 2013.

Smith, Bruce R. *The Acoustic World of Early Modern England: Attending to the O-Factor*. Chicago: University of Chicago Press, 1999.

Sobchak, Vivian. *Carnal Thoughts: Embodiment and Moving Image Culture*. Oakland: University of California Press, 2004.

Stanfill, Mel, and Anastasia Salter. "Avatar Bodies That Matter: The Work of 'Realism' in Gendered Representation." *EA Sports FIFA: Feeling the Game,* edited by Raiford Guins, Henry Lowood, and Carlin Wing, 67–86. New York: Bloomsbury, 2022.

Star, Susan, and James Griesemer. "Institutional Ecology, 'Translations' and Boundary Objects: Amateurs and Professionals in Berkeley's Museum of Vertebrate Zoology, 1907–39." *Social Studies of Science* 19, no. 3 (1989): 387–420.

"sport, n.1". [Etymology and Def. 4a]. *Oxford English Dictionary*. March 2019. Oxford University Press. http://www.oed.com/viewdictionaryentry/Entry/187476 (accessed April 7, 2019).

Statista. "Global Sports Market—Total Revenue from 2006 to 2015 (in Billion U.S. Dollars)." Statista. http://www.statista.com/statistics/194122/sporting-event-gate -revenue-worldwide-by-region-since-2004/.

Steam. "Steam Charts." Accessed March 18, 2025. https://steamdb.info/app/2669320 /charts/#48h.

Stein, Abe. "Playing the Game on Television." In *Sports Video Games*, edited by Mia Consalvo, Konstantin Mitgutsch, and Abe Stein, 115–137. Cambridge, UK: Routledge, 2013.

The Story of Rubber [With illustrations]. London: Educational Productions Ltd. in collaboration with Dunlop Rubber Company, 1957.

Stuchtey, Benedikt. "Colonialism and Imperialism, 1450–1950." *European History Online*, January 24, 2011. http://ieg-ego.eu/en/threads/backgrounds/colonialism-and -imperialism.

Sweet, Sam. "501 N. Mednik." *All Night Menu Vol. 3*, 2016.

Sweet, Sam. "Remembering El Rebote." *Racquet Magazine*, June–August 2017.

Swink, Steve. *Game Feel*. Boston: Morgan Kaufmann Publishers, 2009.

Sylla, Edith Dudley. "Jacob Bernoulli and the Mathematics of Tennis." *Nuncius* 28, no. 1 (January 1, 2013): 142–163.

Takahashi, Dean. "How Females in FIFA Led to a Diversity Movement at EA." *VentureBeat*, July 9, 2019. https://venturebeat.com/2019/07/09/how-female-s-in-fifa-led -to-a-diversity-movement-at-ea/.

Taladoire, Eric. "The Architectural Background of the Pre-Hispanic Ballgames." In *The Sport of Life and Death: The Mesoamerican Ballgame*, edited by E. Michael Whittington, 96–115. New York: Thames and Hudson, 2001.

Tannahill, Devon. "Rise of the Machine: The Making of the Video Game Industry and Military Simulation." Unpublished master's thesis. 2014.https://knowledgecommons .lakeheadu.ca/handle/2453/476.

Tarkanian, Michael J. "3,500 Years before Goodyear: Rubber Processing in Ancient Mesoamerica." Unpublished bachelor's thesis, Department of Materials Science and Engineering, Massachusetts Institute of Technology, Cambridge, MA. 2000.

Tarkanian, Michael J. "Prehistoric Polymer Engineering: A Study of Rubber Technology in the Americas." Unpublished master's thesis, Barker Engineering Library, Massachusetts Institute of Technology, Cambridge, MA, 2003.

Tarkanian, Michael J., and Dorothy Hosler. "America's First Polymer Scientists: Rubber Processing, Use and Transport in Mesoamerica." *Latin American Antiquity* 22, no. 4 (2011): 469–486. https://doi.org/10.7183/1045–6635.22.4.469.

Tarkanian, Michael J., and Dorothy Hosler. "An Ancient Tradition Continued: Modern Rubber Processing in Mexico." In *The Sport of Life and Death: The Mesoamerican*

Ballgame, edited by E. Michael Whittington, 116–121. New York: Thames & Hudson, 2001.

Taube, Karl A. *Olmec Art at Dumbarton Oaks*. Washington, DC: Dumbarton Oaks Research Library and Collection, 2004.

Taussig, M. T. *The Devil and Commodity Fetishism in South America*. Chapel Hill: University of North Carolina Press, 1980.

Taussig, Michael. *Shamanism, Colonialism, and the Wild Man: A Study in Terror and Healing*. Chicago: University of Chicago Press, 1987.

Taylor, T. L. "Internet and Games." In *The Handbook of Internet Studies*, edited by Mia Consalvo and Charles Ess, 369–383. Oxford, UK: Wiley-Blackwell, 2011.

Taylor, T. L. *Raising the Stakes: E-Sports and the Professionalization of Computer Gaming*. Cambridge, MA: MIT Press, 2012.

Taylor, T. L. *Watch Me Play: Twitch and the Rise of Game Live Streaming*. Cambridge, MA: MIT Press, 2018.

Tennyson, Charles. "They Taught the World to Play." *Victorian Studies* 2, no. 3 (1959): 211–222.

Thelia, Jerome (dir.). *Bounce: How the Ball Taught the World to Play*. Bounce Group and Wrecking Crew. 2015.

Thomas, Frank, and Ollie Johnston. *Disney Animation: The Illusion of Life*. New York: Abbeville Press, 1981.

Todd, T. *The Tennis Players: From Pagan Rites to Strawberries and Cream*. Guernsey, UK: Vallencey, 1979.

Tompkins, Eric. *The History of the Pneumatic Tyre*. Produced by the Dunlop Archive Project. Lavenham, UK: Eastland Press, 1981.

Tran, Tony Ho. "Man with Brain Implant Challenges Neuralink's Monkey to 'Pong' Game." *The Byte*. May 19, 2021. https://futurism.com/the-byte/man-brain-implant -challenges-neuralinks-monkey-pong.

Tucker, Ross. "Physiology of Football: Profile of the Game." *The Science of Sport*. June 12, 2010, updated October 26, 2013. https://sportsscientists.com/2010/06 /physiology-of-football-profile-of-the-game/.

Tully, John A. *The Devil's Milk: A Social History of Rubber*. New York: Monthly Review Press, 2011.

Turkle, Sherry, ed. *Evocative Objects: Things We Think With*. Reprint ed. 2007; Cambridge, MA: MIT Press, 2011.

Turner, Fred. *The Democratic Surround: Multimedia and American Liberalism from World War II to the Psychedelic Sixties*. Chicago: University of Chicago Press, 2013.

Turner, Victor Witter. *The Anthropology of Performance*. New York: PAJ Publications, 1986.

Vasquez, Alexandra T. "Americas." In *Keywords for Latina/o Studies*, 10–11. New York: New York University Press, 2017.

Victory through Air Power. DVD. Directed by James Algar, Clyde Geronimi, Jack Kinney, and H.C. Potter. 1943; Chicago: International Historic Films, 2005.

The Vladar Company (dir.). *When Games Went Click: The Story of Tennis for Two*. Produced by Raiford Guins, Kristen J. Nyitray, and Peter Takacs. Stony Brook Universities. 2013.

Wardrip-Fruin, Noah. "Gravity in Computer Space." *RomChip* 1, no. 2 (December 2019). https://romchip.org/index.php/romchip-journal/article/view/91.

Warren, Jamin. "Virtual Reality Will Save Games from Itself." *Killscreen*, published 8/26/14. https://killscreen.com/previously/articles/virtual-reality-will-save-games-itself/.

Weaver, Warren. "Some Recent Contributions to the Mathematical Theory of Communication." In *The Mathematical Theory of Communication*, edited by Claude E. Shannon and Warren Weaver. Urbana: University of Illinois Press, 1949.

Weisberg, David. *The Engineering Design Revolution*. Originally at http://www.cadhistory.net/toc.htm, and now published on the Sharpr3D Blog, https://www.shapr3d.com/blog/history-of-cad?utm_campaign=cadhistorynet.

Wells, Paul. *Animation, Sport and Culture*. New York: Palgrave Macmillan, 2014.

Whannel, Garry. "Television and the Transformation of Sport." In *The End of Television? Its Impact on the World*, edited by Elihu Katz and Paddy Scannell, 205–218. *Annals of the American Academy of Political and Social Science*. Los Angeles: Sage Publications, Inc., 2009.

Whittington, E. Michael, ed. *The Sport of Life and Death: The Mesoamerican Ballgame*. New York: Thames & Hudson, 2001.

Wiener, Norbert. *The Human Use of Human Beings: Cybernetics and Society*. Garden City, NY: Doubleday, 1954.

Williams, Serena. "Serena Williams: I'm Going Back to Indian Wells." *Time*, February 4, 2015. https://time.com/3694659/serena-williams-indian-wells/.

Wilson, Suzanne Lemieux. *Rowland B. Wilson's Trade Secrets: Notes on Cartooning and Animation*. New York: Focal Press, 2012.

Wing, Carlin. "The Ball." In *100 Greatest Video Game Characters*, 15–16. Edited by Jamie Banks, Robert Mejia, and Aubrie Adams. Washington, DC: Rowman & Littlefield, 2017.

Wing, Carlin. "Episodes in the Life of Bounce." *Cabinet*, 56 (2015). http://cabinetmagazine.org/issues/56/wing.php.

Wing, Carlin. "Hitting Walls: Saving a Temple of Handball." *Racquet Magazine*, June–August 2017, 39–47.

Wing, Carlin. "Hitting Walls (v.XXVIII): Captured Play." *Games and Culture: A Journal of Interactive Media* 9, no. 6 (2014): 391–405.

Wing, Carlin. "Hopelessly Devoted: Why We Watch Sports." *Public Books*. May 7, 2013.

Wing, Carlin. "Swerve and Return." In *Tech/Know/Future: From Slang to Structure*, edited by Charlotte Kent and Tom Leeser, 82–103. Montclair, NJ: Montclair State University Galleries, 2021.

Wing, Carlin. "True Bounce: Stories of Dunlop and the Rise of Vulcanized Play." In *Sports, Society, and Technology: Bodies, Practices, and Knowledge Production*, edited by Mary McDonald and Jennifer Sterling, 17–40. Singapore: Palgrave-MacMillan, 2020.

Wingfield, Major Walter Clopton. *The Game of Sphairistike or Lawn Tennis*. London: Harrison and Sons, 1874.

Wingfield, Major Walter Clopton. *The Major's Game or Lawn Tennis*. London: Harrison and Sons, 1873.

Winnicott, D. W. *Playing and Reality*. Routledge Classics edition. 1971; New York: Routledge, 2005.

Winthrop Young, Geoffrey. "The KULTUR of Cultural Techniques: Conceptual Inertia and the Parasitic Materialities of Ontologization." *Cultural Politics*10, no. 3(2014), Duke University Press, 385. https://doi.org/10.1215/17432197-2795741.

Witkowski, Emma, and Rune K. L. Nielsen. "Let's Take a FIFA!": Football and the Free-Time Practices of At-Risk Youth under Remand." In *EA Sports FIFA: Feeling the Game*, 181–196. Edited by Raiford Guins, Henry Lowood, and Carlin Wing. New York: Bloomsbury, 2022.

Wolf, Mark J. P. *The Video Game Explosion: A History from PONG to Playstation and Beyond*. Westport, CT: Greenwood, 2007.

"Women's Quarterfinal Match, Serena Williams vs Jennifer Capriati," *2014 US Open*, CBS, September 7, 2004.

Young, Iris Marion. "Throwing Like a Girl: A Phenomenology of Feminine Body Comportment Motility and Spatiality." *Human Studies* 3, no. 2 (April 1, 1980): 137–156.

Zeavin, Hannah. *Mother Media: Hot and Cool Parenting in the 20th Century*. Cambridge, MA: MIT Press, 2025.

Index

Note: Page numbers in *italics* refer to illustrative matter.

Aberth, O. G., 156

Action, defined, 277n53

Adams, Charles, 154–155, 156

Adorno, Theodor, 139

Advanced Research Projects Agency (ARPA), 143

African American tennis players, 101–104

After Dark (screensaver), 146, 273n9

Agamben, Giorgio, 294n15

Agassi, Andre, 39

Aguilar-Moreno, Manuel, 209, 211, 219

Ahmed, Sarah, 32, 197

AJUPEME (Associación De Juego de Pelota Mesoamerica), 29–30, 181, *204*, 205–207, 221, 224, 230–231, 237, 285n2, 290n52, 290n54

Alcorn, Al, 165–167, 277n69

Ali, Muhammad, 9, 251n10

All Watched over by Machines of Loving Grace (Curtis), 170

Almeter, Gary, 148–149

"America's First Polymer Scientists" (Tarkanian and Hosler), 212–213

Amer Sporting Company, 264n45

Amiga Boing Ball, 146, *147*, 186, *187*

Amiga Soccer (video game), 188, 282n32

Anáhuac, 207, 286n5, 288n34

"Andamos Armados" (song), 205

Anglo-Peruvian Amazon Rubber Company, 81

Angry Birds (video game), 281n10

Animation. See also *Names of specific companies;* Video games; Virtual bounce

balls in, overview, 112–116, 271n61

of bodily movements, 119–125

of gestural semiotics, 126–133

Pixar's *Luxo Jr., 110,* 111–112

relationship of sound and image in, 134–139

sound and image relationship in, 133–139, 278n72

squash and stretch in, 27, 112, *128,* 130, 131–132, 271n60

Animation, Sport and Culture (Wells), 116

Anstey, Peter, 62

Anta Sports, 264n45

Appadurai, Arjun, 20, 254n49, 261n13

Apparatus, 122, 125–126, 160, 162, 244, 294n15

Apple (company), 113, 267n5

Aristotle, 50, 53

ARPANET, 143–144

Arrizón, Alicia, 218

Art du Paumier-Racquetier et de la Paume (M. de Gersault), *55*

Art of conjecture, 62–63

Art of Conjecturing (Bernoulli), 63
Ashe, Arthur, 103
Associación De Juego de Pelota Meso-
americao. *See* AJUPEME
Association for Computing Machinery's
Special Interest Group on Computer
Graphics and Interactive Techniques.
See SIGGRAPH
Atari, 146, 160. See also *Pong* (video
game)
Ayrton Paris, John, 117–118
Aztec, 76, 217, 286n5, 288n32

Bad Call (Collins, Evans, and Higgins),
99–100
Baer, Ralph, 160, *161,* 162–168
Bali, 56–57
Ballcourts
court tennis, *46,* 62
Mesoamerican, 208, *209,* 246, 286n6
tennis, 41, 88, 158, 259–260n76
*Ball Ellipsis (Beached Balls, Si'an Kaan,
Mexico), 249*
*Ball Ellipsis (Globe Stress Ball, Fun Express
Globe Stress Ball, and SAGE 50 Years
Earth Stress Ball), 10*
Ball-making, *67,* 220
Ball play. See also *Names of specific
sports*
author's study of, 7–8
gender and, 30–33, 58, 193–203, 22,
229–231, 242
history of, 8–10
study of, 11–13
Ballplayer figurine, *210*
Ball Portrait (Global Beach Ball), 33
Balls. *See also* Tennis balls
in digital animation, 112–116, 271n61
for dogs, 69, 70
as objects, 11
rubber, 74–85, 123
for *ulama,* 207–208, 212, 224, *225,*
290n56

"Balls" (commercial), 18–19
Ball-tracking. *See* Hawk-Eye Innovations
ball-tracking technology
Bang Goes the Theory (television show),
171, *172*
Barad, Karen, 239, 253n33, 256n23
Barthes, Roland, 29
Basillio, Enriqueta, 218
Basketball, 1–3
Bateson, Gregory, 191
Batey, 214. See also *Ulama*
Becerra, Hector, 232
Benjamin, Walter, 22, 253–254n48
Bentham, Jeremy, 56, 240
Beretta, Marco, 11, 50, 52–53
Bergson, Henri, 124–125, 127
Bernoulli, Jacob, 9, 41, 62–65, 260n81
Berry, David, 101
Betting, 56–58, 258n56, 259n68
Beyond a Boundary (James), 4, 101
Bicycle tires, 81, 263n24
Bindle, Thomas, 12
Bizzarie di Varie Figure (Bracelli), *43*
Black hair, 200, 285n76
Black identity and tennis, 101–104
The Black Jacobins (James), 4
Blatter, Sepp, 284n57
Bodily movements, imagery of,
119–126, 193–202, 269n39, 269n41
Body techniques, as term, 242
Bogost, Ian, 168
Boluk, Stephanie, 56
La bonde, 44–45, 48, 55
Boulding, Elise, 248–249
Bounce, as concept, 11–13, 18–20,
106, 119, 240–241. *See also* Ping,
as concept; *Pok ta pok; Pong* (video
game); Ricochet, as concept; Rub-
ber bounce, as concept; Squash and
stretch, as technique; True bounce;
Virtual bounce
Bounce, software programming of, 115,
146–147, 153–160

Bounce: How the Ball Taught the World to Play (film), 31
Bouncing Ball (computer demonstration), 155–157, 159–160
Boundadou, 44
Boundaries, 20, 95–105
Boundary objects, 20
Bounding logics, 28–30
Bourdieu, Pierre, 242, 247
Bowker, Geoffrey, 20
Bowman, Nick, 141
Boyle, Robert, 59, 60, 66, 67, 72, 259n66
Brazil, 196
Bright Signals (Murray), 94–95
Brockedon, William, 77
Brookhaven National Laboratory, 157
Brunet, Jacques-Charles, 51
Burnett, D. Graham, 12
Burnham, Van, 155, 162, 166, 275n33
Burstyn, Varda, 21
Burt, Jules, 186–188
Bury, Thomas Talbot, 118

Canales, Jimena, 119, 125
Caoutchouc, 76, 211
Capriati, Jennifer, 73, 96, 104
Cárdenas, Lázaro, 218
Carillo, Mary, 97
Carlos, John, 219
Carpenter, Loren, 169–170, 172, 279n90
Carpenter, Rachel, 169–170, 279n90
Cartoons, 17, 133–134. *See also* Animation
Cartwright, Lisa, 120, 122
Casement, Roger, 81, 126, 262n23
Castilla elastica tree. *See* Rubber trees and plantations
Caucheros, 80
Cavanaugh, John, 103
Celluloid, 6, 11, 78, 115, 119, 129, 145, 270n52, 273n6

CGI (computer-generated imagery), 106–107
Chanson du ricochet, 48
Charles II, 60
Charles V, 41, 213–215, *216*
Chase lines, 39, *46*
Chase rules in court tennis, 46–48, 55, 256n24, 256n26, 257n27, 258n57
Château de Fontainbleu, France, *42*
Chavez, Marco Antonio, 224
Chelsea (football team), 18–19
Chicano/a, as term and identity, 235–236
Chichén Itzá, 246
Child development, 239–241
Chronophotography, 120
Cinematic realism, 30. *See also* Realism
Citizen (Rankine), 102–103, 266n73
Civilizing Process and Courtly Culture (Elias), 54
Classical Probability in the Enlightenment (Daston), 64–65
Coba, 246
Cockfighting, 56
Coefficient of restitution, 154
Cold War, 144–145, 274n26
Collections, 23–24, 69–70
Collective participation, 106, 266n80
Color television. *See* Television broadcasts
Commodore, 146, 186–187
Common sense, as concept, 29, 184–186, 190, 197, 202
"The Computer as a Communication Device" (Licklider), 144, 146, 173
Computer demonstration programs, 155–160
Computer Space (video game), 165, 277n69
Computer technology development, 143–144
"Computing Science" (Hayes), 143, 146–147

Congo, 81, 126

Consalvo, Mia, 21

Consumer Technology Association, 186

*Contemporary Absurdities, Existential
   Crises, and Visual Art* (Kent and
   Guinness), 7

Contreras, Sheila Marie, 236

Control Machete (band), 205

Cooperative models, 172–175

Copernicus, Nicolaus, 42

Cortés, Hernán, 213–214, 215

Cosmologies, 207–212, 217

Court games, 24–25. See also *Names of
   specific sports*

Court-paume, 44

Court tennis, 37–40, 65–67. *See also*
   Tennis
   Bernoulli on, 62–65
   game of ricochet in, 43–45, 48–49
   as reflective model of the universe,
      41–43
   as reflective model of the world, 55–62
   rules of, 47–48, 258–259n62,
      258nn58–60
   scientific explanations and, 49–55

Crafton, Donald, 127, 129–130

Cricket, 4, 9, 84, 96, 101, 120, 261n13,
   293n12

Csíkszentmihályi, Mihály, 23

Curtis, Adam, 170, 171, 172, 278n87

Daedaleum, 116

Dance cards, 194, *195*, 196, 283n50

Daston, Lorraine, 64–65

Data Feminism (D'Ignazio and Klein),
   105

Dead rubber, 261n15

De Anima (Aristotle), 50

"The Death of John Cavanaugh"
   (Hazlitt), 1

Deep play, 56–57

Descartes, René, 51–52, *52*

Dick, Leslie, 247

Digital animation. *See* Animation

Digital balls, 112–116, 271n61. *See also*
   Balls; Virtual bounce

D'Ignazio, Catherine, 105

Ding Huan, 115, 267n8

Disney, Walt, 129, 136, 270n55

Disney Studios, 27, 111–112, 129–133,
   136–139, 270nn54–55, 272–273n3

Dog balls, 69, 70

Donald Duck Joins Up (Shale), 133

Donner Analog computer, 157

Doyle, Jennifer, 31, 199–200

Du Cros, Arthur, 84–85

Du Cros, Harvey, 81, 83, 84

Dunlop (company), 81–82, 263n24,
   263nn26–27, 263n32

Dunlop, John Boyd, 81–82, 263n24

Durán, Diego, 213

Duran, Miguel, *207*

Dvorak, Robert, Jr., 158

EA Sports FC, 179, 201–202. See also
   FIFA International Soccer (video
   game)

EA Sports FIFA: Feeling the Game (publication), 202

École Normale Supérieure, 124

Edgeworth, Maria, 118

Edwards, Paul, 44

Elastic self, 268n24

Electronic Arts (EA), 29, 179, 207

Elias, Norbert, 54

Elliot, Lise, 240

Embodiments, 277n59

Enclosure, 52–53

English Football Association, 196–197

Enrigue, Álvaro, 246

E-sports, 21

Even odds, 56–57

Everett, Robert, 152

Fable du ricochet, 48

Female athletes. *See* Gender

Ferenz, Ramona, *153*
Fernandeño Tataviam Band of Mission
    Indians, 206, 221
Fetish theory, 262n23
Feynman, Richard, 126
FIFA International Soccer (video game),
    7–8, 25, 29, 179–184
    body representation in, 193–202
    computer graphics in, 184–190,
    282n40
    gender and, 30, 31
    licensing negotiations of, 284n54
    movement of bodies in, 190–193
    revenue from, 280n3
    Ultimate Team (play mode), 201
FIFA World Cup, 9, 16–17, 196
Fitton, William Henry, 267n12
Fleischer, Dave and Max, 135
Flusser, Vilem, 94
Football (also known as soccer).
    See Soccer
Forrester, Jay, 143–144, 152, 274n26,
    275n27
Fort/da (game), 14, 15, 111, 241
Foucault, Michel, 243, 294n15
Fox, John, 212, 224, 254n53
France, *42*, 44–45, 48–49
French Open, 90
Freud, Anna, 252n22
Freud, Ernst. *See* Halberstadt, Ernst
Freud, Sigmund, 14, 15, 247, 262n23
Full Hair Tech, 200

Gaboury, Jacob, 185
Galileo, 9, 51, 52
Game design, 56
Game feel, 29, 185, 189
#GamerGate, 31
Games and media, 18–23
Games and Sport in Everyday Life (Perin-
    banayagam), 24–25, 37, 95
The Games Ethic and Imperialism (Man-
    gan), 83–84

Garcetti, Eric, 231
Garrison, Zina, 103
Geertz, Clifford, 56, 186, 240, 248
Gender
    ball play and, 30–33, 58, 225,
    229–231, 242
    transgender and nonbinary athletes,
    283n48, 290–291n61
    video game representation and,
    193–203, 284n61
General Instruments, 168
Gere, Charlie, 150–151, 274n17
Gestural semiotics, 126–133
Gibson, Althea, 103
Gillmeister, Heiner, 14, 44–45,
    49, 75, 84, 252n20, 255n16,
    258n60
Gilmore, John, 154–155, 156
Gitelman, Lisa, 21
Glitched bodies, 29
Glitch Feminism (Russell), 203
Goalball, 267n82
Goffman, Erving, 54
Goodyear, Charles, 77
Gramsci, Antonio, 186
Grand Slam Championships, 89–90
Griesemer, James, 20
Guins, Raiford, 21, 180, 183
Gunning, Thom, 118–119

Habitus, 242
Hair, 285n76
Halberstadt, Ernst, 14–15
Hall, Stuart, 186
Hamley Brothers, 144, 162
Hamm, Mia, 31, 200
Hammons, David, 1, *2*, 3
Hancock, Thomas, 77–78, *79*, 116
Handball, 232–239, 291n68
Handicapping rules, in court games,
    58, 64
Hanson, Christopher, 98
Harrison, Bill, 161–162

Haudenosaunee Confederacy, 248,
     294n24
Hawk-Eye Innovations ball-tracking
     technology, 26, 73, 96–105, 106
Hawkins, Paul, 96
Hawkins, Sherman, 44
Hayes, Brian, 12, 146–147, 152, 154
Hazard end, *46, 47*
Hazlitt, William, 1, 103
"Heartbeats" (song), 18, 253n34
Heights Casino, New York City, 5
Heliocentrism, 42
Henry V (Shakespeare), 49
Henry VIII, 41
Herrera, Raul, xix, 223–224, 225, 231,
     237, 289n47
Higher Goals (sculpture by Hammons),
     1, *2,* 251n3
Higinbotham, William, 157, 168
Hinton, Geoffrey, 143
The History of the Pneumatic Tyre (Dun-
     lop), 82
Hobbes, Thomas, 41, 59, 60, 66, 72
Hosler, Dorothy, 212–213
House of Gabriel Giolito de Ferrari and
     Brothers, 50
How to Play Baseball (animated films),
     132, *133,* 141
Huastec, *210*
Husserl, Edmund, 32
Huygens, Christiaan, 63
Hyperrealism, 148. *See also* Realism

ICC Cricket World Cup, 9
Identity, 17–18
*The Illusion of Life: Disney Anima-
     tion* (Thomas and Johnston), *128,*
     129–131
"I'll Ping You" (Almeter), 148–149
India-rubber, 78, *80,* 269n38. *See also*
     Rubber balls
Industrial Revolution, 77
Injury, 226–229

Instant replay, 73, 96, 98, 99, 103,
     265n59, 265n62. *See also* Hawk-
     Eye Innovations ball-tracking
     technology
International Real Tennis Professional
     Association, 259–260n76
International Tennis Federation. *See* ITF
     (International Tennis Federation)
Into the Universe of Technical Images
     (Flusser), 94
"Introduction: A Non-Zero-Sum Game"
     (Suparak and Kashmere), 179
Invensys, 263n32
Iroquois Confederacy. *See* Haudeno-
     saunee Confederacy
Italy, 41, 197
Italy, Classical, Historical, and Picturesque
     (Brockedon), 77
ITF (International Tennis Federation),
     73, 88, 91, 95

Jackson, Wilfred, 136, 139
James, C. L. R., 4–5, 101
Jeffries Royal Lawn Tennis set, *75*
Jeu de bonde, 44, 45, 48, 256n24
Jeu de paume, 41, *42,* 44, *51, 216*
Jhally, Sut, 21
Johnston, Ollie, 129–131, 136
Jones, Cleon, 12
Jones, Jan-Erik, 62
Journal d'un bourgeois de Paris (publica-
     tion), 41

Kahl, Milt, 137
Kashmere, Brett, 179, 283n52
Katz, Elihu, 20
Kelly, Martin Campbell, 168
Kelly, Peter, 32
Kepler, Johannes, 51
Kew Gardens, 80
Kik, 211
Kittler, Friedrich, 20
Klein, Lauren F., 105

Klein, Melanie, 252n23
Knowing and passing parable, 14

Lab testing of tennis balls, 88–89.
  See also Tennis balls
Lacan, Jacques, 247–248
La Condamine, Charles Marie de,
  288n33
Lake, Robert, 54, 55
Landscape with David and Bathsheba
  (met de Bles), *216*
Latour, Bruno, 61, 259n73
Laughter (Bergson), 124–125
Law, John, 187
Lawn tennis, 37, 74–75, 261n2. *See also*
  Tennis
Lederach, John Paul, 248
Lefebvre, Henri, 243–245
Lemieux, Patrick, 56
Leopold II (king), 81
Letter to a Friend on Sets in Court Tennis
  (Bernoulli), 63
Leviathan and the Air Pump (Shapin
  and Schaffer), 59, 60, 66
Lewis, Mark, 187
Licklider, J. C. R., 143, 145, 173, 279n97
The Listener (publication), 93
London Times (publication), 86
Longing, 32
Lord's Cricket Ground, *63, 85*
Los Angeles Plays (art installation),
  24, 71
Louis XIV (king), 63
Lowood, Henry, 21, 167, 180, 184,
  281n13, 282n36
Luxo Jr. (animated short), 111, 140,
  267n2
Luxo lamp, 267n1

Macaloon, John, 23
Machismo, 291n63
MacIntosh, J. J., 62
"Made for TV," 72, 91, 261n6

Magnavox Odyssey, 160, 162–164, 168,
  278n82
Maher, Jimmy, 187
Malaya, 82
Manchester United, 18–19
Mangan, J. A., 83–84
Manovich, Lev, 113
Maravilla Handball Court, 30, 231–237,
  285n2, 292n76
Maravilla Historical Society, 232–233,
  236
Marey, Étienne-Jules, 119–122, *127,*
  269n34, 269n39
Mártir de Anghiera, Pedro, 213
Marxism, 4
Marza Animation Planet, 140
Mauss, Marcel, 241–243
Maya, 8, 76
McDonald, Mary, 21
McEnroe, John, 97–98
McLuhan, Marshall, 21–22
Media
  ball play and, 9
  balls and, 72–73
  games and, 18–23
  spectacles, 266n73, 280n5
  television broadcasts of games, 18–19,
    91–95
Media of interpersonal communication,
  as concept, 21–22
Mediation, as concept, 15, 19–20, 47, 59
Mediation of bounce, 18–20
Mesoamerican ballcourts, 208, *209,* 246,
  286n6, 287n8, 289n45, 291n68
Mesoamerican ballgames, 8–10, 29, 76,
  78, 207–211, 286–287n7, 292n71.
  See also *Ulama*
Mestizaje, as term, 289n38
Metagames, 56
Mexican Revolution (1910), 217, 220
Mexico, 197, 246, 271n62. *See also*
  Mesoamerican ballgames
Mia Hamm Soccer 64 (video game), 200

Miles, Eustace, 85–86

Mirra, Helen, 232

MIT (Massachusetts Institute of Technology), 143, 152–153, 155–157

Mixtec, 8, 76, 211

Modern tennis, 70. *See also* Tennis
   balls of, 71–74, 85–91
   boundaries in, 95–105

Montaigne, Michel de, 9, 41

Montfort, Nick, 168

Moonbounce technique, *151*

Morgan, Roger, *216*

Morris, Barbara, 98

Motion through devices, 116–119

Le Movement (Marey), 120–124

Movement physics, 186. *See also* Bodily
   movements, imagery of

Mummy tyre, *82*

Murray, Sue, 94–95

Muscular Christianity, 73, 83, 118, 134

Museo Galileo, 50

Muybridge, Eadweard, 120, 127, 129,
   134, 269n34

Mythic, games as, 29

Nahuatl, 29, 211, 287n15

Naka, Yuji, 141

National Chicano Moratorium, 235–236

Navigation logics, 186

Navratilova, Martina, 39

Needham, Joseph, 115

Netscape Navigator, 113

Newbolt, Henry, 74, *85*

New England Patriots, 254n52

Newton, Isaac, 9, 41, 51–52

Newton's experimental law, 154

New York Mets, 3

Nintendo, 139, 179

Nishiyama, Michi, 234, *235*

Nishiyama, Tommy, 234, 236, *235*,
   292n76

Nonbinary athletes, 283n48, 290n61

Nunn, Nigel, 283n46

Objective truth, 99–100. *See also* Hawk-
   Eye Innovations ball-tracking
   technology

Off/on the ball, as phrase, 190–193,
   283n43

Olmec, 8, 76, 207–208, 211, 286n5,
   286n6, 287n21

Olympic Games, 9, 218, 251n7

"On the Ball" roundtable, 11–12

On the Nature of Things (Lucretius), 42

Optical toys, 116–119

Oscilloscopes, 120, 150–152, *153*,
   275n27

Oshima, Naoto, 141

Osorio Uscanga, Armando, 207, 221

"Pandora's Vox" (humdog), 69

Panopticon, 22–23

Paralympics, 258n58

Parker Brothers, 144, 162, 164

Parkour, 259n72

Pascal, Blaise, 63

Penang Rubber Estates, 262n21

Pennington, Michael, 196

A People's History of Tennis (Berry), 101

Perez, Amanda, 231–232, 235–237

Perinbanayagam, R. S., 24–25, 37, 40,
   54, 95

*Personal Narrative of the Origin and Pro-
   gress of Caoutchouc or India-Rubber
   Manufacture in England* (Hancock),
   78, *79*

Peru, 81

Peters, John Durham, 20

Petit, Emmanuel, 16

Phenakistoscope, 116, 118

Phillips, Amanda, 194–195

*Philosophy in Sport Made Science in Ear-
   nest* (Paris), 117–118

Photorealism, 119, 125. *See also* Realism

Ping, as concept, 28, 145, 148–153,
   273n15

Ping-Pong (game), 144

Pixar, 111–112, 170
Planetary motion, 51–52
Plantations, 80–85
Plateau, Joseph, 118
Play, 15–16, 240–241
Players, 22–23
Pneumatic objects
  balls, 6, 71, 72, 90
  experiments with, 60, 227
  tires, 81, 263n24
Pok ta pok, 29–30, 209, 237. See also
  Ulama
Pok ta Pok (video game), 220
Pong (video game), 7, 28, 91, 112, 148,
  164–169, 171, 172, 174, 276n48.
  See also Video games
Ponnusamy, Suresh, 88–89, 90
Popol Vuh, 211
Port Royal Logic, 63
Premier League football, 18–19
Priestley, Joseph, 76
Probability, 62–65
Project Diana QSL card, 151
Puc, Reyna, 221
Pung, 149. *See also* Ping, as concept
Putamayo, 126
Putzel, Pete, 69

Quasi-action, 27, 114
Quasi-objects, 16, 25, 27, 40, 106, 113,
  114, 122, 252–253n30
Quasi-selves, 32
Quasi-subjects, 16, 106, 113, 192
Queer Phenomenology (Ahmed), 32
Quevedo, Francisco de, 246
Quiq, 211

Racial identity, 101–104, 285n76
Racing the Beam (Montfort), 168
Racquet and Tennis (R&T) Club, 38
Radiation Laboratory (MIT), 157
Ramos, Maria Isabel, 208–209, 211,
  229–230

Rankine, Claudia, 102–103, 266n73
Realism, 119–126, 148, 157, 190, 195
Rebote, 233–234, 237
Refereeing, 99, 104, 265n62
Renaissance, 41–43
Ricochet, as concept, 39–40, 43–44,
  48–49, 67
Right Stick Switching, 282n40
Rochín, Roberto, 219–220
Roman, Easy, 69
Roman, Roxy, 69
Rowe, David, 21
Rubber balls, 74–85, 123, 212–213,
  219–220. *See also* Balls
Rubber bounce, as concept, 8, 29, 213.
  See also Bounce, as concept; *Pok ta
  pok*; True bounce; Virtual bounce
Rubber trees and plantations, 80–85,
  207, 219–221, 231–232, *238*,
  293n65
Rule-bound frameworks, 28–29, 95–105,
  244–245, 258nn58–62
Rusch, Bill, 161–162
Russell, Legacy, 29, 203
Rutter, David, 198, 200

Sahagún, Bernardino de, 213
Sanders Defense Associates, 161–162,
  168
San Fernando *ulama* team, 206, *207*,
  221–222, 230–231
Scaino, Antonio, 26, 41, 49–50, 53–57,
  83, 257n34
Schaffer, Simon, 59–61, 66, 259n73
Schechner, Richard, 244
Schiebinger, Londa, 216
Scientific Revolution, 50, 165
Scoring system, 58–59, 211–212
Scott, Katie, 198
Screensavers, 112, 113, 146, 273n9
Scroggins, Michael, 171
SEGA, 139–140, 141. See also *Sonic the
  Hedgehog* (video game)

Sekula, Allan, 4, 251n4
Semenya, Caster, 31
Semi-Automatic Ground Environment
    (SAGE) project (MIT), 143
Sensible Soccer (video game), 188
Serres, Michel, 16–17, 25, 40
Servomechanism Lab, 274n24
Seventh Generation principle, 248,
    294n23
Shakespeare, William, 9, 41, 49
Shannon, Nick, 197, 200
Shapin, Steven, 59–61, 66, 259n73
Sharapova, Maria, 103
Sheppard, Samantha, 102
Shot Spot. See Hawk-Eye Innovations
    ball-tracking technology
Siegert, Bernhard, 20
SIGGRAPH, 111–112, 169, 170
Sinclair QL, 146
Sito, Tom, 166
Slavery, 81
Smith, Tommie, 218
Sobchak, Vivian, 149
Soccer, 16–17, 179–180, 191–192, 193,
    207, 280n5. See also FIFA Interna-
    tional Soccer (video game)
Sonar, 12, 18, 149, 150, 274n19
Song Cartoons (animated series), 135
Sonic the Hedgehog (film), 140
Sonic the Hedgehog (video game),
    139–140, 141–142
Sound and image, relationship in ani-
    mations, 134–139, 278n72. See also
    Animation
Spain, 42
Spectatorship
    Benjamin on, 253–254n48
    betting and, 56–58
    longing and, 32
    panopticon of, 22–23
Speculation, 3
Sphairistike (game), 74, 77, 116
Spherical studies, 42–43

Spinning beach ball of death, 113, 114,
    267n5
Sport as a cultural practice, 4, 22–23,
    244–245, 259n72, 261–262n17
Sporting Blackness (Sheppard), 102
Sports-media complex, 21
Sports video games. See Video games
Squash
    author's positionality on, 1, 5
    description of, 5–6
    gender and, 31
    tournament of, xi, xii, xiii
Squash and stretch, as technique, 27,
    112, 128, 130, 131–132, 271n60.
    See also Animation
Star, Susan Leigh, 20
Steam (gaming platform), 179
Steamboat Willie (film), 136–137
Stein, Abe, 21, 181
Sterling, Jennifer, 21
Story of Dunlop, 82–83
Sudden Death (Enrigue), 246
Suparak, Astria, 179, 283n52
Super Bowl, 9
Super Mario Bros (video game), 139
The Super Mario Bros. Movie (film),
    140
Sweet, Sam, 233–234, 235
Swink, Steve, 29
Sylla, Edith Dudley, 11, 65, 260n81,
    260n87

Table tennis, 6, 144
Tarkanian, Michael J., 212–213
Tataviam Fernandeño Band of Mission
    Indians, 221
Taylor, T. L., 21
"Techno Tribe" (Carpenter), 170
"Television, Gaming, and Training
    Apparatus" (patent), 161, 162
Television broadcasts, 9, 18–19,
    91–95, 251n10, 265n49. See also
    Media

Tennis, *13,* 25, 26, 254n51. *See also*
   Court tennis; Modern tennis; Ten-
   nis balls
study of, 7, 14, 255–256n16
Tennis balls. *See also* Balls; Tennis
of 19th-early 20th c. games, 72,
   74–85
collections of, 69–71
Made for TV, 92–95
of modern era, 71–74, 85–91,
   261n4
Tennis for Two (computer demonstra-
   tion), 157–160
Tennyson, Charles, 84, 254n53
Thaumatrope, 116, 117, 118,
   267–268n12, 268n14
"They Taught the World to Play"
   (Tennyson), 84
Thomas, Frank, 129–131, 136
Thompson, Robert William,
   263n24
"Throwing Like a Girl" (Young), 29,
   23, 242
Tippet, Krista, 248
Tlaxcala, 76
Toltec, 211
Tom Brown's School Days (Hughes),
   118, 293n12
Tosi, Alessandro, 50
Toys, 116–119
"Training Tennis Players through Natu-
   ral Philosophy" (Beretta), 50
Transgender athletes, 283n48,
   290–291n61
Transparency, 100, 129, 259n73,
   265n65
Treatise on the Game of Ball (Scaino),
   49–50, 53–56
Trinquetes, 215
True bounce, 6, 70–71, 88, 105, 215.
   See also Bounce, as concept; Virtual
   bounce
Turner, Fred, 139

Ulama, 7, 25, 29–30, 181–182, 205–232,
   287nn8–9, 289n48, 289nn44–45.
   See also Mesoamerican ballgames
Ulama balls, 208, 212, 225, 291n56
Ulama Project, 219, 289n44
Ullamaliztli. See Ulama
Umpiring, 99, 104, 265n62
Uniform of *ulama* players, 222–223,
   290n51. See also *Ulama*
United States Tennis Association
   (USTA), 88–91, 264n43
US Open Championships, 38, 70, 73,
   89–90, 96–105, 225
USTA. *See* United States Tennis Associa-
   tion (USTA)
USTA Billie Jean King National Tennis
   Center, 70, 90
USTA National Championships, 101
US Women's National soccer team,
   193–194

Vasconcelos, José, 218
Vauban, Marquis de, 48–49
Video games, 9–10, 21, 150, 160,182
   280n7. *See also* Animation; *Names of
   specific games*
"Video Imaging Method and Apparatus
   for Audience Participation" (Carpen-
   ter), *169*
Virtual bounce, 27, 113–114. *See also*
   Animation; Bounce, as concept;
   Digital balls
development of, 144–148
ping, 148–153
software programming of, 146–147,
   153–160
Visual realism, 148, 157. *See also*
   Realism
"Vitaï Lampada" (Newbolt), 74, *85*
Vulcanization of rubber, 77–78, 116,
   219–220. *See also* Rubber balls
Vulcanized play, as phrase, 71, 77, 119,
   262n18, 263–264n32

Wardrip-Fruin, Noah, 186
Weaver, Warren, 274n22
Weiditz, Christoph, 214–215
Weisberg, David E., 154
Wells, Paul, 27, 116, 118, 129
"What Computers Should Be Doing?"
    (Pierce et al.), 173
Wheel technology, 81
When Games Went Click (Dvorak),
    158
Whirlwind computer, 151–152, *153,*
    155, 274n22, 274n23, 275n27. *See
    also* Virtual bounce
Whiteness, 103–104, 255n2
Wickham, Henry, 80–81
Williams, angel Kyodo, 266n80
Williams, Serena, *13,* 73, 96–98,
    101–103
Williams, Venus, 103
Wilson, Rowland B., 144, 272–273n3
Wilson Sporting Goods, 264n45

Wimbledon Championships, 86, 90,
    92–93, *93,* 94, 95, 101, 265n57
Wingfield, Walter Clopton, 74–77,
    116, 162
Winnicott, D. W., 15–16, 252n23
Witkowski, Emma, 21
Wittgenstein, Ludwig, 60, 61, 253n41
Wolsey, Thomas, 41
"The Work of Art in the Age of Mechan-
    ical Reproduction" (Benjamin), 22
World Cup, 194, 196, 197, 284n57
World Series, 3, 12
World Squash Federation (WSF), 6

Ximenez, Francisco, 211

Young, Iris Marion, 29, 230, 242–243

Zidane, Zinedine, 16
Zoetrope, 116, 267n8
Zoopraxiscope, 116

Publisher contact:
The MIT Press
Massachusetts Institute of Technology
77 Massachusetts Avenue, Cambridge, MA 02139
mitpress.mit.edu

EU Authorised Representative:
Easy Access System Europe, Mustamäe tee 50,
10621 Tallinn, Estonia
gpsr.requests@easproject.com

Printed by Integrated Books International,
United States of America